IEE POWER SERIES 17

Series Editors: Professor A. T. Johns
J. R. Platts

High voltage engineering and testing

High voltage engineering and testing

Edited by
H M Ryan

Peter Peregrinus Ltd. on behalf of the Institution of Electrical Engineers

Published by: Peter Peregrinus Ltd., on behalf of the
Institution of Electrical Engineers, London, United Kingdom

© 1994: Peter Peregrinus Ltd.

British Library Cataloguing in Publication Data

A CIP catalogue record for this book
is available from the British Library

ISBN 0 86341 293 9

Printed in England by Short Run Press Ltd., Exeter

Contents

Contributors

Professor D.J.Allan
Director & Technical Manager
GEC Alsthom T & D Transformers
Lichfield Road
PO Box 26
Stafford ST17 4LN

Dr.N.L.Allen
Department of Electronic & Electrical
Engineering
The University
Leeds LS2 9JT

S.Armstrong
New Contracts Manager
Electricity Division
East Midlands Electricity
PO Box 4, North PDO
398 Coppice Road
Arnold
Nottingham NG5 7HZ

Dr.I.A.Erinmez
Technical Audit Manager
National Grid
National Grid House
Kirby Corner Road
Coventry CV4 8JY

S.M.Ghufran-Ali
Switchgear Specialist
Merz & McLellan
Amber Court
William Armstrong Drive
Newcastle upon Tyne NE4 7YQ

**Professor Dr.-Ing.
E.Gockenbach**
Schering-Institut
Universität Hannover
30167 Hannover
Callinstraβe 25
Germany

J.Graham
Technical Manager
(Development)
Bushing Reyrolle Company
Hebburn
Tyne and Wear NE31 1UW

R.C.Hughes
Consultant
2 Childs Hall Drive
Great Bookham
Leatherhead
Surrey KT23 3QL

T.Irwin
Projects Manager
NEI Reyrolle Technology
Hebburn
Tyne and Wear NE1 1UB

Professor G.R.Jones
Department of Electrical
    Engineering and Electronics
University of Liverpool
Brownlow Hill
PO Box 147
Liverpool L69 3BX

J.A.Lapworth
National Grid
Technology & Science Laboratories
Kelvin Avenue
Leatherhead
Surrey KT22 7ST

F.J.Liptrot
Technical Director
Allied Insulators
PO Box 17
Leek New Road
Milton
Stoke-on-Trent ST2 7EE

C.J.Morgan
Technical Audit Unit
National Grid
National Grid House
Kirby Corner Road
Coventry CV4 8JY

Professor H.M.Ryan
School of Advanced Technology
University of Sunderland
Edinburgh Building
Chester Road
Tyne and Wear SR1 3SD

M.Simmons
Pirelli Cables
PO Box 6
Leigh Road
Eastleigh
Hampshire SO5 5YE

A.White
Engineering Manager-Power
GEC Alsthom Transformers
PO Box 26
Lichfield Road
Stafford ST17 4LN

Introduction

D.J. Allan

The 1990s are a time of change for the Electric Power Industry; new trading blocks are being formed, transnational mergers of manufacturers are being consolidated and key European utilities are introducing unbundling programmes to split the generators from the transmission network and distributors, often by privatisation of nationalised undertakings. The European Community has introduced the Single Market and the Utilities Directive is changing the way that purchasers and manufacturers can make commercial contracts within the Community.

The widespread use of power electronics to control large items of power plant, and to replace conventional high voltage devices, is revolutionising the control of high voltage networks, allowing operators to redirect power flow through under-utilised parts of a network in order to maximise power flow in a system whilst minimising capital investment. At the same time new testing techniques are evolving using microprocessors to undertake on-line manipulation and enhancement of recorded signals to provide better information regarding equipment under test.

During the 1980s there was an apparent reduction in popularity of power engineering amongst students, and many high voltage laboratories at universities and polytechnics closed down — the equipment was sold and the space converted to provide laboratories for light current engineering. In the United Kingdom the industrial base shrank due to mergers and closures, and many industrial high voltage laboratories were closed or mothballed. Few young engineers saw a future in the area of high voltage engineering and the remaining high voltage laboratories are largely staffed by older although well experienced engineers.

There is a clear need to redress the situation. Industry needs young engineers to work in the design offices and in high voltage research, development and testing laboratories. Few British universities are able to offer undergraduate courses with a substantial high voltage engineering content. Young graduates in industry need further training in this area, and with the ramifications introduced by European Community Directives, courses of this type are needed to ensure the future of the Electric Power Industry.

This book, and the vacation school on which it is based, provides essential information for the engineer engaged in high voltage engineering and testing, and for those wishing to start a career in that area. It addresses changes in practices and procedures and the introduction and adoption of new technical advances. It is essential that the industry is able to progress at a fast rate, and to achieve this aim it is necessary to encourage and develop the expertise of engineers and to nurture and support the flow of new ideas.

I.1 History of high voltage engineering

Serious high voltage engineering began with the induction coils constructed in 1836 by Nicholas Callan, a priest-scientist who was Professor of National Philosophy at Maynooth College, near Dublin. Callan's Great Coil induced voltages of above 100 kV, but it was of academic interest only. In 1882, Lucien Gaulard and John Dixon Gibbs patented the first systems using alternating current, and in 1886 Karoly Zipenowski, Miska Déri and Otto Blathy, engineers from Ganz in Hungary, patented the first transformer. The invention of the transformer was the key that opened the door leading to efficient and effective power systems working at high voltages.

High voltage power systems were installed throughout the world, based on Zipenowski's transformer invention. Systems were introduced in the United Kingdom at 132 kV (in the 1920s), 275 kV (in the 1950s), 400 kV (in the 1960s). Higher voltage systems were introduced overseas at 750 kV (in the 1960s) and 1100 kV (in the 1970s). The development of equipment was rapid up to UHV levels. Unfortunately, equipment at the higher voltages was introduced before the principles were fully understood. The first equipment installed at 132 kV was prone to fail due to lightning impulse activity, the first transformers installed in some 500 kV systems failed due to part winding resonance, and the failure rate of transformers installed at 750 kV is twice that for transformers operating at 400 kV.

The lightning impulse failures in the 1920s led to a better understanding of transient voltage distributions within windings and to the introduction of lightning impulse tests. The analysis of later failures led to the introduction of new dielectric tests in order to achieve acceptable levels of reliability and dependability of the installed equipment.

New causes of failure are being investigated at present, including the effect of Very Fast Transients (VFT) initiated by disconnect switches in Gas Insulated Systems (GIS). VFTs with front times measured in nano-seconds lead to highly non-linear voltage distributions in power transformers, often causing winding failure in service.

In order to demonstrate that equipment would be reliable in service, manufacturers and purchasers had formed third party bodies to develop standards. In Britain, the British Standards Institution was formed to undertake this work and developed a series of internationally accepted British Standards. Similar bodies were founded in other countries, and as early as 1906, the International Electrotechnical Commission was founded in Geneva in order to prepare the common International Standards that were needed to facilitate international trade.

The IEC prepared standards based on current practice. In a fast moving technology, it was shown to be inefficient to carry out the pre-standardisation work within IEC, and in 1921 the International Conference on Large High Voltage Electric Systems was founded to promote free discussion on the research and current investigations into the application of developing technologies, as a feeder organisation to IEC. CIGRE has since developed into a large, well structured organisation with 15 Study Committees, each charged with steering research and investigation on an international scale into the design, building and operation of large high voltage power systems. However, a key section of the CIGRE mission is still to provide the pre-standardisation input to IEC in the field of power generation and transmission systems.

It became apparent that even CIGRE was not the appropriate body to organise basic research in high voltage engineering. In 1972 the first International Symposium on High Voltage Engineering (ISH) was held. ISH linked basic research in universities with the equipment-based research in industry. Formal links exist between the Technical and Sub-Committees of IEC and Study Committees of CIGRE, and between the CIGRE Study Committee responsible for high voltage techniques and ISH.

Regional Standardisation Groups have also been formed to rationalise standardisation within geographical areas.

Within Europe the European Committee for International Standardisation (CENELEC) was established in 1973 to take over the work started in 1959 by CENELCOM. CENELEC Standards are based, wherever possible, on IEC Standards but exist as Harmonised Documents (HDs) or European Norms (ENs) which are common standards adopted by CENELEC countries. ENs, in particular, are strong mandatory documents, enshrined in European Law.

In North America, the equivalent regional group is ANSI, which generally adopts standards prepared by IEEE Technical Committees. In the past, there have been major differences between ANSI and CENELEC standards, but in the 1990s it is expected that these differences will be gradually harmonised, as all Regional Standardisation Groups endeavour to adopt IEC Standards as the basis of Regional and National Standards.

I.2 High voltage power networks

IEC Standards are performance standards, which include the description of dielectric tests to be carried out on high voltage equipment; the test methodology is prescribed together with selection charts relating test levels to system highest voltages.

Most high voltage networks operate at 330-500 kV with a much smaller number operating at 700-800 kV. Where power is to be wheeled between networks operating at different frequencies, or operating at different phase angles, d.c. links are a successful method of providing a channel for power flow. The link may be hundreds of kilometres in length, with intermediate booster stations between the rectifier and converter terminals, or it may be a short 'back-to-back' link between adjacent networks.

Where power is required to flow through a specific channel in an a.c. network, phase-shifting transformers, quadrature booster transformers or other equipment controlled by power electronics are being proposed to force power to flow along paths that would not be taken without the intervention of these devices. The technology covers Flexible a.c. Transmission Systems (FACTS) and employs power electronics devices to provide dynamic load brakes, modular series capacitors, static VAr compensators - all at high voltage.

During the early 1990s, the utility responsible for generation and transmission in England and Wales was privatised. The generation and transmission areas of CEGB were unbundled and two generation companies sold into private ownership. The twelve distributors, previously the Area Boards, were also sold into private ownership and the transmission system formed yet another company, owned by the distributors.

Previously, the CEGB operated one of the largest interconnected power systems in the world, including generation and transmission in Scotland and Northern Ireland within their control. Post-privatisation, the separate generators, distributors and the transmission network operate independently with their own design teams, purchasing specifications and technical specifications. Whereas these previously existed as common technical specifications to cover all equipment, the separate companies now follow independent technical requirements.

Whereas CEGB was a technically strong organisation, the new companies exhibit strong commercial attributes. Problems concerning high voltage plant were previously handled by a single team; each successor company now has its own technical experts. The result could easily lead to divergencies in technical requirements and a lack of interactive discussion concerning common problems. This book will address the situation by providing tutorial information to those engineers who are taking new assignments involving high voltage equipment and will encourage discussion of common problems within a learned society framework.

It is important to differentiate clearly between standards and specifications. Standards may be mandatory, or voluntary, in which case the purchaser may nominate them in a contract. Specifications are the corpus of technical requirements adopted by a purchaser and are supplementary to the standards.

The difference can be clarified by examining the triangle in Fig. I.1, representing the hierarchy of needs. The triangle represents all the technical requirements for a contract; the requirements become more specific to a particular contract towards the apex of the triangle. The band at the base of the triangle represents those horizontal standards that apply to all electrical power plant, eg. insulation co-ordination, EMC and safety. The next band represents the vertical standards that apply to a particular range of equipment, e.g. the transformer or switchgear standards issued by IEC, CENELEC, BSI or ANSI. The next band represents the customers' specifications for a particular range of equipment, e.g. the transformer or switchgear specifications of NGC, YEB, EDF or ENEL. The top band represents particular site requirements included in customer specifications, e.g. voltage and power ratings, installation limitations or seismic requirements.

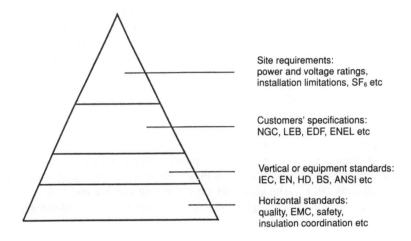

Site requirements:
power and voltage ratings,
installation limitations, SF_6 etc

Customers' specifications:
NGC, LEB, EDF, ENEL etc

Vertical or equipment standards:
IEC, EN, HD, BS, ANSI etc

Horizontal standards:
quality, EMC, safety,
insulation coordination etc

Fig. I.1   *A hierarchy of needs — standards and specifications*

I.3 EC Directives

European standardisation is part of Community law, and must provide the framework and detail to verify that directives issued by the EC are followed. The Utilities Directive, following the introduction of the European Single Market in 1993, had specific requirements that must be followed by utilities in placing orders for all electrical equipment above a certain value. The intention is to remove barriers to trade between member states.

In order to achieve a "level playing field", it is essential that each member state adopts the directives in the same form, and ensures that they are followed with even policing in the same manner. It appears at present that the playing field is far from level; different member states are interpreting the directives differently and the way in which they are followed varies between countries.

The main directives associated with high voltage engineering are:

—   The Utilities Directive
—   The Electromagnetic Compatibility Directive (EMC)
—   The Certification Directive

The EMC Directive will be adopted in 1996, but the situation is far from clear. It is difficult to see how it can apply to high voltage electric power systems or to railway systems. The requirements of the Directive will be that electromagnetic emission from the system should not affect the working of other equipment, and the power system should be immune to the electromagnetic interference from other equipment. Clearly, there is a problem where power systems with overhead lines and power electronic equipment are concerned. The Department of Trade and Industry advises that the Directive will cover electric power systems; other European member states have advised utilities that they will be excluded.

The Utilities Directive, linked to the Public Procurement Directive, includes requirements concerning quality assurance certification of manufacturers. Certification of a manufacturer includes the operation of its high voltage test laboratory, and at present it is necessary to validate the calibration of the instruments used for third party testing (on behalf of other manufacturers) and will require the validation of test systems, including the *in situ* calibration of high voltage dividers used to measure a.c., d.c., lightning or switching impulse voltages.

Again, there are substantial differences in the ways that these requirements are being interpreted in different member states. Some require much stronger qualification than others, and the result may well be a barrier to trade rather than a "level playing field". The goalposts are constantly moving and it is important that all engineers engaged in high voltage engineering, or equipment design, have a strong, up-to-date grasp of the legislation in place.

I.4 The future of HV engineering

In some countries there is an evident change in demand for the type of power station envisaged. The ordering of new large coal-fired or nuclear stations is in decline due to the long manufacturing and building times. More Combined Cycle Gas Turbine (CCGT) power stations are being ordered and embedded in lower voltage sectors of the electricity networks. CCGT stations have become viable in Europe since the EC changed rules to allow gas to be burned in power stations. However, the move towards gas fired stations has stretched gas supplies and the price of gas may make CCGT stations uneconomic to operate.

It is certain that in the future there will still be a demand for large power stations operating at high voltage. High voltage transmission lines will still be required, operating under both a.c. and d.c. conditions to take power from where it was generated to the centres of load, perhaps many hundreds of kilometres away. Components and plant will be needed for the new stations and transmission lines, and to replace ageing equipment. All the high voltage components and equipment will require to be validated by testing.

At the same time, new test techniques are being developed and new tests proposed. It is important that as much information as possible is extracted from the test results, and that tests give a true insight into the reliability and dependability of the equipment manufactured and tested. The progress in power electronics will allow new concepts in equipment design to be trialled and installed. The use of microprocessor-based instrumentation in testing equipment in the factory, and in monitoring the condition of equipment in service, will offer new opportunities for high voltage engineers to develop novel instrumentation, new techniques and to change and replace existing test procedures with new methods that are both searching and successful in establishing the quality, reliability and dependability of the components and equipment used in the high voltage electric power networks.

There has never been a more opportune moment for engineers to take up high voltage engineering as a career. There is a shortage of engineers qualified in high voltage technology in the industry, and the age spectrum of those engineers in place indicates the need to have newly trained engineers ready and able to replace many of them within the next decade. The opportunities are available, and with the surge in new computer-controlled instrumentation, there will be a most interesting career available that will require engineers with a bias to both heavy and light current engineering, and with experience in both high voltage engineering and electronics. The situation will demand the emergence of both managers and technical experts to undertake the research, development and industrial testing of equipment and of high voltage power systems for the future.

Chapter 1

Electric power transmission and distribution systems: Part 1

I.A. Erinmez and C.J. Morgan

1.1 Introduction

1.1.1 Nature of transmission and distribution systems

In all countries in the world that utilise electricity as a source of energy, there exists a transmission and distribution system of some form. Although both systems carry current at different voltages, the functionality of the two systems provides a clear separation.

A transmission system can be likened to the motorway system in that it interconnects the major load and generation centres within a large geographical area, thereby allowing for the efficient bulk transfer of power. The distribution system, on the other hand, is concerned with the delivery of electricity from the load centres to places of actual demand in much the same way as the local road system enables you to travel from a major centre to your final destination.

The transmission systems operate at voltages of 220 kV and above with other standard voltages of 275 kV, 330 kV, 400 kV, 500 kV and 765 kV. Generators connect onto the transmission system via their generator transformers and directly feed power into the grid of transmission lines (see Fig. 1.1). The major load centres referred to earlier are points on the transmission system where the demand for a smaller specific geographical area is marshalled together. It is at these points that the distribution system starts.

The distribution system is built in a layered structure, each layer decreasing in level of power supplied. The primary distribution system is usually 132 kV, with other standard distribution voltages at 66 kV, 33 kV, 11 kV, 3.3 kV and 240/110 V at the customer terminals. The distribution system, at its highest voltage level, is directly connected to the transmission system via large transformers. The distribution system also differs from the transmission system in terms of the complexity of interconnectivity between load centres. The distribution system

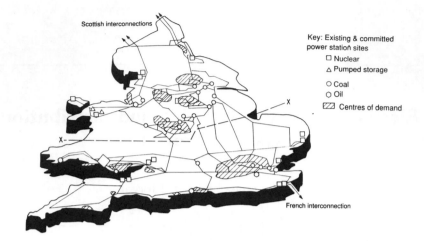

Key: Existing & committed
power station sites
□ Nuclear
△ Pumped storage
○ Coal
○ Oil
▨ Centres of demand

Scottish interconnections

French interconnection

X

X

Fig. 1.1   Major power station sites and centres of demand

tends to be structured as radial outlets from the transmission grid sites. At the end of these radial outlets the lower voltage distribution system, operating progressively at lower voltages, delivers power to the customer via transformers rated to customers needs.

1.1.2 Early developments (1880-1930)

The first large scale use of electricity was initiated by Thomas Edison in 1882 in what became known as "Edison's Illuminating Companies". The most famous of these was the Pearl Street System which was a system comprising of a load of approximately 400 lamps each consuming about 83 W, which provided electric lighting for Lower Manhattan.

At about the same time in England, the Holborn Viaduct Generating Station provided approximately 60kW of power for customers who had mixed loads (i.e. not just lighting loads). It is worth noting that both these systems relied on direct current (d.c.) generation with the power being distributed by underground or surface laid cables.

Throughout the first decade of the age of electric lighting the demand for electric lighting grew at a phenomenal rate, but towards the end of the century came the revolution in the way in which electricity was generated and transmitted. Up to this point in time the direct current generators had been powered by steam engines or complicated mechanical linkages to water wheels. This resulted in the situation whereby the majority of industry dependent on electricity was sited physically close to riverbanks.

There had been much debate on the merits of direct current generation and transmission, but with the invention of the transformers, the advocates of alternating current (a.c.) generation and transmission prevailed and the gradual development of an a.c. transmission system took place.

Gradually, in towns and cities all over the industrial world, generating stations were developed to a point that they were able to supply the demand within their own area. This resulted in the situation whereby each electricity authority defined their own parameters of supply, such as frequency and voltage.

From the turn of the 20th Century, electricity has been seen as a service or product which should be available to all. Its generation and distribution has always been the subject of debate, both by the authorities who produced and sold electricity and by Parliament who laid down the laws for the industry.

There had also been much debate on the reserves of fossil fuel available to the electricity industry and on the effects of pollution, which was excessive in conurbation areas. In 1918, the Williamson Committee reported to the President of the Board of Trade that "the cheap supply of electricity, on town conditions in particular, would be most marked. The reduction of pollution by smoke would result in a lower death rate from bronchial diseases". The Committee also recommended that the concentration of larger generation units in fewer and bigger power stations was the only solution to reducing the costs of industrial power to an absolute minimum. In addition, they recognised the need for more efficient power stations to help conserve fossil fuel reserves.

Along with this report and many others produced for the Government of the day, it was soon realised that something was going to have to be done to the electricity industry to ensure its efficient, economical and effective development. Lord Weir, a leading entrepreneur in his time, was invited to head a committee for the newly elected Conservative government of the early 1920s. His remit was to investigate the problem of electrical energy in the UK. Weir completed his report in four months and considered three ways in which the electricity supply industry might operate. These three suggestions were as follows:

(1)   Power could be bought and sold by a Transmission Board, buying only from the cheap suppliers and allowing inefficient plant to close down;
(2)   The Transmission Board would act as a carrier for electrical power and leave the buying and selling negotiations to the producer and purchaser;
(3)   All generation to be brought under the control of the Board.

All of these suggestions were rejected (although it is interesting to note that the present structure of the UK Electricity Supply Industry is in fact based on a combination of the first and second options).

The recommendation of the Committee which was finally adopted was the establishment of a "GRIDIRON" of high voltage power lines covering the whole country.

1.1.3 Development of the grid concept

In 1926, an Act of Parliament was passed which set up the Central Electricity Board whose responsibility was to build and operate such a Gridiron which would connect the lowest cost generating stations together. For the first time in the UK, customers benefitted from the use of the cheapest power generation stations supplying their electrical energy needs. Between the Electricity Supply Act of 1926 and the Second World War, the idea of public ownership or "nationalisation" of the Electricity Supply Industry gained favour. In 1934, the Fabian Society produced a document entitled "The Socialisation of the Electricity Supply Industry". The principles advocated by this document, whilst causing concern amongst the municipal companies and being regarded as radical, were not adopted until the 1947 Electricity Act. The multiple municipal generation stations were formed into a single Generation Board, whilst the municipal distribution companies were formed into the Electricity Boards.

The first Transmission Grid System, operating at 132kV, was planned and built in the early 1930s and interconnected major generating sites in England and Wales by 1938. Over the first decade or so of Nationalisation, the central planners of the industry had many claims of poor project management laid at their feet for the inability to build new power stations. Demand was growing so fast that proposals were put forward to divide the Grid System into three sections.

Arguments for and against this were many, but the then President of the Institution of Electrical Engineers, T.G.N. Haldane, suggested an increase in the grid voltage. Economic studies carried out then (and still applicable today) showed that it was cheaper to build power stations on local fields and transmit the electricity via high voltage lines. Sense prevailed and plans for the construction of the 275kV Supergrid [1] were put forward in the early 1950s and actioned. Bigger generation stations were built and these tended to be concentrated around the Nottingham and Yorkshire coal fields and the River Trent, from which their cooling water was obtained.

By the early 1960s it became evident that even the 275kV system would not be capable of carrying the predicted power flows and plans for a further increase in the Supergrid voltage to 400kV were considered and accepted [2]. Diagrams showing the development of the grid concept can be seen in Figs. 1.2 - 1.6.

In all this time, when there were vast changes to the transmission system within the UK, the distribution system changed very little in structure. It still essentially took power from the major load centres and distributed it to individual customers. The biggest change that the distribution system has seen has been in the number of customers now connected to an electricity supply. In 1948, one quarter of the domestic properties in the UK were connected to a supply of electricity, (approximately 9.5 million domestic premises). By 1989 that figure had risen to 20 million domestic premises [3]. In the same period, domestic electricity consumption has risen from 11GWh to almost 80GWh.

1.1.4 Current position

Today the physical structure of the Electricity Supply Industry in the UK is basically the same as that set out in the Weir Committee Report of 1925 and as amended by the Fabian Society Report of 1934. Essentially, large power stations

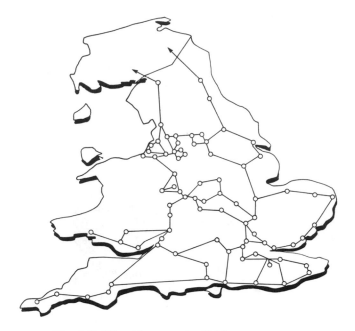

Fig. 1.2   The 132 kV grid system in 1934

are physically sited next to their primary fuel source and the power generated is transmitted to load centres via a high voltage transmission network where it is then distributed to customers via a large and extensive network of overhead lines and cables.

The private generation companies sell their product either directly to a customer or to a central pool where large customers or Regional Electricity Companies can buy power in bulk.

As the owner and operator of the high voltage transmission system in England and Wales, the National Grid Company (NGC) operates within a demanding framework of statutory and regulatory requirements. It has a dual obligation under the Electricity Act:

- To develop and maintain an efficient, co-ordinated and economical transmission system; and
- To facilitate competition in the generation and supply of electricity.

These obligations are specified by the transmission licence, which requires NGC to:

- Schedule and despatch available generation in merit order to meet demand;
- Plan and operate the transmission system to meet defined security and quality of supply standards, including statutory levels of system frequency and voltage;

Fig. 1.3   *The 132 kV transmission system in 1948*

- Administer the settlements process, particularly the complex computer systems needed to calculate payments due as a result of daily trading in the electricity market or "pool".

It was recognised from the outset that freedom of access to the transmission and distribution systems would be the key to effective competition in generation and supply. Major national and international debates have taken place and continue about the practicalities of open transmission access, particularly from the point of view of technical co-ordination and security of supply.

In England and Wales, the overriding aim has been to provide a "level playing field", meaning that would-be entrants to the market, irrespective of location or size, are able to use the transmission system and that old, established players have no privileged rights of access.

Accordingly, NGC must be operationally independent of both generators and suppliers and must be seen to be so by all users of its system. It must be able to earn sufficient revenue from the operation of its energy transport system to ensure that it can invest in new connections and reinforcements, maintain existing assets to required standards and earn a return for its shareholders.

Transparency of information is essential to create the right conditions for open access to the transmission system. NGC makes available to existing and

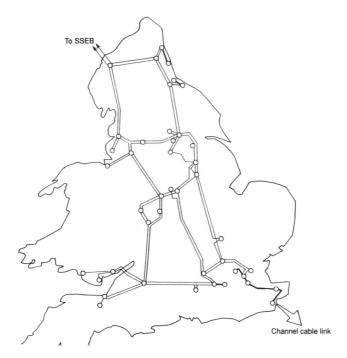

Fig. 1.4   275 kV system for 1961

prospective users alike detailed information including:

*   **Annual Statement of Charges** for connection to and use of the transmission system;
*   **Seven Year Statement** — NGC's snapshot of the development of the system in each of the seven years ahead, which provides essential background information for investment decisions by generators and suppliers;
*   **The Grid Code** — comprehensive technical and operational requirements for all plant connected to NGC's system which help to guarantee access for all users without discrimination.

NGC is obliged by its licence to provide access to its system to all parties who agree to abide by the Grid Code and to pay appropriate charges for their use of the transmission system. Devising equitable charges which accurately reflect the costs to NGC of providing the transmission system has been a major challenge, bearing in mind that before the 1989 Electricity Act transmission and generation were operated as an integrated system, the costs of which were recovered through a single, uniform tariff.

NGC's use of system charges seeks to reflect the investment costs of providing a system which can accommodate the bulk power transfers which result from

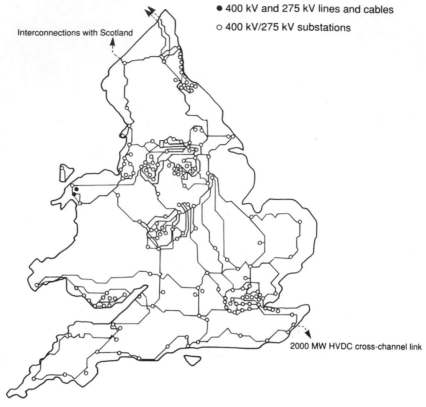

● 400 kV and 275 kV lines and cables
○ 400 kV/275 kV substations

Interconnections with Scotland

2000 MW HVDC cross-channel link

Fig. 1.5   The National Grid System 1990

imbalances between generation and demand in England and Wales.   Zonal differentials in the charges aim to create financial incentives for generators to locate new plant near to load centres where there is inadequate generation, and conversely for demand to locate in zones where there is currently a surplus of generation. The transmission charging principles are regularly reviewed to ensure appropriate financial messages to the users.

Although NGC provides both technical and financial messages to the users on the optimum areas of the system for the location of new generation and demand it is, nevertheless, obliged by its licence to provide connection to, and use of, the system to any applicant, irrespective of their choice of location.   Furthermore, technical and financial terms for connection to the system need to be provided within three months of application by a prospective user — a dramatic change from past practice.   The new commercial environment is a demanding one, requiring multi-disciplinary teamwork bringing together engineers, lawyers, accountants and economists to ensure a timely and effective response to customer needs.

Fig. 1.6   *The Supergrid system*

Now that transmission and generation in England and Wales are no longer centrally co-ordinated, NGC must plan and operate its system, without full knowledge of the intentions of its customers. By contrast with new connections, which are of course signalled to NGC some years in advance, closure of older power stations can take place at short notice. This new background of uncertainty, together with regulatory controls on income from connection and use of system charges, means that the need for new investment in the transmission system is subjected to more rigorous internal scrutiny than ever before and that more effort is being focused on making existing assets work harder. Where the need for new investment is established, procurement practices have changed significantly towards greater flexibility and innovation in transmission system engineering.

As an example, NGC is increasingly investing in static var compensators and quadrature boosters to help control the power flows arising from the changing pattern of generation and demand. The application of these and other flexible a.c. transmission system (FACTS) technologies will enable certain system reinforcements to be deferred, or even removed.

Wherever there is a clear cost benefit transmission line reconductoring, live line working techniques and on-line capability monitoring are utilised to improve

availability of transmission circuits enabling more efficient and flexible use of the
transmission system.

The announcement by the Government that ownership of generation in England
and Wales was to be separated from control of the transmission system gave rise
to concern that the security of the system might be undermined.   Since the
restructuring of the industry (see Fig. 1.7), however, the new arrangements have
been dramatically tested on a number of occasions which have shown that there
has been no reduction in standards:

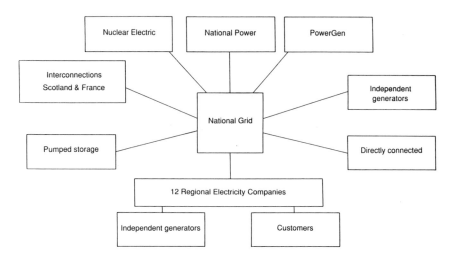

Fig. 1.7   *The current industry structure*

•   Following the televised World Cup semi-final in July 1990, the system
       experienced a record-breaking demand surge of 2,800 megawatts which
       was successfully met by scheduled reserve generation (see Fig. 1.8);
•   In December 1990, severe blizzards and high winds caused widespread
       damage to the transmission and distribution systems, resulting in almost
       800 NGC circuit trips in just 40 hours — as many as we would expect in
       a whole year.   Co-operation between NGC, generators and Regional
       Electricity Companies ensured that loss of supply to customers was
       minimised;
•   In September 1991, a plant failure in France resulted in the loss of the full
       2,000 megawatts import capacity of the cross-Channel link — double the
       instantaneous loss which NGC's system is required to sustain operationally.
       System frequency was restored to normal within two minutes of the
       incident, without loss of supply to customers.

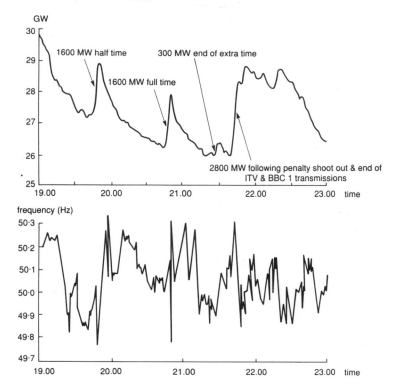

Fig. 1.8   *The effect of screening the World Cup semi-final, England vs. West Germany, in 1990*

System security has been maintained despite the unbundling of the industry because:

- All users of the transmission system, as a condition of access, agree to abide by the technical and operational rules set out in the Grid Code;
- NGC continues to plan and operate the transmission system according to the standards of security observed by the CEGB;
- As a condition of its licence, NGC purchases from generators the ancillary services — reactive power, black start capability and reserve — needed for the maintenance of system security and stability.

However, the most important factor is that all companies — generators, suppliers and network operators — have a vested interest in ensuring high quality of supply to customers.

1.2 Structure of transmission and distribution systems

Electricity, unlike other forms of energy, has several factors which set it aside in the way its infrastructure is built and operated.

Firstly, electricity cannot be stored. The electricity generated must continually meet the level of continually varying demand, plus losses. Failure to maintain this balance would resort in a fall in frequency, which could in turn result in demand being lost to large areas. It is the role of the control engineer to continuously meet the demand by scheduling sufficient generation to maintain the system voltage and frequency. This would not be difficult in an ideal world where demand did not vary on a minute to minute basis, but an indication of the complexity of the problem can be seen from Fig. 1.9 which shows typical demand curves for England and Wales during both summer and winter minimum and maximum demand.

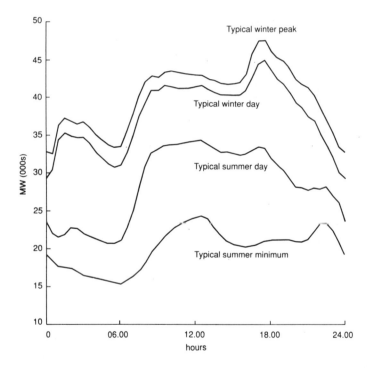

Fig. 1.9   Typical summer and winter demands

Secondly, the worldwide demand for electricity is growing (see Fig. 1.10). Although in some parts of the world electrical energy consumption may have slowed down or reversed into decline, the overall trend still shows an increase. This means that the utility planners have to anticipate this growth several years in advance since generation projects can have long lead times. Identification of the sites of demand growth is also required as distribution systems have to be extended firstly to provide a supply of electricity and secondly to ensure the network is capable of carrying the increase in demand.

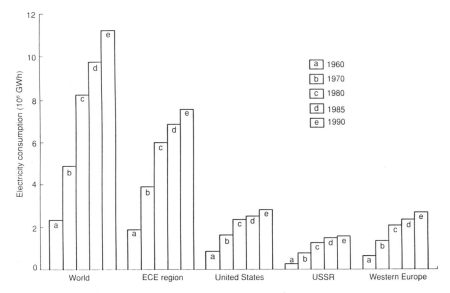

Fig. 1.10   World and regional consumption of electricity, 1960-90. ECE = United Nations Economic Commission for Europe. Source: Annual Bulletin of Electric Energy Statistics for Europe, 1992 (United Kingdom)

Finally, since electricity is generated from a primary fuel, the nature and location of this fuel is an important factor in influencing the structure of a transmission and distribution system.

As already described, electricity generation from coal tends to be carried out very close to the actual coal fields themselves. This is due to the fact that it is more economic to transport energy in the form of electricity over high voltage lines, as opposed to the transportation of coal. Large power stations also have a requirement for large amounts of water for use within the steam cycle and for this reason most power stations tend to be sited either on the coast or by large rivers.

The same principle applies to nuclear generation in that there is normally a wish to keep the nuclear reactors away from population and hence demand centres. Given this situation, it is not surprising to find that the majority of demand is in fact situated distant from generation. This situation will, in all probability, remain since one further factor has come into play from the early 1970s. There has been a gradual emphasis on the environmental effect of power generation and transmission and distribution systems.   This has resulted in the call for undergrounding of overhead lines and for more environmentally friendly generation methods.

1.2.1 Typical characteristics of transmission and distribution systems

A transmission system by definition is a system whereby the bulk transfer of power can take place via a high voltage network which interconnects the major load centres in a country with the main generating stations. Distribution is concerned with the movement of power at lower power levels to the final customer. Generators normally generate electricity at voltages in the region of 11 - 25 kV at a frequency of 50 - 60 Hz. This voltage is then increased in magnitude to the transmission voltage. In the UK, this transmission voltage is 400 kV in the main, but 275 kV is also used in large conurbations (e.g. London, Manchester etc.) whilst the frequency of generation is 50 Hz.

The main transmission system covers the whole country and at strategic points, power is taken via transformers from the transmission to the radial bulk distribution operating at 132 kV (see Fig. 1.11). Although most of the generating stations are connected at transmission voltage levels, some generation still exists at 132 kV and lower voltages.

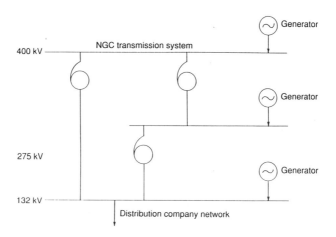

Fig. 1.11   The grid and Supergrid

It is at this point that the real distribution system starts. Voltages, transformed down to 66 kV or 33 kV, carry power to towns or sections of cities. Transformed once again down to 6.6 kV or 11 kV, the power is now distributed to actual load centres, be they housing estates, office blocks or factories.

One of the most important characteristics of a transmission or distribution system is the security of supply, which by implication means a supply of electricity which is continuous and of the required quantity and quality, especially in terms of frequency and voltage. This in turn means that the generation, transmission and distribution systems have sufficient inbuilt flexibility to maintain supplies under conditions of plant breakdown or weather induced failures for a wide range of demand conditions.

The frequency of the system is governed by the speed at which the generating plant operates. The quality of the frequency generated is conditional on the transmission system operator scheduling sufficient generation to meet the demand plus the losses of the transmission and distribution system. This quality is further assured by holding generation in reserve for immediate availability in the event of plant losses. In this way, the level of the frequency can be maintained to predetermined levels. In the UK, the statutory frequency is 50 Hz with ±1 % permitted variation.

The other important characteristic of a transmission and distribution system is the voltage which is also governed by the generating plant. Whilst the frequency of a system will be constant at all points, the voltage levels will differ at different locations on the system. This difference will be governed by the capacitive and inductive characteristics of the system. In low power flow situations, the capacitive effect dominates and causes the voltage at the end of transmission and distribution lines to rise. At high power flows, the inductive effect dominates and the voltage tends to fall along the length of a line. By utilising the effect that causes the high or low voltages it is possible, through the installation of capacitive or inductive compensation plant termed "Reactive Compensation", to negate the wide variations in voltage.

Statutory voltage variations permitted on transmission and distribution systems are usually around ±5 % and ± 10%, respectively.

The above technical principles and characteristics are applicable to virtually any transmission and distribution system around the globe. The only difference between transmission and distribution utilities is related to the organisational structures of the companies which control the various systems.

1.2.2 System organisational structures

In essence, there are four main headings under which electricity supply industries around the world can be classified:

(1)   Vertically integrated;
(2)   Horizontally separated;
(3)   Privately owned;
(4)   Publicly owned (nationalised).

In a vertically integrated industry, everything from generation, through transmission to distribution is controlled by one authority. This authority may be publicly or privately owned, but essentially will be able to carry out integrated generation, transmission and distribution planning and construction within a defined geographical area. Within the UK, the only vertically integrated systems exist in Scotland and Northern Ireland. Electricité de France and ENEL (Italy) are also typical examples of a vertically integrated system.

England and Wales currently has a horizontally separated structure where generation, transmission and distribution companies are separated from each other. This method allows for the private ownership of all, or parts, of each function. However, the transmission system tends to be the essential mechanism by which a horizontally separated system operates in a secure and economic manner. The pivotal role of a transmission company is to provide a transmission and operational infrastructure by which power can be securely transported from one site to another and hence allow for trading between generators and distributors. This type of system allows for the distribution companies and other large customers to buy the cheapest power available from the generators, who can include other interconnected electricity utilities outside the geographical area. Another typical example of a horizontally separated structure is the Spanish Electricity Industry.

Interconnection between two transmission systems is an extension of the philosophy of obtaining the cheapest power available. Here, two systems, which may be physically close together or distant and which may or may not possess the same characteristics (e.g. voltage and frequency), can connect each other together. This connection may be in the form of a high voltage a.c. overhead transmission line, as in the case of the England/Scotland Interconnection, or it may be a d.c. link as in the case of the cross-Channel interconnector with France.

Although three phase a.c. overhead lines is the usual method of interconnection between utilities, in cases where long transmission distances and/or sea crossings are involved, High Voltage Direct Current (HVDC) transmission is preferred despite the relatively higher costs of convertor terminal equipment. HVDC transmission is also utilised for interconnecting utilities with unequal frequency of supplies, e.g. 50 Hz and 60 Hz and in cases where an asynchronous link is required.

A combination of all the above organisational structures exists in the United States and Germany. Here, each geographical area (e.g. a state or city) is served by its own power company which will tend to be of a vertically integrated nature, but tie lines or interconnectors exist between power utility companies and/or groups of companies forming energy trading pools to enable them to take advantage of a more economic supply of electricity.

1.3 Design of transmission and distribution systems

Transmission and distribution systems are designed to have three basic principles:

(1)   Economy;
(2)   Quality;
(3)   Reliability.

An economic system is one where adequate and cheap sources of electricity exist and the ability to transmit power to the customer can be carried out effectively and efficiently. The quality of a system is defined as a measure whereby standards are maintained and the customer is continually satisfied. Finally, the supply of electricity has to be reliable, i.e. it must be continuous and available when needed.

The quality and reliability of a power system can be summarised in one phrase: security of supply.

1.3.1 Security of supply

Security means providing the customer with a supply of electricity which is continuous (i.e. uninterrupted except in exceptional circumstances), and is of the required quantity and of defined quality (i.e. frequency and voltage). This in turn means that the generation, transmission and distribution system must have sufficient flexibility to maintain supplies under conditions of plant breakdown or weather-induced fault failures for a wide range of demand conditions. Interruption of supply can also result from insufficiency or unavailability of generation, transmission or distribution capacity.

Security of supply is potentially at its greatest when the source of power and voltage is close to the demand that it supplies. However, other factors may require that supplies are provided over the transmission system from a more distant point. To provide security of supply in such cases, the main transmission system is designed so that the transmission capacity between any two parts of the system is sufficient to withstand the loss of any one or two circuits at time of peak demand depending upon the particular utility practice.

1.3.2 Transmission system capability

Since October 1938, the national grid has been utilised to allow generation surpluses in one part of the country to supply demand in other parts of the country where there is a generation deficit. In assessing the capability of the system to meet this task, the system designer splits the system into predominantly importing or predominantly exporting areas. The transmission circuits connecting such areas together tend to constitute the weakest 'links' in the system and thus reflect the thermal capability of the system to accept bulk power transfers. These circuits which link areas together constitute system boundaries. As an example, NGC adopted a zonal allocation of plant and transmission capacity comprising 14 "tariff" zones with nine system boundaries. Fig. 1.12 shows these tariff zones with the cost of connecting 1000MW of demand in each zone.

Three factors can limit the capability of the transmission system to transfer power across a system boundary, namely thermal ratings, voltage and stability, and these are discussed in turn.

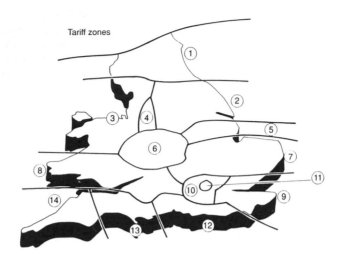

Fig. 1.12   *The 14 tariff zones (with effect from April 1st 1993)*

1.3.2.1 Thermal capability

The amount of power which can be transferred across a boundary on the system is limited by the **rating** of the individual circuits and the way in which the power transfer is shared between them. The "firm" thermal capability (i.e. the capability after the loss of either one or two circuits, dependent upon the chosen security criteria) is usually less than the sum of the individual ratings of the remaining circuits; this is because one circuit usually reaches its rating limit before others due to the resulting unevenly distributed power flows.

1.3.2.2 Voltage

At times of winter peak demand it is sometimes necessary to restrict power transfers to a level lower than the firm thermal capability in order to ensure that satisfactory voltages can be maintained in the importing area, especially under contingency outage conditions. However, at other times the firm thermal capacity will usually apply.

1.3.2.3 Stability

Power transmission capability between two areas or between a major generating station and the system can be limited by considerations of electrodynamic stability, rather than by the thermal capability of the connecting transmission lines. Two stability regimes are usually defined: *transient stability* following severe disturbances, for example a network fault, and *steady state stability* which concerns the response to small disturbances such as the normal random load fluctuations.

1.3.2.3.1 Transient stability. Generators will remain transiently stable if — following a large disturbance such as a nearby fault — each generator settles down to a new steady operating condition, i.e. continues to operate at the same mean speed. During the fault the electrical output of each generator will be substantially less than the mechanical power input from the turbine. The excess energy will cause the generator rotor to accelerate and start to electromechanically oscillate (swing) against other generators. Provided that the faulted circuits are disconnected quickly (80-100 milliseconds), and adequate transmission remains, generator voltage and speed controls will respond quickly and steady operation will be achieved. If, however, the fault persists or inadequate transmission remains, then large cyclic exchanges of power between the generators on the system will occur. This is likely to cause extensive damage to the generator and system break-up through the forced operation of the system protection.

1.3.2.3.2 Steady state or dynamic stability. Large disturbances are relatively infrequent, but small disturbances arise frequently as a result of normal load variation and switching operations. The damping of such oscillations is a function of the transmission system design, generator excitation control design and generator and demand technical characteristics. Problems with undamped oscillations, which can lead to loss of the synchronism between generators, have been experienced in the UK and elsewhere recently.

System transient stability is assured through use of:

- Fast protection system which ensure faults do not persist;
- Braking devices which assist deceleration;
- Series compensation devices which improve the connection adequacy of the transmission system;
- Shunt compensation devices which improve voltage levels throughout the system;
- Fast generator excitation systems which rapidly restore the ability of the generator to transmit power following a large disturbance;
- Use of Power System Stabilisers on generator excitation control systems and shunt compensation devices which improve damping performance;
- Adjustments to system configuration.

1.4 Operation

1.4.1 Introduction

Grid system management, or system operation, is concerned with the routine day to day operation of a power system. Within the sphere of grid system management, four distinct timescales can be identified which are essential to the safe and economic operation of a power system. These timescales are as follows:

- Operational planning, extending from several years (say 3 to 5) down to a few days ahead;

- Extended real time, extending from a few days to say a few hours ahead;
- Real time, extending from a few hours ahead to an hour or a few hours after the event;
- Post-operational event, covering collection and analysis of data from actual operation, which may extend from a few hours to a few months after the event.

Different organisational structures often exist to carry out the functions of planning and operating a power system, but sometimes these functions overlap. For example, operational planning engineers may be involved in the design of the power system — especially when some projects can be operational in a very short timescale, an example being the connection of CCGT power stations which can be constructed and be operational within 24 months. On the other hand, the operational planning engineers prime interest will be ensuring the security of a system given constraints such as generation outages and circuit maintenance outages.

As stated above, there are four main timescales in which a system is operated. These timescales will be described in turn giving the main objectives of operating a major power system.

1.4.2 Operational planning

The task of operational planning can be considered in terms of the primary objective — to achieve secure cost operation of the power system at minimum cost. Such an objective involves many decision areas, as summarised below.

1.4.2.1 Demand forecasting

Demand forecasting is fundamental to all aspects of predictive system operation work. Together with the forecasting of plant availability, these provide the basis for all decisions on resource needs.

Demand estimates are required over the whole of the operational planning timescale (several years to one or two days lead times). Power and energy estimates will be needed for the whole utility with a geographical breakdown of the power estimates (probably to individual bulk supply points) and time profiles, or at least demand duration histograms. Estimates of demand power factor or reactive power, with geographical and time profiles, will also be needed.

1.4.2.2 Plant availability forecasting

This is complementary to demand forecasting and with it provides the basis for such decisions as plant maintenance programmes, level of plant scrapping, fuel requirements and distribution, and trading. Forecasts are required for the whole operational planning timespan. As far as known, estimates are based on judgement from past operating results, levels of maintenance expenditure, experience on similar plant, age of plant and other commercial/economic factors.

1.4.2.3 Generation and transmission outage planning

Generation and transmission outage plans are required to programme manufacturers' and utility maintenance resources, as input to loading simulation and network security studies, as input to studies with shorter lead times and finally for the preparation of outage and switching schedules in real time. Programmes will be required for the whole of the operational planning period, in outline for the longer lead times and in detail for say lead times of one year and less. These programmes will need to include any transmission construction related outages on the system. Computational techniques used will include pre- and post-fault loading simulation (using both d.c. and a.c. analysis, where the dc option is often used as a filter to select only "severe" contingencies for ac assessment), dynamic analysis (including detailed excitation, power system stabiliser and governor modelling) and possibly some form of mathematical optimisation which helps place outages at minimum operating cost. In general, the generation programme will be the dominant factor, with the transmission outages arranged to have minimum or zero impact on the economic operation of the remaining generation.

An important end product of operational planning is to provide advice to the control staff on expected operating conditions with supporting information. Typically, this will include for the day(s) ahead:

- Generation incremental costs/merit order;
- Outage programmes;
- Expected available generation;
- Expected peak and minimum demands;
- Preferred network configuration including constraints on power flows and remedial switching in the event of faults;
- Preferred voltage profile and reactive sources;
- Any special situations;
- Demand forecasting, loading simulation and network analysis.

Operational planning is also concerned with providing the control engineer with data which is of a longer term nature.

1.4.2.4. Fuel allocation and energy modelling

A fuel allocation model may be needed to ensure that fuel is distributed at minimum source plus transport costs to meet the station's needs overall. This is a transportation type problem and is normally encountered in an organisation which controls both generation and transmission systems (e.g. a vertically integrated utility).

1.4.2.5 Protection settings

Although types of protection are likely to have been settled at the planning stage, the normal settings will need to be determined together with any changes necessary to meet given outages and phases of any transmission construction and maintenance programme. This often requires a more detailed analysis than that necessary at the planning stage.

1.4.2.6 Automatic system protection

This term is used to cover such post fault features as automatic reduction of generation or circuit switching to reduce network loadings, disconnection of demand by low frequency and/or rate of change of frequency relays. The need for some of these may have been identified at the planning stage but others may have to be introduced to accommodate changes in construction and maintenance programmes. In some cases, adequate time may be available for remedial actions to be initiated by the operators, in which case operational instructions will have to be prepared.

In these cases, network analysis will be mainly used, and for the low frequency studies dynamic analysis extending over several seconds.

1.4.2.7 Preparation for abnormal situations

It is sometimes necessary to prepare contingency plans for possible periods of abnormal operation, for example interruption in the supplies of fuel, loss of communications, exceptionally severe weather, restoration of supply following a large scale loss of supply.

The emphasis in such studies will be on maintenance of supply when faced by shortage of resources. Demand prediction, loading simulation (perhaps with different objectives) and network analysis will all be needed. Quite abnormal operating states may need to be analysed.

1.4.2.8 Operational standards

One of the functions of operational planning is to periodically review past operating experience and in the light of this, of the expected development of the system, and of the policies of neighbours, to decide whether any changes are needed in the operational standards of security and quality of supply. In order to analyse the impact of any changes, sensitivity studies covering ranges of systems conditions over a number of years using the methods and programs of the sections above will be required. Probability analysis may also be used (e.g. in assessing running spare requirements).

1.4.2.9 Operational memoranda and procedures

It is essential that operational memoranda and procedures are updated in line with system and organisational developments. This is likely to be done jointly by operational planning and control centre support staffs. It is unlikely to involve much computational work.

1.4.2.10 Facilities for operational planning or real time control

Experience shows that Energy Management System (EMS) and System Control and Data Acquisition (SCADA) systems have been updated, perhaps replaced, at intervals of approximately 5 to 15 years. The lead time from first thoughts to commissioning is likely to be between 4 and 8 years. Grid system management will provide a main input to the user and functional specifications.

In the authors' experience, facilities for operational planning are also likely to need enhancement more frequently than the real time facilities. Again, grid system management will provide a main input to the specifications.

1.4.2.11 Computational tasks

It is evident from the above that in addition to demand prediction, the main computational tasks are loading simulation and network analysis. The former is, in one form or another, an essential component of all predictive studies, e.g. for economic studies to estimate operating costs and for network security studies — to estimate plant outputs and, together with demand estimates, transfers at substations.

The information required will be the operating state and output of each generating unit at specified times of the day for one or more days, or at specified load levels. Invariably some form of optimisation will be involved, even if this is only summation down a merit order to meet a given demand at lowest cost. The problem will also usually be constrained, for example by network capability, system reserve needs, generation response limits (in increasing order of detail and importance).

Network analysis will in one form or another be an essential component of most predictive studies. There will be three main purposes — to confirm that the proposed operating conditions will provide adequate security of supply, to determine the capability of the network and hence any constraints on system operation (or demand) imposed by the network, and finally to assess transmission losses and hence any adjustments of generation to achieve minimum cost operation.

1.4.3 Extended real time analysis

Extended real time is the period which runs from one or two days prior to real time to about one hour ahead. Its function is to complement the operational planners by modifying the information received to take account of changes that have occurred (e.g. in demand level, plant availability and outage conditions), and detailing it to a very high level.

The four main tasks of extended real time analysis are demand forecasting, scheduling of generation, trading and network security analysis. In each case, the level of data required for each task will increase significantly. For example, demand forecasting data is required on a half hourly basis, generator loading data requires synchronising/desynchronising times with outputs over each half hour and network analysis data requiring to know the present network configuration and contingencies for certain fault conditions.

1.4.4 Real time operation

The objectives of real time operation are:

- To ensure that a supply of electricity of adequate security and quality at minimum cost is provided to consumers;
- To provide necessary access to plant and system for maintenance, repair and new construction;
- To minimise the effects of disturbances.

These tasks are achieved with the help of the following systems.

1.4.4.1 SCADA systems

To run any power system in real time requires knowledge of actual network topology and power flows. This is provided by SCADA systems. The essential SCADA functions will be the acquisition and display of current system information, normally with a cycle time of a few seconds. The essential telemetered data will include equipment states, flows, voltages, frequencies, alarms for status changes, protective gear operations and possibly operating variables outside limits. A hierarchy of displays will be used from block diagrams of the system showing basic operating quantities in geographical areas to system diagrams, substation and circuit operational diagrams and sometimes substation and circuit safety diagrams. Alpha-numeric displays of status changes and alarms will be provided, with provision for acceptance.

Monitoring of the values of the telemetered quantities is essential to check whether any operating quantities are outside limits. This will be additional to and more comprehensive than alarms generated in the substations.

Not essential, but often very desirable, will be the ability to telecommand the operation of equipment from the control centre. This may include adjustment of starting up of emergency generation, circuit breaker operation, tap changing, changing of protective gear status, demand disconnection and reconnection.

A specific and increasingly discussed topic nowadays is alarm handling and fault analysis, i.e. to determine from the alarms given what system events have occurred, noting that there is likely to be more detail than needed in the alarms, that some may be incorrect, and that although alarms are correct some of the secondary equipment (e.g. protective gear) operations initiating these may have been incorrect.

With the increasing use of remote control and demanning of substations, the control engineer may no longer be able to complement his or her indications by discussion with the substation attendant which, although time-consuming, has proved invaluable in many situations in the past.

1.4.4.2 Automatic generation and tie line frequency control

Utilising a control signal from the deviations of frequency from nominal and tie-line flows from target values, this control is utilised to automatically adjust generation outputs such that the frequency and tie-line flows between utilities are kept within specified limits. Although this process is usually required to be automatic, if the specified limits are within a narrow bandwidth it is possible to implement the process to a wider bandwidth of limits by manual dispatch.

1.4.4.3 Generation scheduling and economic dispatch
This is the process of determination of output of all plant, whether synchronised or unsynchronised, to minimise the total cost of generation. It is a multiple constraint minimisation problem although many systems even now only minimise the cost of generation subject to meeting the total system demand. In the case of mixed hydro-thermal systems, it has been common practice to assign to the hydro-plant whose running is optional (i.e. storage plant) an incremental cost which causes this plant to be operated so as to use the water run-off determined from longer term studies. This process should include system and plant constraints to achieve an "optimum" solution and is usually achieved through dynamic/integer programming combined with heuristic methods.

1.4.4.4 State estimation
This is the process of determining a complete and consistent set of variables representing the state of all transmission system components which best fit (e.g. by a least squares criterion) the telemetered data. It will include as a first stage some form of network configurator which will attempt to determine a network configuration most likely to be that of the actual system.

Some form of network configuration and state estimation are essential if contingency evaluation is to be implemented.

1.4.4.5 Contingency evaluation
This is a desirable function in which, hopefully, the effects of all potentially critical outages classified as credible contingencies are evaluated. It will comprise at least a series of load flows one for each of the outage cases, with the resulting flows and voltages checked against limits and, if outside these, alarmed. The contingency analysis will include a procedure to check system fault levels.

1.4.4.6 Trading and accounting
The control engineer must operate his or her system so as to fulfil longer term contracts and probably take advantage of any differences in generation costs between his/her and neighbouring systems to undertake opportunity trading. This will require knowledge of marginal cost levels, and network flows and capability, derived from the economic dispatch, monitoring and contingency evaluation facilities.

1.4.4.7 Load management
This is the ability to control demand, by prior agreement with the consumer, so as to reduce the operation of very expensive generation or to avoid low frequency operation and/or forced disconnection of demand. The main decisions in implementation of load management are when and how much, which are essentially simple decisions in economic and computational terms.

1.4.4.8 Containment of disturbance and restoration of normal conditions
An essential part of a control engineer's work is to restore the system to a viable steady state after a disturbance. In rare cases, the disturbance may have developed

to the extent that demand is disconnected, or the system split or large amounts of generation lost. In this event the first priority will be to stabilise or secure the situation (e.g. to relieve overloads), the second priority to restore transmission preparatory to restoring demand, and finally to restore demand as the build up of system capacity permits.

The task of the operator will be significantly eased by auto-reclose facilities on individual circuits and, in a few utilities, by automatic restoration of circuit paths through a network. Otherwise, the operator must at present largely rely on previously evaluated strategies and procedures.

1.4.5 Post-operational events

1.4.5.1 Post-event tasks
The statistics of past operation are the most important source of data for estimating future commitments and requirements. They also provide the raw data by which operational and general management can monitor the efficiency with which many aspects of the utility's work is being done. A further area of work will be the analysis of system performance. In abnormal events, such as major loss of supply it may be possible to determine recommendations to avoid similar recurrences. The data collected, and in particular the analysis done, will be particular to each utility but a representative sample is given below.

1.4.5.2 Routine on-line data
The on-line data collected may include those listed in Table 1.1, most or all of which will be logged automatically via the SCADA System.

Table 1.1   On-line data collection and analysis

	Quantity	Analysis
(1)	Frequency	No. of times and duration outside limits
(2)	Tie line flows	Deviations outside limits
(3)	Circuit flows	Flows outside limits, histogram of loadings
(4)	Voltages	Voltages outside limits
(5)	Generator outputs	Deviations from instructed values, plant flexibility
(6)	Demands met	Maximum and minimum demands, geographical distribution, profile, sensitivity to weather, frequency and voltage
(7)	Power and energy traded	Trading, evaluation of losses and billing.

Table 1.2   Operational data and assessment for performance analysis

	Quantity	Analysis
(1)	Demand forecasts ⎫	Errors in forecasts; errors in estimated
(2)	Generation forecast ⎭	margins
(3)	Assessed total system cost	Computed from generator outputs and merit order
(4)	Generation schedule versus actual operation	Out of merit operational costs due to plant breakdown
(5)	Transmission constraints and costs	The cost of out of merit operation due to transmission constraints
(6)	Actual generation and transmission outages	Comparison with programmed outages; generator and circuit availability
(7)	Various forms of ideal operation	Efficiency of the scheduling and dispatch process, including control centre and station performance
(8)	Pumped storage operation	Loading cycles on plant, savings from pumped storage
(9)	Two-shifting of plant	Plant flexibility
(10)	Load management instructions ⎫	Quality and reliability of supply;
(11)	Frequency and voltage ⎬	and planning standards
(12)	Interruptions of supply ⎭	

Table 1.3   Data collected and assessment of abnormal situations

	Quantity	Analysis
(1)	Protective gear performance	Reliability, maintenance needs, type or installation problems
(2)	System response to generation losses	Time response of the system to sudden generation-demand imbalances
(3)	Low frequency relay ⎫	Reliability of supply, adequacy of planning
(4)	Instructed reductions in demand ⎬	and operational planning margins, adequacy of capital programmes

1.4.5.3 Performance analysis

Objectives of performance analysis will include estimation of the efficiency of operation and accuracy of forecasting, plant performance and characteristics, and reliability and quality of supply. The data used will generally be from operating statistics and predictive assessments, see Table 1.2.

1.4.5.4 Abnormal occurrence data

The data for the monitoring and analysis of abnormal situations will come from on-line sources and staff reports; the main purpose being to obtain data on system performance in such conditions for future design work and to assess what plant, control facility or organisational changes would be valuable, see Table 1.3.

A special case will be the analysis of conditions leading up to and during a major disturbance. This will require the collection of system wide data — telemetered, manually logged and for example transient recorders — followed by detailed analysis of all aspects — load flows, transient and dynamic stability, protective gear operations. Time tagging of telemetered data is particularly valuable in these circumstances.

1.4.5.5 Summary of operational tools

It can be seen that the efficient and secure operation of a power system is a complex and demanding task. Many tools, often common across the operational and planning timescales, have been developed to aid the decision-making process to aid the engineer to complete his/her objectives.

These tools include calculation methods for:

- Demand prediction;
- Generation scheduling (including dispatch);
- Economic dispatch;
- Fuel and energy modelling;
- Assessment of generation/operating costs;
- Assessment of alternative operation methods;
- Load flows and short circuit levels;
- Network configuration and state estimation;
- Transient stability and protection performance;
- Steady-state stability behaviour;
- Longer term dynamic behaviour.

Usually most of these tools are utilised in computing platforms in an off-line mode. However, increasingly tools like state estimation, load flow and transient stability prediction are being utilised on an on-line basis with a clear trend towards achieving real time simulation.

1.5 Future developments

1.5.1 Introduction

In recent years, there have been many developments in the electricity supply industries of various countries throughout the world. These developments have resulted in independent power producers and consumers demanding easier access to transmission systems to take advantage of larger markets and cheaper electricity. This access, known as "Third Party Access" (or T.P.A.) has resulted in transmission companies radically rethinking their strategy with regard to development of their systems.

Environmental considerations have also resulted in severe restrictions in the building of new overhead lines, with calls being made for the undergrounding of existing overhead lines. Transmission companies are being forced to find ways of working their assets harder and finding methods of utilising the "wires" to a much higher extent. This has resulted in the development of refurbishment techniques for overhead lines and substations as well as Flexible a.c. Transmission Systems (FACTS) devices.

Managerial structures are also changing from centrally planned, vertically integrated systems to horizontally structured systems, with public ownership of many, or all, parts of the structure. With this changing structure, transmission companies are quickly becoming the vehicle for competition in electricity supply and are increasingly subject to legislation to ensure Third Party Access is allowed on a fair and transparent basis.

1.5.2 Flexible a.c. transmission systems

Since alternating current transmission systems won the debate over direct current systems over a century ago, engineers have made efficient use of the a.c. system to move large amounts of power over great distances with relatively small losses. However, they have been plagued with the problem of controlling the power flow, especially during transmission outage periods, as well as in connecting new generation on the system.

1.5.3 Present status of FACTS devices

At present, many utilities across the world are participating in the development and use of early forms of FACTS devices. Examples of these devices include:

- *Shunt compensation* either in the form of shunt capacitance or shunt inductance to support the voltage at a node (capacitance will raise the voltage whilst inductance will reduce the voltage). These devices tend to be controlled by mechanical switches, and their effect is seen in blocks of discrete levels of compensation.

- *Static var compensators.* These devices are similar to shunt compensation devices, except that the inductance and capacitance are controlled by thyristors. They can, therefore, provide any amount of reactive compensation within their design limits at a very fast rate and have the potential to vary depending upon system conditions. For example, if installed on the end of long, heavily loaded transmission lines, SVCs will help to maintain the voltage at those nodes between preset limits for a range of operating conditions, including post fault voltage recovery. Examples of MSC/SVC installations on the NEC system are shown in Fig. 1.13.
- *Quadrature boosters.* These devices are similar to transformers in nature, but have one major difference. A second winding is wound on each limb. This winding is energised from another phase, thus giving a quadrature voltage injection into the three phases. This allows the control of power flow along circuits and power flow sharing between parallel circuits.
- *Series Capacitors.* These devices are large capacitors, sited along the length of long transmission lines which help to reduce the series impedance of a line, thus allowing more power flow through that line.
- *HVDC technology.* This technology is utilised where non-synchronous ties between different utilities or parts of the same transmission system are required. The most obvious cases of this are where the two systems have different frequencies or voltage levels, or are separated by seas or oceans. HVDC is also a tool that can be used within an a.c. transmission system to help influence the power flows between two zones since a.c. overhead lines have the capability to be converted to d.c. lines and thus no new circuit or cable is required.

1.5.4 Future developments in FACTS technology

New ideas for the control of a.c. power systems are emerging year by year. An indication of the types of equipment we may see in the future can be seen below:

- *Series capacitor technology* can be modified to include thyristor switching. This offers the ability to enhance system damping and avert the occurrence of system resonance conditions at sub-synchronous frequencies.
- *Thyristor/GTO converter technology* gives rise to controlled components within power systems that may inject a series voltage into a transmission line in much the same way as a quadrature booster, but also provides shunt compensation at the same time without utilising inductive and capacitive elements.
- *Thyristor/GTO technology* also offers the possibility of introducing faster electronic switches to replace mechanical switches controlling existing shunt compensation, quadrature boosters and series capacitors thereby improving the response of the transmission system and its controllability.
- *Optical fibre technology* offers the power system engineer the ability to provide fast signalling channels from the control room to the substations

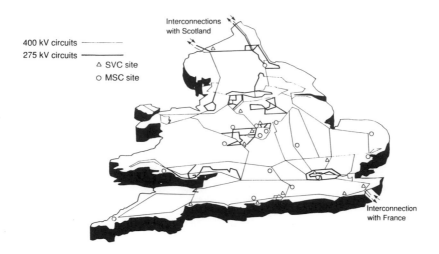

400 kV circuits ————

275 kV circuits ————

△ SVC site

○ MSC site

Interconnections with Scotland

Interconnection with France

Fig. 1.13   Existing and authorised SVCs and MSCs

for protection, plant control and communications.   These signalling channels will take the form of optical fibres embedded within the earthwire of the transmission line.

1.6 References

1   SAYERS, D.P., FORREST, J.S. and LANE, F.J.: "275 kV Developments on the British Grid System". *Proceedings IEE 1952*, **99**, **Pt. II**, p.582

2   BOOTH, E.S., CLARKE, D., EGGINGTON, J.L. and FORREST, J.S.: "The 400 kV Grid System for England and Wales". *IEE Proc. A*, 1962, **109**, (48), p.493

3   "Handbook of Electricity Supply Statistics, 1989" (Electricity Association, London)

4   "Electrical Transmission and Distribution Reference Book" (Westinghouse Electric Corporation)

Chapter 2

Transmission and distribution: Part 2

T. Irwin and H.M. Ryan

2.1 Introduction

The international standard covering insulation coordination is IEC publication No. 71. According to the IEC, insulation coordination comprises the selection of the electric strength of equipment and its application, in relation to the voltages which can appear on the system for which the equipment is intended and taking into account the characteristics of available protective devices, so as to reduce to an economically and operationally acceptable level the probability that the resulting voltage stresses imposed on the equipment will cause damage to equipment insulation or effect continuity of service.

The operation of a transmission system will present the system insulation with various types of voltage stress from the continuous voltage at which it must operate to various forms of overvoltage.

Overvoltages can be classified into two groups. The first group is externally generated overvoltages from lightning which manifests itself in two ways; by direct strikes to the lines or substations and by strikes to the towers or overhead earth wires from which back flashover voltages from the towers to the conductors can occur. The second group comprises internally generated overvoltages due to switching. These include the switching of capacitive and electromagnetic loads and travelling waves due to the energising of transmission lines. Also taken into consideration are power frequency overvoltages which are caused either by load rejection or by voltage changes on the two healthy phases when a single phase fault occurs.

Up to about 300 kV experience indicates that the highest voltage stress arises from lightning. For transmission systems above 300 kV the switching overvoltages increase in importance so that at about 500 kV the point has been reached where they are equivalent to that of lightning overvoltages.

Power frequency voltages are important since they affect the rating of protective arresters or coordinating gaps which provide a means of controlling the overvoltages for the purpose of insulation coordination.

The complexity of insulation coordination may be visualised from the flow diagram shown in Fig. 2.1 which indicates some of the factors involved and their inter-relationship. The procedure takes an iterative approach where several scenarios for insulation coordination may be under consideration. Evaluation of the dielectric stresses imposed on the insulation may be carried out using a range of computer models determined from the system characteristics to enable the selection of the insulation requirement. To assist with the selection of insulation levels IEC 71 has "Standard Insulation Levels" an extract from which is shown in Fig. 2.2 and Table 2.1 for system voltages in the range 300 kV to 525 kV.

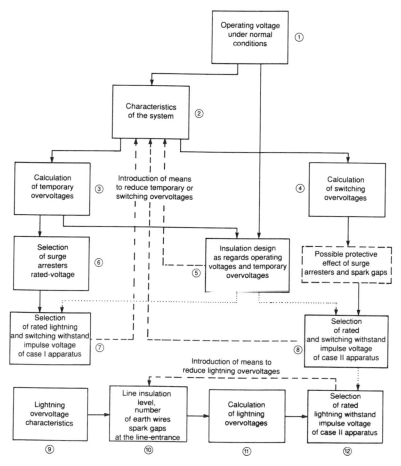

Fig. 2.1   Block diagram of insulation coordination and design

2.2 Classification of dielectric stress

The following classes of dielectric stresses (see Fig. 2.3) may be encountered during the operation of the transmission system:

(1) Power frequency voltage under normal operating conditions;
(2) Temporary overvoltages;
(3) Switching overvoltages;
(4) Lightning overvoltages.

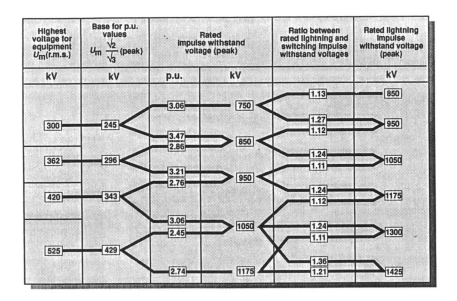

Fig. 2.2   Standard insulation levels

2.2.1 Power frequency voltage

Under normal operating conditions the power frequency voltage can be expected to vary somewhat but, for the purpose of insulation coordination, is considered to be constant and equal to the highest system voltage for the equipment.

2.2.2 Temporary overvoltages

The severity of temporary overvoltages is characterised by amplitude and duration. This overvoltage is a significant factor when considering the application of surge arresters as a means of surge voltage control. Although the surge arrester is not

Table 2.1　　Standard insulation levels from IEC 71

Highest voltage for equipment U_m	Standard switching impulse withstand voltage			Standard lightning impulse withstand voltage
kV (r.m.s. value)	longitudinal insulation (+) kV (peak value)	phase-to-earth kV (peak value)	phase-to-phase (ratio to the phase-to-earth peak value)	kV (peak value)
300	750	750	1.50	850 950
	750	850	1.50	950 1050
362	850	850	1.50	950 1050
	850	950	1.50	1050 1175
420	850	850	1.60	1050 1175
	950	950	1.50	1175 1300
	950	1050	1.50	1300 1425
525	950	950	1.70	1175 1300
	950	1050	1.60	1300 1425
	950	1175	1.50	1425 1550

used for the control of temporary overvoltages it results in a considerable increase in the resistive component of leakage current in a metal oxide gapless surge arrester. This results in a temperature rise within the surge arrester and the possibility, if left long enough, of arrester failure.

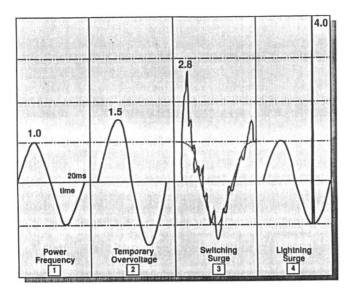

Fig. 2.3   Classification of dielectric stress

Temporary overvoltages arise from (i) earth faults, (ii) load rejection and (iii) resonance and ferroresonance.

When considering earth faults, the voltage on the healthy phases will rise during the fault period to values approaching line voltage dependent on neutral earthing arrangements. Typically for 400 kV systems a value of 1.5 times normal system phase voltage can result for periods up to 1 s by which time the fault is usually cleared.

Load rejection can, on long uncompensated transmission lines, produce voltages of 1.2 times nominal system voltage, due to the Ferranti effect, at the substation end of the line which is disconnected from the source (i.e. remote end). The Ferranti effect is caused by the current flow through the line shunt capacitance which produces a voltage across the source and line inductance which increases with distance from the source end of the line.

Temporary overvoltages due to resonance and ferroresonance conditions generally arise when circuits with large capacitive elements (transmission lines, cables etc) and inductive elements (transformers and shunt reactors) having non-linear magnetising characteristics are energised or through sudden load changes. Parallel line resonance can also occur during de-energisation of one circuit of a double circuit transmission line which has shunt reactive compensation. The energised line feeds the resonance condition through the intercircuit capacitance and voltages as high as 1.5 times nominal system phase voltage have been recorded on 400 kV systems [1]. This voltage will remain at this level until the line is energised or until the compensating reactor is switched out. Magnetic voltage transformers can also produce ferroresonance conditions but usually these are sub-third harmonic and the resultant voltage is close to nominal system voltage [2]. This resonant voltage will not present any problem for the insulation but the VT primary current will be many times the nominal current and the resultant heat generated in the primary winding would be of prime concern for the VT insulation. However, fundamental ferroresonce can also occur with VTs and voltages in excess of 2 times nominal system voltage have been reported. If ferroresonant conditions are indicated during a system study then design modifications may be considered or steps can be taken to avoid the switching operations that cause them or to minimise the duration by selection of an appropriate protection scheme.

2.2.3 Switching overvoltages

Switching overvoltages according to IEC 71 can be simulated by a periodic waveform with a front duration of hundreds of microseconds and a tail duration of thousands of microseconds. The waveform shown in Fig. 2.3 may be typical of switching surges which have in practice a decaying oscillatory component superimposed on the power frequency waveform. Of major importance are the switching surges produced by line energisation and re-energisation which cause travelling waves on the transmission line and are most severe at the end remote from the switching point. This will be discussed later in more detail.

2.2.4 Lightning overvoltages

Lightning overvoltages according to IEC 71 can be simulated by an aperiodic wave with a front duration of the order of one microsecond and a tail duration of the order of several tens of microseconds. The lightning surge voltage is produced by a strike to the tower or earth wire which either induces a voltage in the phase conductors or by back flashover of the line coordinating gap injects current into the phase conductor. A direct strike to the phase wire can also occur if the phase wire is not well shielded by the earthwire. The voltage wave front arriving at the substation can be significantly affected by the line termination and with cable terminations the wave front may be drastically reduced to the extent that it more resembles a fast switching surge. This will be discussed later in the chapter.

2.3 Voltage-time characteristics

In practice the waveform of lightning overvoltages varies considerably, but for the purpose of testing equipment there is an internationally agreed wave shape used which has a rise time of 1.2 μs and a time of decay of 50 μs to 50 % of the maximum amplitude.

Switching overvoltages vary from oscillatory voltages of several tens of thousands of cycles per second to travelling waves with a rise time of up to

1000 μs. It is now generally accepted that the withstand strength of air insulation is a minimum for a wave with a rise time of approximately 250 μs and with a decay time to 50 % of 2500 μs, although there is a tendency for this minimum to occur at longer wave fronts as the air gap increases. Similarly, the minimum under oil withstand strength occurs with a wave front of approximately 100 μs.

Gas insulated substations (GIS) have a virtually flat voltage-time (*V-t*) characteristic from power frequency through to the switching surge range and show an upturn near the 10 μs point.

Fig. 2.4 shows a typical *V-t* characteristic for 400 kV GIS along with a range of dielectric stresses. From the standard insulation levels (IEC 71) 400 kV equipment can have a 1425 kV lightning impulse withstand level (LIWL) with a corresponding 1050 kV switching surge level (SIWL) based on a rated voltage of 420 kV (r.m.s.) i.e. phase voltage peak of 343 kV.

$$\frac{420 kV}{\sqrt{3}} \times \sqrt{2} = 343 kV$$

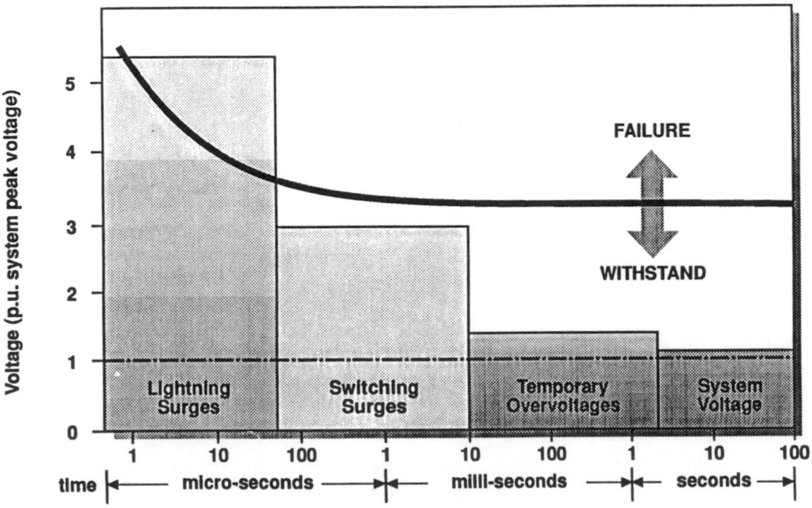

Fig. 2.4   Voltage-time characteristics

The voltage-time characteristic of insulation used in equipment must be carefully assessed when comparing performance with lightning and switching surges. Also the *V-t* characteristics of protective devices must be considered; for example, coordinating gaps in air will not offer practical protection of GIS against switching or lightning surges because the *V-t* characteristics are totally incompatible.

2.4 Factors affecting switching overvoltages

When energising or re-energising transmission lines severe overvoltage can be generated [3]. The overvoltage magnitude is dependent on many factors including the transmission line length, the transmission line impedances, the degree and location of compensation, the circuit breaker characteristics, the feeding source configuration and the existence of remnant charge from prior energisation of the transmission line.

The magnitude of these switching transients is the main factor determining the insulation levels for EHV and UHV transmission systems; consequently reduction in their severity has obvious economic advantages. The two methods used to achieve substantial reduction in transmission line energising overvoltages are resistor insertion [4] and controlled closing of the energising circuit breaker close to voltage zeros [5]. There have, however, been instances where both methods have been combined for overvoltage reduction in a UHV system [6]. Alternatively, for systems up to 500 kV, gapless metal oxide surge arrestors can also be used to reduce the phase-to-earth switching surge overvoltages to below 2.4 p.u. of system nominal phase voltage peak.

2.4.1 Source configuration

Source networks can very crudely be split into two types: (a) those with purely inductive sources, i.e. where no lines are connected to the energising busbar, e.g. a remote hydro station; and (b) those with complex sources, i.e. with lines feeding the energising busbar or a mixture of lines and local generation.

Much work has been done over the years on both types of sources [7, 8, 9]. Trends are more easily defined with the simple lumped source and normally slight increases in overvoltages are obtained with increase in source fault level except where resonance conditions are approached with very low source fault levels. With the complex source configuration no general trends exist due to a large number of interacting parameters in the source network.

2.4.2 Remanent charge

The remanent charge on a transmission line prior to its reclosing has a significant effect on the overvoltages produced. The value of trapped charge is very much dependent on the equipment permanently connected to the line, as this determines the decay mechanism (see Fig. 2.5).

If no wound VTs, power transformers or reactors are connected, the line holds its trapped charge, the only losses being due to corona and leakage and thus the decay is very much weather dependent. In good weather conditions the time constant of the decay is of the order of 10-100 s so that no appreciable discharge will occur in an automatic reclosure sequence.

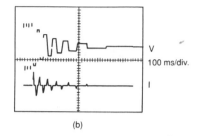

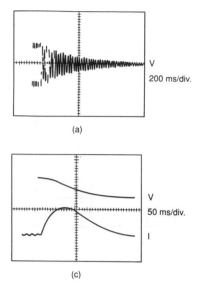

(a)

(b)

(c)

Fig. 2.5   *Decay of trapped charge by:*
(a) reactor
(b) power transformer
(c) voltage transformer

With a power transformer connected to the transmission line the trapped electromagnetic energy oscillates within the RLC circuit formed by the line and transformer. The transformer oscillates between the saturated and unsaturated state resulting in a decay in some 5-50 cycles of the supply frequency. Fig. 2.6 shows a computer simulation of the 3 phase decay of trapped charge with initial overvoltage when the circuit breaker opens due to mutual coupling effects.

The decay mechanism with a wound magnetic voltage transformer is similar to the power transformer but the damping is much more effective due to the high winding resistance, of the order of several tens of k-ohms, of the voltage transformer. Most of the stored energy on the line will be dissipated in the first hysteresis loop of the voltage transformer core mainly due to copper loss, although conditions have been encountered with long transmission lines and wound VTs in SF$_6$ insulated substations where complete dissipation requires 5 cycles of the supply frequency.

When a shunt reactor is connected to the line, on de-energisation an oscillation exists determined by the line capacitance and the reactor similar to the power transformer. In this case, however, there is no saturation and the oscillations are slowly damped due to high reactor Q value. In addition, beat effects are introduced due to mutual coupling effects. The damping time constant is of the order of seconds thus only slight decay in the trapped charge occurs within a practical high speed auto-reclose time sequence.

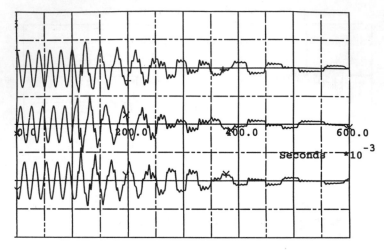

Fig. 2.6   Decay of power transformer voltages (with saturation)

2.4.3 Transmission line length

The length of transmission line being energised affects the overvoltage magnitude in that the longer the line the greater is the steady state open circuit receiving end voltage (Ferranti rise) on which the high frequency transients, the frequency of which is determined by the line length, are superimposed.

The frequency of the switching surge can be approximated by:

$$f = \frac{1}{4T}$$

Where T is the transit time of the line ≈ {line length (in km)/0.3} μs assuming the surge travels along the line at the speed of light.

2.4.4 Compensation

Shunt reactor compensation has a two-fold effect when situated at the transmission line receiving end, both aspects of which contribute to a reduction in the severity of the energising overvoltages. The reactor reduces the magnitude of the Ferranti rise along the line by negating the effect of a portion of the line shunt capacitance and presents a line termination other than an open circuit to any travelling waves from the transmission line sending end. For a 400 kV line the capacitive line charging current is approximately 1 A per km (0.25 MVA per phase). A 200 km line would typically require 40 MVAr of shunt reactive compensation per phase depending on the system operational requirements.

2.4.5 Circuit breaker pole scatter

Circuit breakers will seldom produce a simultaneous close onto a three-phase transmission line for two reasons:

(1)   The circuit breaker mechanism closes the contacts at high speed and mechanical tolerances will give a spread of closing times between the three phases. Typically this may be of the order of 3-5 ms between the first and last pole to close.

(2)   Depending on the point-on-wave at which the circuit breaker initiates a close, the phase with the highest instantaneous value of power frequency voltage will pre-arc first, just before contact touch.

The pole scatter effect produces voltage through mutual coupling from the first phase to close on the other two phases. This pre-charging effect then produces a voltage greater than phase voltage across the contacts of the other two phases of the circuit breaker. This in turn forces a greater than 1 p.u. step voltage to be applied as the other phases close. Certain critical points can be reached depending on the pole scatter time and point-on-wave of closure, for example when the second pole to close occurs at $2T$, $4T$ etc for the transmission line i.e. the point at which the switching surge on the first-phase to close, returns to the sending end of the line. By studying the various combinations of pole scatter, points of maximum/near minimum overvoltages can be determined. With Transient Network Analyser (TNA) studies, pole-scatter diagrams can be created showing the effect of incremental changes in pole scatter and the maximum overvoltage position located. With computer analysis, using programs such as EMTP, a statistical approach is normally adopted using random point-on-wave (uniform distribution) with gaussian distribution for pole-scatter. From work previously carried out with TNA using 500 operations with random point-on-wave and circuit breaker pole scatter determined from typical distribution curves, the 2 % probability value (i.e. the overvoltage value which will be exceeded for 1 in 50 operations) was approximately 15 % below the maximum value derived using the conventional (maximum) method.

2.4.6 Point-on-wave of circuit breaker closure

The magnitude of the transient voltage is very much dependent on the instantaneous value of power frequency voltage at which the circuit breaker closes. If all three poles of the circuit breaker closed at voltage zeros then only a very small transient voltage would occur.

2.5 Methods of controlling switching surges

On systems of 400 kV and above, the energisation of long transmission lines (200 km and greater) is commonly required. Voltages above 4 p.u. of the normal system phase voltage peak have been shown to occur by T.N.A. and computer studies. Methods therefore have been employed to reduce these overvoltages to 2.5 p.u. or less in order to achieve economic design of the transmission line and substation. At 400 kV the overvoltages can be well controlled using metal oxide surge arresters at the send and receive end of the line. At 500 kV, circuit breaker pre-insertion resistors have been used with great effect but result in complicated contact arrangements on the circuit breaker with appropriate increase in maintenance. For 500 kV systems with line lengths below 300 km, metal oxide surge arresters can give an acceptable voltage profile along the line length, with the maximum voltage occurring at the line mid-point. With lengths near 300 km and above, additional surge arresters can be placed at the line mid-points. However, in all cases the substation voltages can be adequately controlled with the surge arresters at the line ends only. Dependent on the line design and the acceptable risk of failure for the 500 kV line, the mid-point surge arrester may not be required even with 300 km lines.

2.5.1 Circuit breaker pre-insertion resistors

2.5.1.1 Single stage resistor insertion

Energising a transmission line through single stage resistors results in the waves transmitted along the line being reduced in magnitude and hence the overvoltages at the receiving end being less severe. Resistors used in this way must also be removed from circuit and the removal of the resistors also initiates travelling waves which create overvoltages. Fig. 2.7 illustrates the severity of the overvoltages produced in the initial energising operations through single stage resistors and the subsequent resistor shorting operations indicating that there is an optimum insertion resistor value where the overvoltages produced by the initial energisation and the subsequent resistor removal are equal.

The optimum resistor value varies with different system conditions but is typically in the range 300-500 ohms.

2.5.1.2 Insertion time and pole scatter

The overvoltages generated when energising transmission lines are relatively insensitive to the resistor insertion time. However, if the insertion time is less than the circuit breaker pole scatter plus twice the transmission line transit time an increase in the overvoltage results, especially in the cases when remanent charge exists. Fig. 2.8 illustrates the results from a series of studies to investigate the relationship between pole scatter and insertion time. They show that there is significant increase in the overvoltage if one phase is energised through its resistor and the resistor shorted out before another phase has been energised for the first

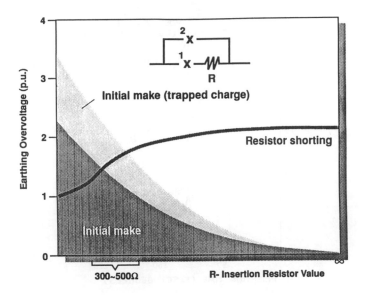

Fig. 2.7   Overvoltage reduction with closing resistors

time as the damping effect of the resistor on the mutually induced voltage is ineffective. The effect is most pronounced in the regions of small insertion resistor values where the mutually coupled transient components are greater.

2.5.2 Metal oxide surge arresters

2.5.2.1 Selection of surge arrester rating

The temporary overvoltage level and duration must be carefully considered before selecting the rating of the surge arrester (MOA). The surge arrester must be capable of withstanding, from thermal constraints, the temporary overvoltage (TOV) which in most circumstances determines the surge arrester rated value. From the rated value stems the protective or voltage limiting characteristic of the surge arrester — the higher the rating, the higher the limiting or residual voltage the arrester will have.

Thermal constraints are very important with MOAs since if the rating is too low, temporary overvoltage may cause excessive heating resulting in thermal instability with a runaway condition being produced and subsequent failure. Fig. 2.9 shows typical TOV capability for MOAs. The energy capability of the surge arresters are usually expressed as kJ/kV of arrester rating. The maximum continuous operating voltage is considered as 80% of the rated voltage. Typically for 400 kV systems an arrester rating of 360 kV will be used which gives a maximum continuous operating voltage of 1.25 p.u. of nominal system voltage.

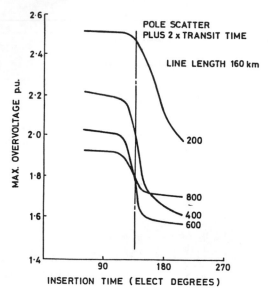

Fig. 2.8   Overvoltage variation with resistor insertion time

From the curve given in Fig. 2.9 a temporary overvoltage of 1.5 p.u. can be withstood for approximately 20 minutes with a 360 kV MOA. The surge arrester voltage-current characteristic exhibits an extremely non-linear relationship once the "knee" point voltage has been exceeded, which causes large increases in current for small voltage increase (see Fig. 2.10). Typically for a 444 kV rated arrester (TOV capability of 1.7 p.u. for 10 s on a 500 kV system) a residual voltage change from 860 kV to 1220 kV produces a current change from 1 kA to 40 kA. This arrester will also limit switching surges at the substation to approximately 2.4 p.u. on 500 kV systems. The rise time of the switching surge is relatively slow in comparison to the transit time of the substation busbars allowing surge arresters at the line end only to give complete substation protection.

2.5.2.2 Application of MOAs for switching surge reduction
For the 400 kV transmission system shown in Fig. 2.11 the maximum switching surge is reduced from 2.15 p.u. to 2.03 p.u. (see Fig. 2.12) by locating a 396 kV MOA at the line end. This is not surprisingly, a very small voltage reduction for a level of voltage which, even without MOA application, is well within the SIWL (3.0 p.u.) and would not justify the cost of surge arrester application. For this condition, the surge arrester draws a peak current of 190 A with a pulse width of 0.5 ms. Smaller current pulses are also evident for the other oscillations at the power frequency voltage peaks.

For the same transmission system Fig. 2.13 shows the maximum overvoltage with trapped charge which is well in excess of the substation SIWL. For this condition a reduction from 4.28 p.u. to 2.35 p.u. in the switching overvoltage can

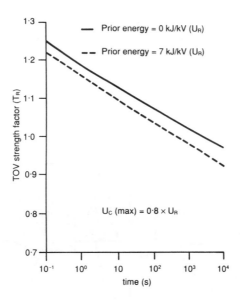

Fig. 2.9   *TOV capability for typical surge arrester, expressed in multiples of $U_R(T_R)$*

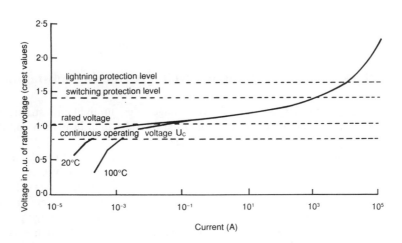

Fig. 2.10   *Typical voltage-current characteristics for ZnO arresters*

be achieved with a 396 kV rated MOA. For this case the surge arrester draws a 2 kA current pulse with a 0.5 ms width. By comparison of the two conditions, i.e.

with and without surge arresters, not only is the amplitude significantly reduced but the oscillations are much more quickly damped. For the trapped charge (worst case) condition a safety factor (SIWL/protection level of MOA) of 1.27 has been achieved (IEC 71 recommends a minimum of 1.15).

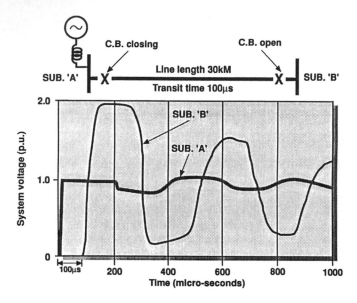

Fig. 2.11   Switching surges — travelling waves

2.6 Factors affecting lightning overvoltages entering substations

The magnitude and rate of rise of overvoltages due to lightning strikes on transmission lines is an important consideration for substation insulation and the strategy adopted for limiting these overvoltages.

Having determined the insulation required for the line, it is usual to find that the lightning withstand level is in excess of commercially available LIWL levels of the substation equipment.   Thus, unless precautions are taken, overvoltages entering the station can cause undue insulation failure.   Surge arresters can be situated at the line entrance but consideration must be given to the voltage profile as the surge travels through the substation.   Alternatively, consideration may be given to using rod gaps, set to operate marginally below the station LIWL and SIWL levels, and fitted to the first 3 or 4 towers.   However, consideration must be given to the voltage-time characteristic of the substation equipment in comparison to that of line gaps.   For example, 132 kV and 400 kV GIS cannot be adequately protected by line coordinating gaps.   Fig. 2.14 shows a comparison of

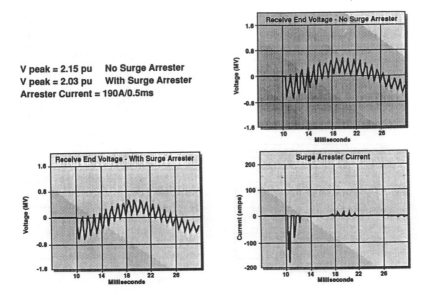

Fig. 2.12   *Control of switching surges using surge arresters (low level surge)*

line gaps and a metal oxide surge arrester in relation to the standard insulation levels for a 400 kV GIS and shows that the 2 m line gap is totally ineffective as a method of reducing the incoming surge voltages.

The number of strikes to transmission lines is generally accepted to be related to the isoceraunic level which is defined as a number of days in a year in which thunder is heard at a given location. Assumptions are made in relating this isoceraunic level to the number of strikes to towers and earth wires; the Transmission Line Reference Book [10] provides methods of calculating the number of line flashes — see Fig 2.15. The number of strikes is directly proportional to the isoceraunic level, i.e. a level of 20 entails twice the number of strikes as a level of 10.

The calculation method is based on the number of ground flashes that would occur to the area of ground shielded by the transmission line. Two possibilities exist for generating lightning overvoltages on the line conductors — the "backflashover" and the "direct" strike. Fig. 2.16 shows a typical 400 kV single circuit tower illustrating the two strike conditions and gives the tower and line parameters.

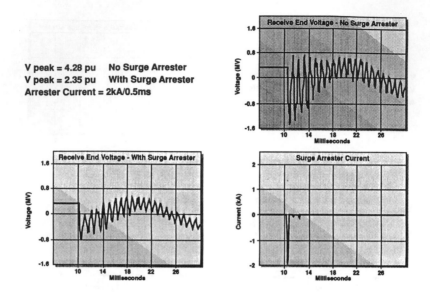

V peak = 4.28 pu   No Surge Arrester
V peak = 2.35 pu   With Surge Arrester
Arrester Current = 2kA/0.5ms

Fig. 2.13   *Control of switching surges using surge arresters (high level surge)*

2.6.1 Backflashover

A backflashover occurs as a result of the tower or shield wire being struck by lightning, the current passes to earth via the tower steelwork causing a voltage difference between the tower cross arms and the line conductors. The magnitude of this current can vary from a few kA to over 200 kA. The statistical data for amplitude and steepness of lightning currents is given in Figs. 2.17 and 2.18 derived from data by R.B. Anderson and A.J. Eriksson [11].

Due to the height of the tower and rate of rise of current, a travelling wave can be set up on the tower. The combination of shield wire and tower surge impedance, see Fig. 2.19, and lightning current impulse will produce a voltage at the tower top which is oscillatory due to successive reflections from the tower base. When the surge current arrives at the base of the tower it dissipates through the tower grounding to earth. An additional voltage is produced at the tower foot which is dependent on the grounding impedance. Fig. 2.20 shows typical voltages with the tower travelling wave voltage superimposed on the tower foot voltage. The first voltage pulse width can be estimated by doubling the tower transit time (typically 0.2-0.4 s). Not all of the tower top voltage will appear across the line insulator because there is some reduction due to the position of the insulator on the crossarm and also voltage will be mutually coupled from the shield wire to the phase wire. So the voltage that appears across the line insulator coordinating gap will be similiar but marginally smaller (85 %) than the tower top voltage.

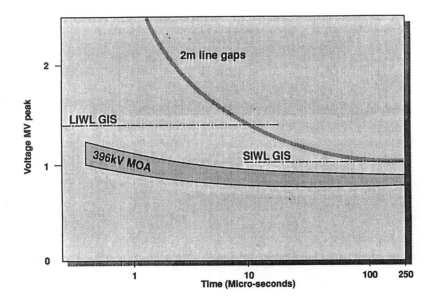

Fig. 2.14   *Comparison of voltage-time characteristics*

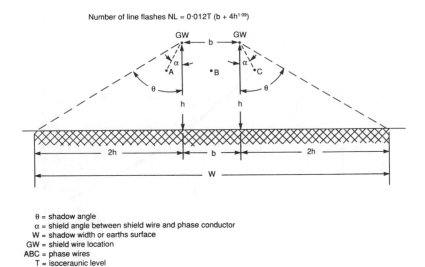

θ = shadow angle
α = shield angle between shield wire and phase conductor
W = shadow width or earths surface
GW = shield wire location
ABC = phase wires
T = isoceraunic level

Fig. 2.15   *Model for line flash calculations*

Depending on the *V-t* characteristics of the line coordinating gap, the backflashover (i.e. from tower to line) may occur near the peak of the voltage pulse or on the surge tail. Test data for line coordinating gaps is limited for

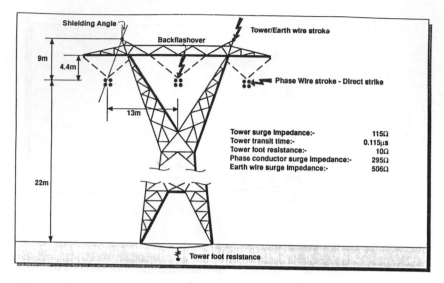

Fig. 2.16   Lightning surges

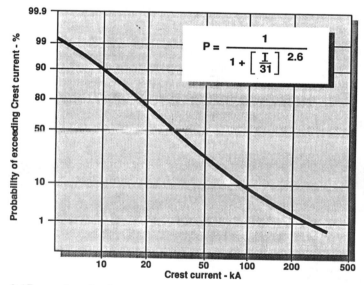

Fig. 2.17   Amplitude of lightning stroke current

"non-standard" lightning impulse voltages. However, work has been done [12] to establish models of the line gap flashover mechanism. Leader progression models have been proposed which can be used to assess the time to flashover for these waveshapes. Fig. 2.21 shows a line gap flashover from a standard 1.2/50 μs lightning impulse voltage and illustrates a "completed" flashover on one of the

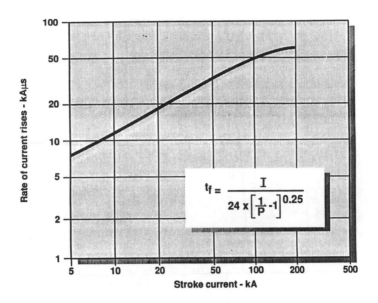

Fig. 2.18   *Lightning stroke current steepness*

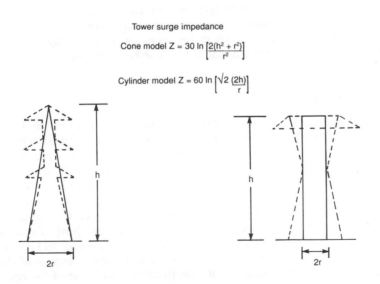

Fig. 2.19   *Tower surge impedance*

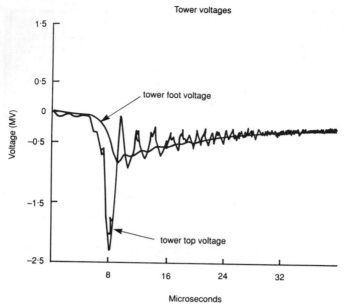

Fig. 2.20   Tower voltages

gaps with leaders only partially bridging the second gap. It is important to note that as the tower foot resistance is increased the more dominant the tower foot voltage will become to a point where for short towers the voltage waveshape across the line gap will approach that of the "standard" impulse.

2.6.2 Direct strike

Most transmission line towers will be equipped with shielding wires. In the tower shown in Fig. 2.16 there are two shield wires. The purpose of these wires is to divert the lightning stroke away from the phase wire and thus provide shielding. Any lightning strike which can penetrate the shield is termed a "direct strike" or "shielding failure". The electrogeometric model proposed in [10] and shown in Fig. 2.22 is a simplified model of the shielding failure mechanism for one shield wire and one phase conductor. As a flash approaches within a certain distance "S" of the line and earth, it is influenced by what is below it and jumps the distance "S" to make contact. The distance "S" is called the strike distance and it is a key concept in the electrogeometric theory. The strike distance is a function of charge and hence current in the channel of the approaching flash. Use of the equation given in Fig. 2.22 requires the calculation of S_{MAX} and S_{MIN} which then relate I_{MAX} and I_{MIN} the corresponding stroke currents. The probabilities for I_{MAX} and I_{MIN} can then be determined (P_{MAX}, P_{MIN}) along with the unshielded width X_s — if $X_s = 0$

Fig. 2.21   Line gap flashover

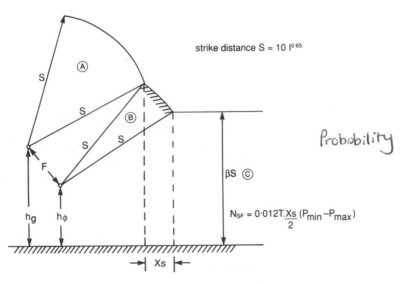

strike distance S = 10 I$^{0.65}$

$N_{SF} = 0.012T.\dfrac{Xs}{2}(P_{min} - P_{max})$

Fig. 2.22   Electrogeometric model for shielding failures

then shielding failure will not occur. The objective when designing the line shielding is to minimise X_s. Fig. 2.23 compares the results of the electrogeometric model calculations for the shielding performance of two towers with identical conductor and shield wire configuration but are of different heights. A corresponding increase in the number of line flashes and shielding failures is indicated with the taller tower. Also the maximum shielding failure current is three times that for the smaller tower.

For the purpose of insulation coordination the direct strike may not warrant further investigation if the transmission line is effectively shielded, particularly in the last 5 km of line approaching the substation. Considering the data from Fig. 2.23 shielding failure rate of 0.2/100 km/year would mean that a direct strike inside the last 5 km of line would occur once in 100 years or a 1 in 3 chance during the life of the substation. When assessing the risk of failure for the substation, however, a sum of all the probabilities for each substation line is required (i.e. 6 lines would give two surges from direct strikes in the life of the substation).

2.6.3 Attenuation of lightning overvoltage

As the lightning surge travels towards the substation from the struck point the wave front above the corona inception voltage will be retarded by corona loss. Skin effect on the line conductors will cause further attenuation due to the high frequency nature of the surge. It is usual therefore to consider lightning strikes that are "close-in" (within 3 km) to the substation when assessing surge arrester requirements and the associated risk of failure of the substation.

2.7 Methods of controlling lightning overvoltages

For well shielded transmission lines, the backflashover condition, close to the substation, is of prime concern for determining the location and number of surge arresters required to achieve insulation coordination of the substation for lightning surges. The risk of a backflashover can be reduced by keeping the tower foot impedances to a minimum, particularly close to the substation (first 5-7 towers). The terminal tower is usually bonded to the substation earth mat and will have a very low grounding impedance (1 ohm). However, the procedure for "gapping" down on the first 3 or 4 towers where line coordinating gaps are reduced in an attempt to reduce incoming voltage surges will increase the risk of a "close-in" backflashover.

2.7.1 Location of surge arresters

Considering the system shown in Fig. 2.24 where the transmission line is directly connected to a 400 kV GIS, a computer model can be created to take into account the parameters previously discussed. A transient study would reveal the level of

	Tower height d.c. 132 kV	
	30 m	60 m
I_{min}	5 kA	17 kA
I_{max}	28 kA	84 kA
Number of line/tower flashes	19/100 km/yr	39/100 km/yr
Number of shielding failures	0.2/100 km/yr	0.65/100 km/yr
Prob. max E/W stroke current	93 kA	126 kA
Annual thunder days	13	13

Fig. 2.23   Lightning performance of transmission lines

lightning stroke current required to cause a backflashover. Then according to the number of lineflashes/100 km/yr calculated for the transmission line and by using the probability curve for lightning current amplitude, a return time for this stroke current can be assessed (i.e. 1 in 400 yrs, 1 in 10 yrs etc.) in say the first km of the line. The voltage then arriving at the substation can be evaluated and compared with the LIWL for the substation equipment. The open circuit breaker condition must be studied here, since if the line circuit breaker is open the surge voltage will "double-up" at the open terminal. Various levels of stroke current can be simulated at different tower locations and the resultant substation overvoltages can be assessed. If it is considered that the LIWL of the substation will be exceeded or that there is insufficient margin between the calculated surge levels and the LIWL to produce an acceptable risk, then surge arrester protection must be applied.

The rating of the MOA will have been assessed from TOV requirements and from the manufacturer's data a surge arrester model can be included in the system model. Repeating the various studies will reveal the protective level of the arrester and from this the safety factor for this system congfiguration can be assessed [13 - 16]. IEC 71 recommends a safety factor of 1.25 for 400 kV equipment (safety factor = LIWL/protective level). The surge arrester current calculated for this condition should be the "worst" case or can therefore be used to assess the nominal discharge current requirement of the surge arrester (5 kA, 10 kA or 20 kA). (IEC 91-1 is the international standard for surge arresters [14] and an accompanying guide is available which contains detailed information on the application of surge arresters).

To make full use of the MOA protective level the arrester should be placed as close as possible to the equipment being protected. In the case of the open line circuit breaker this may well be 10-20 m distance. Dependent on the rate of rise of the surge voltage, a voltage greater than the residual voltage at the surge arrester location will be experienced at the terminals of the open circuit breaker. This must be taken into account when assessing the substation overvoltage. Fig. 2.25 illustrates the surge voltage profile of the GIS with the line circuit breaker

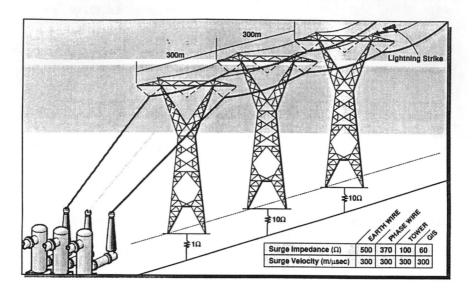

Fig. 2.24   System schematic

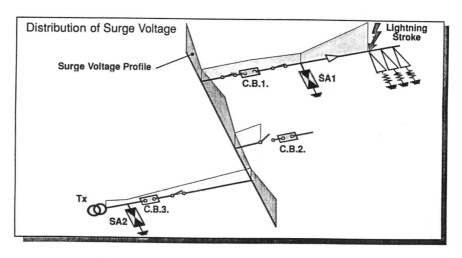

Fig. 2.25   Analysis of lightning surge for gas insulated substation

closed. It shows that additional surge arresters may be required because of the distances involved in the layout of the substation. It then follows that surge arresters have a "protective length" [13] which is sensitive to the rate of rise of the incoming surge voltage and this must be taken into consideration when assessing the lightning overvoltage on equipment remote from the surge arrester.

2.8 Conclusions

This chapter has introduced the important concept of insulation coordination of high voltage systems. It is vital for any engineer working in, or planning to work in, the electrical power industry to be aware of design choices regarding electrical stresses, insulation levels, service performance etc., together with testing procedures and the importance of IEC standards for a wide range of equipment [17]. These aspects will be considered further in later chapters. The topics have also been discussed in several major publications in recent years (e.g. ELECTRA, CIGRE, IEC, IEEE).

2.9 References

1   CLERICCI, A., AL RASHED, S.A. and AL SOHAIBANI, S.N.: "380 kV system from Riyadh to Quassim: studies of temporary overvoltages and methods to prevent them". CIGRE, 1990

2   GERMANY, N., MASTERO, S. and VROMAN, J.: "Review of ferroresonance phenomena in high voltage power system". CIGRE, 1974, Paper 33-18

3   RITCHIE, W.M. and IRWIN, T.: "Limitation of transmission line energising overvoltages by resistor insertion". IEE Conf. Pub. 182, 1979

4   JOHNSON, I.B., TITUS, C.H., WILSON, D.D. and HEDMAN, D.E.: "Switching of extra-high voltage circuits II - surge reduction with circuit breaker resistors". *IEEE Trans.*, 1964, **PAS 196**, p.1204

5   MAURY, E.: "Synchronous closing of 525 and 765 kV circuit breakers. A means of reducing switching surges on unloaded lines". CIGRE, 1966, Paper 143

6   STEMLER, G.E.: "BPA's field test evaluation of 500 kV PCSs rated to limit line switching overvoltages to 1.5 per unit". *IEEE Trans.*, 1966, **PAS 95**, (1)

7   BICKFORD, J.P. and EL-DEWIENY, R.M.K.: "Energisation of transmission lines from inductive sources". *Proc. IEE*, 1973, **120**, (8), pp.883-890

8   BICKFORD, J.P. and EL-DEWIENY, R.M.K.: "Energisation of transmission lines from mixed sources". *Proc. IEE*, 1973, **112**, (5), pp.355-360

9   CATENACCI, E. and PALVA, V.: "Switching overvoltages in EHV and UHV systems with special reference to closing and reclosing transmission lines". *Electra*, 1973, **30**, pp.70-122

10   ANDERSON, J.G.: "Lightning performance of transmission lines", in *Transmission lines reference book 345 kV and above*

11   ANDERSON, R.B. and ERIKSSON A.J.: "A summary of lightning parameters for engineering applications". CIGRE, 1980, Paper 33-06

12   WATSON, W., RYAN, H.M., FLYNN, A. and IRWIN, T.: "The voltage-time characteristics of long air gaps and the protection of substations against lightning". CIGRE, 1986, Paper 15-05

13   IEC 71-1, 1993. "Insulation coordination, Part 1: terms, definitions, principles and rules". Contains the definitions of terms to be found here. It gives the series of standard values for the rated withstand voltages and the recommended combination between these and the highest voltage for equipment

14   IEC 99-1, 1991. "Surge arresters, Part 1: non-linear resistor type gapped arresters for a.c. systems". Specifies the test requirements and maximum protection levels for this arrester type

15   IEC 99-4, 1991. "Surge arresters, Part 4: metal-oxide arresters without gaps for a.c. systems". Specifies the test requirements and maximum protection levels for this arrester type

16   IEC 99-5. "Surge arresters, Part 3: application guide to IEC 99-1 and IEC 99-4". Gives guidance for the selection of surge arresters depending on the overvoltage stress conditions of the system. [Presently under preparation; see 37 (Secretariat) 85]

17   RYAN, H.M. and WHISKARD, J.: "Design and operating perspective of a British UHV laboratory", *IEE Proc. A*, 1986, **113**, (8)

Chapter 3

Overhead lines

F.J. Liptrot

3.1 Introduction

The purpose of this chapter is to provide an overview of the design, manufacture, construction, testing and maintenance of the various components which go to make up overhead transmission and distribution lines. The transmission of electrical energy is carried out over long distances at voltages of 66 kV and above whereas distribution is carried out over short distances at voltages of between 11 and 66 kV using the same technology.

In the early days of electricity transmission, the methods already established by the Post Office, using telephone wires fixed to wood pole supports, were adopted by the electricity supply authorities. This involved the fixing of bare, uninsulated conductors to small porcelain insulators (Fig. 3.1(a)) screwed on to metal spindles which were mounted on to metallic or wooden crossarms supported by wooden poles (see Fig. 3.1(b)).

As transmission/distribution distances increased, it became necessary to increase the system voltages. Consequently, the pin-type insulators were increased in size to provide longer creepage paths to earth; conductor sizes were also increased to allow higher current carrying capacities (ampacities). This resulted in increased vertical loads due to the conductor mass and increased transverse wind loads due to the increased conductor projected area. Insulator spindle diameters were of necessity increased to cater for the greater mechanical loads and similarly single wood poles were replaced by "A" and "H" pole construction, i.e. two poles side by side at 1 or 1.5 m centres and braced together to cater for the higher transverse loads. The simple pin-type insulator designs progressively reached their limit of mechanical strengths under side or transverse loads.

To overcome the problems of the limited mechanical strength of pin-type insulators, the cap and pin insulator was developed (see Fig. 3.2). With this design, the tensile forces applied to the insulator are axial, and it is not required to support side loads, therefore higher conductor loading can be achieved. Short suspension strings of 1 or 2 insulators for 11 kV up to 6 insulator units for 66 kV were used, thus allowing increased conductor diameters and longer spans because

6.6kv
→
11kv
system

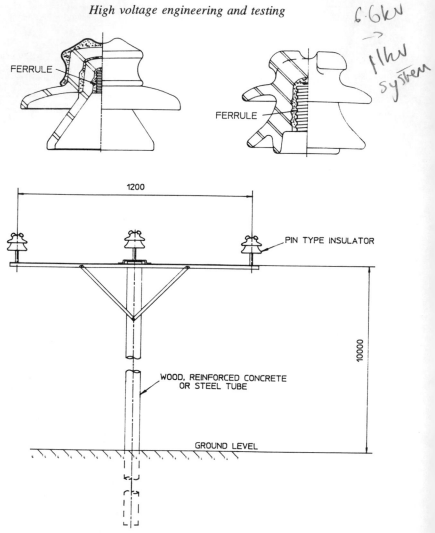

Fig. 3.1   *(a) Section drawing of toughened glass (left) and porcelain pin-type insulators*
(b) Overhead line construction (11 kV)

of the increased mechanical strength of the insulators.

The strength of the wood poles tended to limit any further increases in span lengths and progressively small towers of lattice steel construction were introduced as supports. As voltages increased from 66 to 132 kV, the lattice steel tower became the accepted method of support. Over the years, as voltages have further increased through 275 - 420 - 500 - 765 - 1100 kV, the most frequently adopted method of support remains the lattice steel tower.

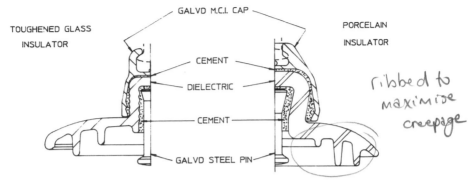

Fig. 3.2   Section drawing of cap and pin insulators

As a substitute for wood poles, concrete poles of pre-stressed concrete of rectangular cross-section — and more recently spun concrete poles — are used with voltages up to about 132 kV.

In recent years, wood pole lines have been developed for 66 to 132 kV using a "trident" form of construction where either line-post or post-type insulators are employed. This type of construction is lightweight and easy to transport and erect. It is increasingly used in areas of high amenity value where the low height construction allows the line to blend in with the background. A number of authorities employ the trident form of construction to provide short lengths of quickly erected lines to bypass faults or substation bays during construction of the substations. A typical "trident" structure is shown in Fig. 3.3.

3.2 Towers and supports

3.2.1 General

The tower or supporting structure is required to carry the overhead line conductors and earth conductors, each of which will be subjected to a variety of forces varying from normal still air load, extreme wind loads, ice loads in some parts of the world, and any additional loads during erection and maintenance of the conductors or insulator sets. The tower must be capable of safely withstanding all the various forces applied to it and at the same time the electrical clearances between the live conductors and the earthed metal must be maintained under all loading conditions.

The tower may be designed to cater for a single 3 phase circuit, double 3 phase circuits, multiple voltage circuits and, with direct current transmission, either a monopolar (with earth return) or bipolar construction.

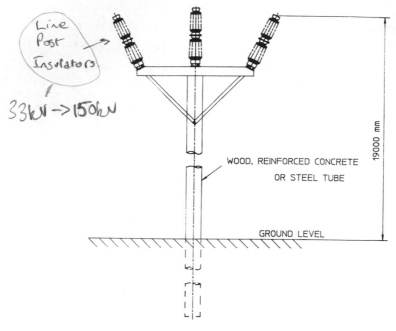

Fig. 3.3   Trident overhead line construction (132 kV)

3.2.2 Self-supporting single circuit towers

Single circuit (SC) towers are generally of one of three forms: horizontal, delta or triangular conductor configuration. Some typical towers are shown in Fig. 3.4.

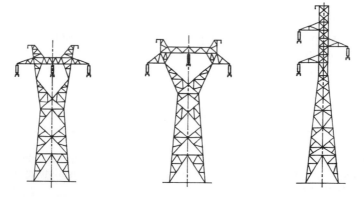

Fig. 3.4   Single circuit steel towers (left to right: horizontal, delta, triangular formations; twin, twin, single earthwire, respectively)

Towers equipped with horizontal configuration conductors require a lower tower height than an equivalent vertical configuration tower but because of the clearance requirements between phases they require a wider strip of ground. If the region in which the line is to be constructed has a high lightning level (isoceraunic level) it is usually necessary to install two overhead earth wires to provide proper shielding of the line conductors. Horizontal configuration results in the capacitance between conductors and earth being approximately the same for each phase but the capacitance and inductance between phases is not equal. To overcome the problem of asymmetry, towers may be constructed with conductors deployed in delta formation and with this arrangement the inductance and capacitance between conductors is virtually the same for each phase.

3.2.3 Self-supporting double circuit towers

A line comprising double circuit (DC) towers can be constructed at a lower capital cost than the equivalent parallel two SC lines. Double current lines are generally 7 to 10% cheaper than the equivalent two SC lines depending on the voltage and conductor sizes. A double current line will occupy less land than the equivalent two SC lines. However, the most frequently adopted design of double current tower with two vertical configuration conductor arrangements is of necessity higher than the equivalent SC tower.

There may be a case for using SC lines when heights must be restricted, perhaps near to an airport or on the grounds of amenity. Generally, two SC lines can assist in this situation but specially designed double current lines can also be employed. A typical "low height" double circuit tower is shown in Fig. 3.5.

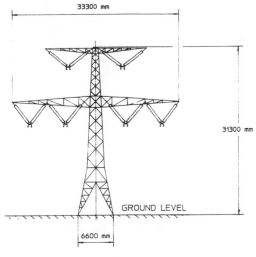

Fig. 3.5   Double circuit low height tower

Probably the main disadvantage of double current lines is the possibility of a double circuit fault arising from a lightning strike, accidental damage resulting from a vehicle or aircraft colliding with a tower, or possibly due to abnormal weather conditions, for example a tornado or very severe icing. Deliberate damage to towers, perhaps by means of an explosive charge or by the removal of bolts critical to the tower strength, is increasingly occurring in many parts of the world and this type of damage will cause more disruption with double current towers than with SC towers. If a reliable interconnection is required then two SC lines spaced as far apart as possible reduces the likelihood of a total interruption of supply due to many of the reasons listed above.

3.2.4 Guyed towers

In addition to self-supporting towers, structures which rely on guy wires to provide longitudinal and transverse support are employed (see Fig. 3.6). The use of guyed towers will result in a saving of about 5 % of erected cost over the equivalent self-supporting towers at 500 kV system voltages. The saving is largely achieved because of the reduced weight of tower steel.

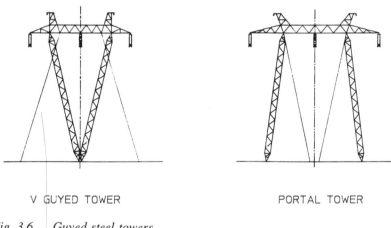

V GUYED TOWER   PORTAL TOWER

Fig. 3.6   Guyed steel towers

Guyed towers can only be employed on relatively flat ground because, on sloping ground, one pair of guy wires can become prohibitively long. In addition, relatively level ground is required if a mobile crane with a 40-50 m jib is to be employed safely for tower erection.

In regions which are densely populated or intensively farmed, the total area of land required by guyed towers makes them less attractive than a self-supporting tower.

3.2.5 Tower design

3.2.5.1 General
Tower design involves a number of distinct stages:

(a)   Derivation of loads to be resisted and clearances to be accommodated by the tower;
(b)   Selection of tower shape resulting from (a);
(c)   Derivation of member loads;
(d)   Design of members.

Stages (c) and (d) will also affect the choice of tower shape and will be repeated by a designer to provide the optimum design.

The various tower types described in the previous section can all by analysed by similar methods. It is not the intention of these notes to provide a complete guide to tower design and the following sections relate mainly to the checking of contractors' designs. However, the basis behind each stage is described so that towers requiring special characteristics can be properly checked.

At each conductor attachment point on the tower, the loads are defined in three axes X, Y, Z more commonly known as transverse, longitudinal and vertical respectively. The derivation of each of these is discussed below.

3.2.5.2 Transverse loads
These comprise those wind-induced loads acting on the conductor and insulators, together with the transverse component of conductor tension, resulting from a deviation in the line for angle towers.

(a) Wind loads

The wind load on a conductor (T_w) may be derived by the formula:

$$T_w = P_c \left(d\, n\, S_w \right) + A_i\, P_c$$

where

P_c = the design wind pressure for cylindrical surfaces
d = the conductor diameter
n = the number of conductors in the bundle (note: no allowance is made for the shielding effect of one conductor on another)
S_w = the effective wind span
A_i = the projected area of the insulator sets.

The wind span is defined as half the sum of the two spans adjacent to the tower and is based on a percentage increase over the chosen economic basic (or standard) span, to allow for errors in plotting and profiling. The increase is usually in the region of 10% over the basic span. To arrive at the effective wind span, the length projected normal to the direction of the wind is taken. This is the true situation for a suspension (straight line) tower, for example tower No. 2 of Fig 3.7. For angle towers, however, due to the many possible permutations of deviation and wind direction on spans S_1 and S_2, it is usual to assume the conductors to be in a straight line for the calculation of S_w.

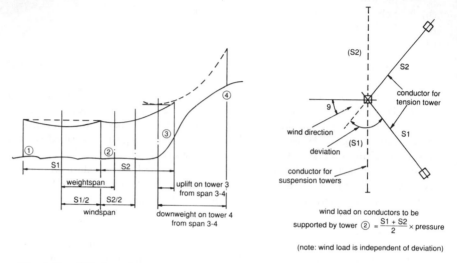

Fig. 3.7   Wind and weight spans

The effective wind span for all tower types may therefore be written as:

$$S_w = \frac{S_1 + S_2}{2} \cos\theta$$

where θ = the line deviation angle

and

$$\frac{S_1 + S_2}{2} = \text{basic span} \times 1.1$$

(b) Deviation loads

For angle towers (and small angle suspension towers) a proportion of the transverse load at a conductor attachment point is due to the tension of the conductors. This load is known as the deviation load (T_d) and is calculated by:

$$T_d = T \, n \, \sin\frac{\theta}{2}$$

where T is the maximum horizontal tower design tension of the conductors which is dependent on the wind pressure. The loads T_d and T_w are therefore considered to act simultaneously on the tower.

3.2.5.3 Longitudinal loads

These are the loads which result from the conductor tension. Suspension towers are not subjected to these design loads except in exceptional cases where a conductor is broken. The incidence of broken conductors is sufficiently infrequent to allow the longitudinal conductor load for the design of suspension towers to be reduced by some factor from the maximum tension so as not to unduly penalise the line in economic terms because of unnecessary longitudinal suspension tower strength. Tension towers on the other hand, which experience the full conductor tension at all times, require an enhanced longitudinal design load to provide the necessary security and capability to resist the associated dynamic loads developed during conductor failure.

3.2.5.4 Vertical loads

These are the loads resulting from the deadweight of conductors, insulators and fittings. The length of conductor to be considered is termed the weight span, the definition being the horizontal distance between the lowest point of the conductors on the two spans adjacent to the tower (see tower No. 2 of Fig. 3.7). The vertical load is therefore:

$$V = \left(S_{wt} \, n \, w \,\right) + W_f$$

where

S_{wt} = Weight span. The design weight span is taken as the basic span increased by a profile factor, usually 2, to allow for undulating terrain, together with an additional factor to allow for errors in plotting and setting out.

n = Number of subconductors.

w = The total mass of conductor inclusive of grease, an allowance for spacers and, if necessary, ice.

W_f = The weight of insulators and fittings.

When the lowest point of the conductors in a span approaches the tower position itself, the effective vertical weight due to the conductors reduces in proportion. Should the "apparent" lowest point move past the tower and into the next span the tower will experience uplift from the former span.

Tower designs must be checked for vertical conditions other than that for maximum weight because minimum weight and uplift conditions are critical for the leg and foundation designs in tensions and uplift respectively. For suspension towers it is usual to consider in the design a minimum vertical condition of zero. This is necessary, although the actual *in situ* weight span is restricted to a positive value which may be, for example, 45 % of the conductor weight in the adjacent spans. This positive load is maintained to ensure that the insulator strings, which are designed to cater for tensile loads only, are not subjected to compressive loads. Tension towers may experience any variation of vertical load from the maximum permitted by the specification through zero to uplift.

The design uplift loads may be defined as either:

(a) $T \tan \phi$, where ϕ is the incoming angle of the conductor above the horizontal and is dependent on the type of terrain

or

(b)   some number, for example "one" or "two", of uplift or negative weight spans. It is recommended this be based on the value obtained from (a).

3.2.5.5 Tower wind loads

Traditionally, tower wind loads have been derived by multiplying the design wind pressure (for flat surfaces) and the projected area of tower members of one face increased by a shielding factor (usually 1.5) to allow for the members on the back face of the tower. For especially large towers, often required for river crossings, the shielding factor should be increased.

The transverse and longitudinal wind loads on towers are derived from:

$$T_t = P_f A_t \cos\theta \, f$$

$$T_l = P_f A_l \cos\left(90 - \theta\right) f$$

where

$\quad T_t$ and T_l = Transverse and longitudinal tower wind loads
$\quad P_f$ = Design wind pressure for flat surfaces
$\quad A_t$ and A_l = Net area of tower members in one face
$\quad f$ = Shielding factor

3.2.5.6 Tower self-weight

These loads are applied at various positions down the tower and shared equally between all four legs. For iced conditions, it will be necessary to consider the weight of members when covered with a stipulated thickness of ice.

3.2.5.7 Loading cases

Each type of tower must be designed to withstand the loads imposed by the conductors and/or ancillaries whether under normal or unbalanced (broken wire) conditions together with the loads imposed on and by the tower itself (wind and self-weight). Each type of tower may be utilised on the line within clearly defined limits, for example a specification may define normal load, unbalanced load, broken wire load and maintenance load.

3.2.5.8 Tower clearances
The second stage of tower design is to determine or check the tower shape. As mentioned previously, the function of the tower is to support and maintain all the necessary clearances required for the earth and phase conductors. It is this latter requirement which determines to a major degree the shape of the tower. The main parameters are:

—   Ground clearance (minimum distance between phase conductors and normal flat ground)
—   Sag of conductors (maximum conductor temperature, still air, standard span)
—   Insulator set length (suspension towers only)
—   Specified vertical or horizontal spacing between phase conductors
—   Specified height of earth conductor above upper phase conductor
—   Minimum groundwire shielding angle

These parameters enable the conductors to be located in space (see Fig. 3.8) and all that remains necessary before the initial shape can be determined is to allocate those areas around the conductors in which the earthed steel may not encroach. This is accomplished by the drawing of wire clearance diagrams which take into account the methods of transferring the live conductors through or past the towers whilst maintaining electrical clearance.

3.2.5.9 Methods of analysis
Traditionally, lattice towers have been designed by static methods employing force diagrams (Maxwell) or hand calculations (Method of Sections). More recently, the computer and finite element space truss programs have allowed elastic techniques to be employed. For the purpose of analysis by both techniques, it is assumed that the transmission towers are pin-jointed structures with axially loaded members.

In the past all types of towers, whether of conventional construction (vertical configuration) or waisted (horizontal formation with central window), were designed employing the static method of analysis. The proof of an adequate design was demonstrated by means of type tests (see 3.2.5.12) where any differences between the predicted and actual failing loads became apparent.

3.2.5.10 Tower weight
There are a number of methods for deriving tower weights for preliminary studies, the most common of which involves the use of the Ryle formula. This was developed originally to cover towers designed with typical United Kingdom 132 kV loading but its use may be extended to cover all tower types by sensible adjustment of the factor K in the relationship:

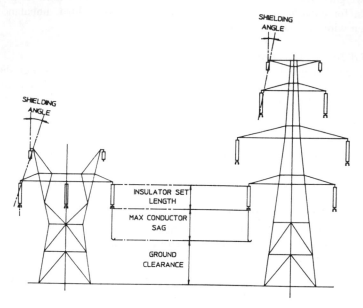

Fig. 3.8   Parameters defining minimum tower shape: SC horizontal (left) and DC (right)

$$W = KH \sqrt{M}$$

where

   M is the maximum ultimate overturning moment due to all loads
        (including wind) in tonne-metres.
   H is the total tower height in metres.
   W is the tower weight in tonnes.
   K is the Ryle factor (approx. 0.014).

The factor K is influenced by a number of parameters, some of which are outlined
below:

—   *Tower material*, whether mild steel only or mild and high yield steel (the
    tower containing HY steel will be lighter).
—   *Tower type*, whether suspension, tension or terminal. The heavier the
    tower, due to the increased complexity of joints and connections, the
    greater the proportion of bolts and plates to the tower members.

—   *Tower formation,* whether "waisted" or "conventional". "Waisted" towers are lighter than "conventional" towers but, due to their reduced height and overturning moment, require an increased K factor.

—   *Range of steel sections available.* The wider the range of sections available the lighter the tower can be designed.

—   *The presence of any restrictions* that have been imposed on the tower so that the design shape deviates from its optimum (i.e. restricted base width).

3.2.5.11 Tower materials

The vast majority of overhead line (OHL) towers throughout the world are constructed from galvanised, hot rolled, mild and/or high tensile steel angle sections. This material is convenient to fabricate by means of numerically controlled presses which crop to length the bars, then punch the many holes required for the various joints. After forming, the steel is galvanised and, when assembled as a tower, should have relatively few pockets and depressions to hold water which could allow corrosion to set up.

Mild steel (MS) with yield strengths of up to about $250 \, \text{N/mm}^2$ is readily available throughout the world and is a very suitable material to provide a reliable tower. High tensile (HT) steel with yield strengths of up to about $400 \, \text{N/mm}^2$ is now used in conjunction with MS. A tower fabricated with HT steel requires a lower tonnage than an MS only tower. However, suitable HT steel is not produced in all parts of the world at the present time.

Steels with yield strengths in excess of about $400 \, \text{N/mm}^2$ should not normally be used in tower construction particularly in those parts of the tower in which vibration, possibly as a result of wind action on the conductors or the tower members themselves, is likely to arise. Cases of fatigue failure of tower members fabricated from very high tensile steel have been reported on a number of occasions although generally in regions where low temperatures (-10 to -20 °C) prevail.

In order to minimise the mass of towers, particularly for areas with difficult access problems, e.g. in swampy ground or in mountainous regions, towers fabricated from aluminium alloy have been employed.

A number of utilities, particularly in the USA and Japan, employ towers fabricated from galvanised steel tube of circular or hexagonal tapering cross-sections. These tubes, usually from 0.5 to 1 m in diameter, are used as a single support. Cross-arms are also formed from similar tubular sections. It has been said that a structure formed from steel tube is aesthetically more pleasing than a lattice steel tower. A tubular steel tower certainly takes up a smaller area of land and it is possible to construct a line in a heavily populated area in which it would not be possible to construct a lattice steel tower line with its comparatively wide base.

Towers for use up to 500 kV constructed from laminated timber bonded with resins under heat and pressure are used in parts of the USA. It is believed that the capital cost of this material is greater than for the equivalent lattice steel towers. An advantage of timber structures is that the timber itself can be considered to provide part of the insulation when calculating the basic impulse level (BIL) of the insulation. This form of construction is often adopted on aesthetic grounds.

3.2.5.12 Tower testing

Generally each contractor will have his or her own methods of analysis, member design and method of detailing of members. The purpose of tower testing is to check that the selected method of analysis (and inherent assumptions) is correct and to ensure that the methods of detailing the connections etc correctly reflect these assumptions. Testing is normally carried out to IEC 652 [1].

Each tower test station usually has its own particular methods of recording and applying the loads. Sometimes it may not be possible to apply the loads exactly in accordance with the design assumptions due to the physical constraints of the test station. In this situation, the test programme should be altered and checked to ensure the loads as applied provide the correct design component loads.

3.3 Conductors

3.3.1 Conductor types

The first transmission lines utilised stranded copper conductors largely because of the good electrical conducting properties (volume resistivity at 20°C, 0.0177 ohm mm^2/m) and resistance to corrosion of the material.

Pure copper has a relatively low ultimate breaking strength of about 250 MN/m^2 and the maximum working tension of conductor manufactured from annealed copper must be limited so that the stress in the conductor does not exceed about 125 MN/m^2. Copper can be "work-hardened", resulting in a material with an improved breaking load of about 420 MN/m^2 at the expense of a slight increase in resistivity. The increase in breaking strength allows higher working loads and longer span lengths. Alloys of copper, particularly cadmium copper, provide a material with an even greater breaking strength of about 630 MN/m^2 but again an increase in resistivity to 0.0217 ohm mm^2/m results. Copper and cadmium copper are expensive and as a result these have largely been replaced by aluminium as a conductor material.

Aluminium has a higher resistivity than copper being 0.0282 ohm mm^2/m. However, if the resistivity is expressed as mass resistivity at 20°C in ohm g/m, because of the lower density of aluminium, copper is seen to have a value of 0.1532 and aluminium 0.076 ohm g/m. Because aluminium has a low breaking strength of only 165 MN/m^2, in order to provide a conductor with an acceptable breaking strength, it is necessary to combine the good conducting properties of aluminium with a material with high mechanical strength. This is done by "laying-up" a centre core of galvanised steel wire with surrounding strands of aluminium, i.e. aluminium conductor, steel reinforced (ACSR). IEC Publication 1089 [2] describes manufacturing and testing requirements for ACSR and for all aluminium conductors, aluminium alloy stranded conductor and aluminium alloy conductors steel reinforced.

Various alloys of aluminium can be produced and formed into conductors and some of these alloys, often containing silicon or magnesium, have virtually twice the breaking or ultimate tensile strength (UTS) of pure aluminium. The gain in UTS is offset to some degree, however, by the increased resistance which many aluminium alloys exhibit. In recent years, big improvements have been made with aluminium alloys and stronger alloys with reduced electrical resistance have resulted. A big advantage of aluminium alloy over pure aluminium is that alloys are in general more resistant to the corrosive effects of saline or industrial chemical attack.

Other methods of increasing the effective breaking strength of both copper and aluminium are employed. Copper or aluminium clad steel are used. This material is formed by "drawing" an ingot of steel which is surrounded by a concentric layer of aluminium or copper, the ratio of the thickness of the coating to the diameter of the steel core in the ingot being the same as the ratio of thickness of the coating to the diameter of the core material of the finished strand. In "drawing" the material through a series of dies to produce the required strand diameter, the steel core is "work-hardened" which increases the strength of the steel. The core wire of clad steel is therefore usually of a higher UTS than the core wire of normal ACSR. The aluminium clad steel core wire can then be used as the load-bearing portion of a conductor with 2, 3 or more layers of standard strands of aluminium wires.

One of the advantages of clad steel is that the material of the conducting strands and the material cladding the core is the same and therefore the possibility of interstrand electrolytic corrosion, due to the dissimilar metals ie zinc/aluminium in an ACSR conductor, is avoided. Clad steel conductors therefore are not required to be provided with greased cores which is usually necessary with ACSR conductors. In addition, there is a slight reduction in resistance since the aluminium of the cladding has a lower resistivity than zinc.

There are various other types of conductor produced for particular situations. ACSR conductors can be produced as "smooth bodied" conductor. To achieve the "smooth body" the final layer of strands is made up of wedge shaped segments. Smooth body conductors are used to reduce the possibility of wind induced conductor vibration which might otherwise occur particularly with long river crossing spans.

A number of producers manufacture a "self-damping" conductor. This is a conductor which is laid-up with an annular air space between the inner aluminium layer and generally the final two layers of aluminium strands. The system of damping is said to operate because if conductor vibration occurs the double layers of conductor strands move relative to the inner conductor and absorb and "damp" the vibrational energy.

A further type of conductor which is perhaps worthy of note, although little used, is of "expanded body" construction. This type of conductor is built around spacers to produce a conductor of larger diameter than the normally laid-up conductor with no increase in cross-sectional area. The intention of this conductor is to provide an increased diameter and hence reduced surface electrical stress at working voltage, resulting in improved radio-interference performance and reduced corona loss. Mention must also be made of a conductor into which optical fibre strands are layered, usually within a protective tube at the centre of the conductor and surrounded by conventional wire strands. The optical fibre is used for the transmission of communication and control information.

Summarising, the conductors most frequently employed at present are either ACSR or aluminium alloy as these provide the most economical type of construction, aluminium being cheaper than copper, and the reduced sag resulting from the higher tensile loadings available from the steel content, or high strength aluminium alloy allows tower heights and hence costs to be minimised.

3.3.2 Clamps and joints

Clamps and joints are very critical components of overhead lines. Joints and dead-ends must be capable of supporting the full line conductor tension, with any current up to full line current (and occasionally fault current) and at any temperature, from minimum ambient to the maximum conductor operating temperature (usually 75 °C). Clamps and joints must also provide a radio noise performance at least equal to that of the line conductor.

For an ACSR conductor, the joint is made in two parts — a galvanised or stainless steel inner cylindrical sleeve, surrounded by an outer cylindrical aluminium sleeve. Fig. 3.9 shows a cross section of midspan joint. The inner and outer sleeves are compressed by means of an hydraulic, motor-driven press. The dies of the press are interchangeable and are selected for the particular diameter of sleeve which is being compressed.

In preparation for fitting the sleeve, the aluminium conductor strands must be cut away by exposing the correct length of steel core wire. The steel strands must not be damaged when the aluminium strands are cut. Most manufacturers require the steel and aluminium to be coated with grease compounds before compression. These compounds generally have corrosion inhibiting properties and may contain lithium based soaps. Some manufacturers require the compounds to be loaded with small abrasive particles which increase the friction developed between strands and the joint body.

A vital detail which must be carefully controlled during fixing of two-part joints is that the outer sleeve must be correctly centred over the inner sleeve. This feature can be controlled by means of the simple jig shown in Fig. 3.9. Small dome-headed aluminium plugs are driven into the two locating holes after compression.

During type testing [3], samples must be subjected to a heating cycle current test amounting to at least 1000 cycles of heating due to full load current followed by cooling to room temperature with zero current. During the test, the conductor must be supported under a tensile load applied between clamp and conductors of 20 per cent of the conductor breaking load. After the application of 1000 heating and cooling cycles the clamps and joints are cut open for examination. There must be no indication of burning, fusing or local heating of the fittings or conductor.

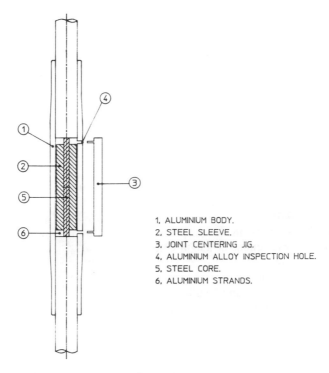

1, ALUMINIUM BODY.
2, STEEL SLEEVE.
3, JOINT CENTERING JIG.
4, ALUMINIUM ALLOY INSPECTION HOLE.
5, STEEL CORE.
6, ALUMINIUM STRANDS.

Fig. 3.9   Midspan joint for ACSR conductor

Conductor dead-end clamps are manufactured using similar designs to mid-span joints and in the past have generally been of the "compression" type. Latterly, wedge-type clamps are being used more frequently but generally only on all-alloy conductors. This is because the compression derived from the wedging action of the clamp can only be applied to the complete conductor. It is not easy to apply a wedge clamping system to the steel load-bearing strands of an ACSR conductor.

It is good line maintenance practice to examine the operational performance of mid-span joints and dead-end clamps. This is done by means of infra-red or heat sensitive cameras either from ground level, or more frequently, by means of a camera fitted into a helicopter. The helicopter allows the camera operator to film each joint and clamp and, by means of simultaneous voice recording, the current being carried by the line whilst each joint is examined can be superimposed on the record. After the flight, the condition of each joint can be monitored in comfort on the ground.

Badly fitted clamps or those of poor design will become heated. Therefore, any clamp showing signs of heating must be replaced soon after being located. Heating of the clamp is caused by poor contact which results in high resistance. This becomes progressively more severe until total burn-out occurs, resulting in the dropping of the line conductor.

3.3.3 Equilibrium of the suspended wire

When a wire or conductor is supported at equal height at its extremities the wire will sag in the centre and assume the form of a catenary. The curve of the catenary can be expressed as

$$y = c \cosh\frac{x}{c}$$

For convenience the origin of the curve is chosen so that:

c = T/W
L =   span length
S =   true conductor length C to B
D =   sag
T =   tension at C
W =   mass per unit length
c =   distance between C and origin.

The diagram below shows the catenary under consideration:

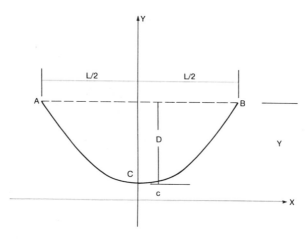

The method of calculating the true conductor length S from A to B is given in Appendix 3.1 together with the derivation of an equation for determining the sag of a conductor in a span. The "change of state equation" which enables the conductor tension to be calculated for a range of temperatures is given in Appendix 3.2.

The method of calculating conductor sag and tension in steeply sloping ground situations is covered in detail in [4].

3.3.4 Conductor creep

When a conductor is strung in a "section" of line ("section" being the name for the length of line between any two angle or section towers) the conductors in each span will form catenaries and the tension throughout the section will be virtually constant provided the conductors are supported on suspension insulator sets which are free to swing. The average or equivalent span throughout a section is calculated using the equation:

$$\text{equivalent span} = \sqrt{\frac{L_1^3 + L_2^3 + L_3^3 + L_n^3}{L_1 + L_2 + L_3 + L_n}}$$

where the span lengths, L_1, L_2, L_n etc, are the horizontal distance in metres between the centre of each tower.

When a new ACSR conductor is tensioned for the first time the conductor is found to change in length in a non-elastic manner. That is, the conductor increases in length due to the applied tension and does not return to its original length when the tension is reduced to zero. This increase in length is termed "creep". The amount of creep which a conductor exhibits is a function of the proportion of aluminium to steel, the number and diameter of strands and the degree of tightness with which the strands are "layered-up" during manufacture.

Manufacturers are required to measure and state the creep of the conductor. This information is required because, when the sag of a conductor is calculated to establish the ground clearance for a particular tower height, any slight increase in conductor length will generally result in a significant increase in sag. Typically, with a 500 m span of "dove" conductor at a tension of 10 kN, a 200 mm increase in true length of the conductor will result in an increase in sag of one metre.

The value of creep can be expressed as an increase in length for a particular tension. A more convenient method is based on the sag and tension curves which a contractor produces as an aid to establishing correct conductor tensions during stringing. The sag and tension curves are a series of curves of conductor sag against tension for a range of all likely ambient temperatures for the region in which the line will be constructed. If the temperature of a conductor is increased, the true length increases reducing the tension and increasing the sag. A convenient method of indicating the creep factor is to relate the change in conductor length due to creep to the increase in temperature which would cause the same increase in length. It is found that with typical 25 mm diameter ACSR conductor the extension due to creep is equivalent to an increase in temperature of 20 to 25 °C.

The contractor is required to provide "initial" and "final" sag and tension charts. The "initial" chart showing the sag and tension of the newly strung conductor and the "final" sag and tension chart showing the sag and tensions which will prevail after about one year in service and for the remainder of a conductor's service life. The amount of creep is high during the first few hours after tension is applied and gradually diminishes with time during the service life of the conductor.

Typically the allowance for the increase in sag which will result from the inelastic extension, due to creep, is equivalent to the increase in sag which would result from raising the conductor temperature by 22°C.

3.3.5 Wind and ice loads on conductor

The effect of wind pressure on transmission line conductors is to increase the conductor tension. The resultant of the transverse wind load and the vertical load due to conductor mass (combined with ice load if appropriate) is applied to the change of state equation (see Appendix 3.2) in order to arrive at the conductor tension under these particular conditions. It is necessary to consider various combinations of transverse and vertical loads when arriving at the final design parameters for both conductors and towers.

The design wind pressure is derived from meteorological records of wind gust velocities. A number of years' records are necessary to provide statistically meaningful information. An approximate empirical formula which relates wind velocity to pressure on cylindrical bodies is:

$$P = \frac{1}{2} \rho C G V^2$$

where:

ρ is the air density which is equal to 1.225kg/m^3 at 15°C and at normal atmospheric pressure.

C is the drag coefficient, typically equal to 1 for most conductors.

G is the gust response factor which varies with terrain roughness, span length and mean height above ground and typically is between 1.5 and 2.0.

V is the design wind velocity in m/s.

If ice loading has to be considered in the design, it is usual to consider that this is of a uniform radial thickness with a density of 900kg/m^3. It is unusual to find a combination of maximum wind and maximum ice loading and various combinations of the two loadings are used in transmission line design. The selection of the combination of loads is heavily dependent on the designer's experience.

3.3.6 Ampacity

When electrical energy flows through a conductor, the conductor gains heat due to I^2R heating and, if the conductor is exposed to the sun, heat is also gained due to solar radiation. The conductor loses heat by convection and radiation.

In order to calculate the current rating of a conductor, a heat balance equation is employed where the heat gained by the conductor is equal to heat lost by the conductor. This empirical equation can be expressed by:

$$I^2R + \alpha \, s'd = H_c + E_s \left[(\theta + t + 273)^4 - (t + 273)^4 \right] \Pi \, d \quad \text{watts/cm}$$

A typical conductor ampacity calculation is included in Appendix 3.3 to this chapter. A more rigorous method of conductor ampacity calculation is given in [5]. The method shows the allowance which is to be given not only for solar and joule heating but also due to heating developed by magnetic effects.

3.4 Dampers and spacer dampers

3.4.1 Introduction

Transmission line conductors frequently vibrate, generally because of the action of wind blowing over the conductors. In this section, the causes of vibration and methods of minimising vibration levels are discussed.

3.4.2 Single conductors

Single conductors exhibit two separate forms of vibration. The first, of small amplitude and high frequency, is known as aeolian vibration. The frequency of aeolian vibration is usually in the range of 5 to 100 Hz and with an amplitude of 0.5 to 2 conductor diameters. The wave motion is largely in the vertical plane. The second type of oscillation is called galloping and this type of violent oscillation is most often, but not always, associated with ice-covered conductors.

Aeolian vibration is generally caused by wind blowing normal or up to an angle of about 75° to the conductor axis with a velocity of 2 km/hr (0.55 m/s) to 32 km/hr (8.8 m/s). When wind blows past a conductor at a velocity greater than the critical velocity, vortices begin to form on the leeward side of the conductor. The conductor, because of the helical form of the outer strands, will be subjected to either "lift" or "depression" and the direction of the force will depend on the "yaw" angle of the wind and on the lay of the conductor. Fig. 3.10 depicts the lift and drag forces developed by the action of wind on a single conductor.

Aeolian vibration was first noted in the 1920s and it was at about that time that the Stockbridge damper was developed to counteract the effects of this type of

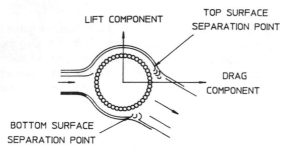

TYPICAL PATTERN OF AIR FLOW

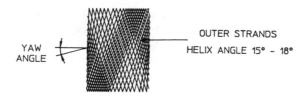

VIEW ON SURFACE OF CONDUCTOR

Fig. 3.10 Aerodynamic lift developed with stranded conductor

vibration. The vibration is developed because of the synchronous shedding of vortices along a length of conductor. This results in alternating lift forces which occur at some natural frequency which is proportional to the wind velocity and inversely proportional to the conductor diameter. If the conductor is assumed to have negligible stiffness a conductor will have a natural oscillation frequency of:

$$f = \frac{1}{2L}\sqrt{\frac{T}{m}}$$

where

f = frequency (Hz)
L = length between node points (m)
T = tension (N)
m = mass (kg/m).

If uncontrolled, aeolian vibration can give rise to fatigue damage to conductor strands adjacent to clamps. The damage may occur to the strands beneath the outer layer and may not become visible until severe damage has been incurred.

The Stockbridge damper provides assistance in controlling aeolian vibration. The dampers are fixed at points where antinodes form during conductor vibration in order to absorb the maximum amount of vibrational energy. This energy is dissipated as heat in the messenger wires of the dampers. An equation to establish the fixing position of dampers with respect to the suspension clamp is as follows:

$$D = \frac{0.0152}{V} \sqrt{\frac{Td^2}{W}}$$

where

D = distance, Stockbridge clamp to conductor clamp centres (m)
V = critical wind velocity (m/sec)
T = conductor tension (kg)
d = conductor diameter (mm)
W = conductor mass/metre (kg/m)

The design of the Stockbridge damper means that each type of damper will have its own natural vibration frequency. The frequency will depend on the mass of the damper weights and the stiffness of the messenger cable. The result is that a damper will absorb vibrational energy in a relatively limited range of frequencies. One method of improving the performance of Stockbridge dampers is to provide weights of two different masses. The result is to produce a "flatter", less peaky vibration response curve which will enable a wider range of vibrational frequencies to be absorbed.

A further method of minimising the possibility of aeolian vibration occurring is to string the conductor with a relatively low tension. With ACSR, it is found that if the tension is kept to less than 20% of the conductor breaking tension then the likelihood of vibration is significantly reduced.

The second mode of conductor vibration, namely galloping, most frequently occurs with ice coated conductors. The conductor is usually found to be coated with ice with a crescent formation which results in a marked change in the aerodynamic properties of the normally circular conductor. Usually, the conductor movement is in the vertical plane and amplitudes of 1 or more metres can occur. In severe cases, interphase flashovers can occur due to the transient reduction in phase clearances. Cases of conductors of different phases clashing are not unknown. Generally wind speeds no greater than 45 km/h are required to induce galloping.

There are a number of recorded cases of conductor galloping which had apparently been induced by short duration wind gusts. In general, gust-induced galloping occurs in very long, usually river crossing, spans. Gust-induced galloping can be largely controlled by the use of segmental, smooth-bodied conductors. The smooth conductor surface providing an improved aerodynamic performance over that achieved with a conventionally-stranded conductor.

Cases have been recorded where galloping has been induced as a result of corona discharges on the conductor. In these cases the conductor has been of relatively small diameter and the system voltage such as to provide a surface stress of greater than 23 kV RMS/cm. This stress is significantly higher than the stresses at which most normal overhead lines are operated. Accordingly, we do not consider corona induced galloping further.

Fortunately, the occurrence of severe galloping is very rare and generally only affects a limited number of isolated spans and then only in particular geographical locations with particular wind conditions. One method of controlling this type of oscillation is to use interphase ties. These are composite insulators capable of withstanding the full phase to phase voltage, with a creepage path and dry arcing distance of at least √3 times the requirements for line to earth voltages. The rod insulators are clamped between the conductor of each phase usually in a delta formation to position the conductors of each phase and eliminate clashing.

3.4.3 Bundled conductors

Twin, triple and quadruple bundled conductors are subject to wind induced aeolian vibration and galloping and, when fitted with spacers, also to sub-span oscillation. Multi-conductor bundles are more prone to vibration than single conductors because any vortices formed by the "up wind" conductor will affect the "down wind" conductor. For example, with twin conductors the leading conductor may initially lift causing the leeward conductor to be depressed and then the leading conductor can "stall" and drop. The action can easily set up a natural rocking action of the conductor.

When first employed some years ago, twin conductors were fitted with Stockbridge dampers to control aeolian vibration. These dampers do control vibration but only at the end of the spans. Vibration of the conductors frequently occurs between the spacers because the Stockbridge dampers fitted to the end of the spans cannot provide any damping action beyond the first spacer. This type of vibration is termed sub-span oscillation. The sub-span vibration generally has a vibrational frequency in the range 0.5 to 2 Hz with amplitudes of up to 2 or 3 times the conductor diameter. The result of much research work was the development of the Spacer damper. Some of the early Spacer dampers employed messenger wires as energy absorbing devices as employed in Stockbridge dampers. Later devices have largely utilised elastomers, loaded in shear or compression, as energy absorbers. Many of the early designs of elastomer spacer provided energy absorption by allowing the arms to swing through a small arc at right angles to the conductor axes. Since sub-conductor oscillation tends to take place in the vertical plane these simple types of spacer damper provide relatively limited amounts of vibration energy absorption. More recent spacer dampers have energy absorption capacity in both the vertical and horizontal planes.

Correctly designed Spacer dampers will provide protection against aeolian vibration, thus dispensing with the requirements for Stockbridge dampers. They will also provide protection against sub-span oscillation.

3.5 Foundations

3.5.1 General

The foundation is the name given to the system which transfers to the ground the various steady state (dead) and variable (live) loads developed by the tower and conductors. Foundations may be variously subjected to compressive or bearing forces, uplift and shear forces, either singly or as a result of any combination of two or three of the forces.

Usually, the limiting design load with transmission line foundations is the uplift load. In this respect, there is a major difference between the design of foundations for transmission lines compared to the design of foundations for most normal civil engineering structures. Accordingly, the amount of literature describing design techniques for overhead line foundations is relatively small compared to the literature available for more traditional civil engineering foundation design practice.

The selected foundation design for a particular tower must provide an economical, reliable support for the life of the line. The foundation must be compatible with the soil and must not lose strength with age. With the progressive increase in transmission system voltages there has been a related increase in foundation sizes and it is worth noting that with a typical quad conductor 500kV line, single leg uplift and compression loads of 70 or 80 tonnes ultimate are usual for suspension towers. With tension towers, ultimate loads of 200 or 300 tonnes are often developed. In ground of poor load-bearing capacity the dimensions of foundations becomes considerable.

In the past, it was often acceptable to "over-design" foundations to allow for uncertainties in the soil characteristics. With the very considerable sizes of present-day foundations for EHV and UHV transmission it is obvious that considerable economies can be made in producing foundation designs to exactly match the soil conditions.

Increasingly, transmission lines are routed through areas of poor ground conditions, often for reasons of amenity. This results in the need for the use of special, generally larger, foundations. The logistical problems of installing large foundations, often in difficult ground conditions, can also be considerable and must be taken into account when considering the design of foundations.

3.5.2 Types of ground

The ground in which the foundations are installed can vary from igneous, sedimentary or metamorphic rock, non-cohesive soils, sand or gravel to cohesive soil, usually clays. Equally, soils with a high organic content, for example peat, can also prevail. Composite soils will also be found and examples of these are sandy gravels and silty sand or sandy peat.

Fundamental to the proper design of foundations is an accurate series of soil tests to determine the range of soil types for which the foundation designs will be required. It is good practice to carry out soil tests at a rate of 1 in 5 tower sites. This is generally sufficient to enable an accurate forecast of the range of soil types to be established. It should be pointed out however, that with large towers having 15 or 20 m square bases, occasionally each of the four legs of a tower may be founded in four different types of ground.

3.5.3 Types of foundation

There are seven basic types of tower foundations:

(i)   steel grillage;
(ii)   concrete spread footing;
(iii)   concrete auger or caisson;
(iv)   pile foundation;
(v)   rock foundation;
(vi)   raft foundation;
(vii)   novel foundations.

3.5.4 Foundation calculations

There are a number of methods of calculation of foundation uplift and bearing capacity. In his paper [6], M.J. Vanner gives some relatively sophisticated methods of foundation uplift design. For the purposes of this chapter however, we will confine ourselves to a simple approach which must be treated with care. Nevertheless, the methods indicated will give reasonably accurate results for the relatively shallow foundations which are normally employed with transmission line towers. A shallow foundation is usually defined as one in which the breadth of the pad is greater than the setting depth.

It is usual to calculate the uplift capacity of a foundation as being equal to the mass of soil contained in the frustum developed between the base of the foundation pad and the soil surface. The angle of the face of the frustum to the vertical is usually designated ϕ and will vary from 35° to 40° in rock, to 25° in good homogeneous hard clay to zero in saturated non-cohesive ground. The soil density will vary from just over 2000 kg/m^3 for homogeneous rock to about 1600 kg/m^3 for soil with normal moisture content to about 800 or 900 kg/m^3 in the case of ground subjected to water uplift. Methods of calculation of uplift capacity are shown on Figs. 3.11, 3.12 and 3.13.

3.6 Insulator design

3.6.1 General

As indicated in Section 3.1 of this chapter it is usual to employ "pin-type" insulators for distribution voltages up to 30 or 40 kV. Above this voltage and up to about 132kV system voltage "line post" insulators may be used or more usually from 40kV to the highest transmission system voltages the "cap and pin" insulator is used.

In Germany and in parts of the world influenced by German design practice, the "long rod" insulator is sometimes used. In recent years, other designs of insulator have been developed, notably rigid insulators utilising a central load-bearing core formed from glass fibre bonded to form a rod, and having plastic sheds. This type of unit is often referred to as a "composite insulator".

3.6.2 Pin-type insulators

The pin-type insulator is so called because in use it is screwed onto a galvanised forged steel "pin" which is mounted vertically on a metal or wooden crossarm.

For low voltage systems, 6.6 to 11kV, it is usual to have a one-piece insulator shed in which the porcelain is loaded largely in compression. A sketch of a typical pin-type insulator is shown in Fig. 3.1(a). Reference to the sketches will show that the top of the porcelain body is formed into a groove into which the conductor is bound by means of wire or fixed with the aid of special clips. Toughened glass pin type insulators require a metal cap; this holds together the "diced" pieces of glass which result if the glass becomes shattered.

For increased working voltages, the number of "sheds" of the insulator unit are increased, additional sheds being jointed together by means of portland cement/sand mortars.

For system voltages above about 33kV, pin-type insulators become quite large and unwieldy devices. For 33kV systems a creepage distance of about 650mm will be required for reasonably clean areas. Accordingly, above about 33kV cap and pin insulators or line post insulators are usually employed. A typical so called "trident" wood pole 132kV line is shown in Fig. 3.3 using line post insulators. An upper limit for systems insulated with line post units would be about 150kV with present day practice.

3.6.3 Cap and pin insulators

The names cap and pin insulators are given because the insulators are fitted with "bell" shaped galvanised, malleable cast iron or forged steel caps and galvanised forged steel pins. The dielectric material can be of toughened glass or porcelain. With both porcelain and glass the shed is formed to produce a smooth upper surface and generally a ribbed lower surface to maximise the creepage path. Typical cross sections of cap and pin insulators are shown in Fig. 3.2.

The load bearing section of the dielectric must be carefully dimensioned to ensure that the material is correctly stressed electrically and mechanically. This is because both toughened glass and porcelain are brittle materials with comparatively limited tensile strengths. The load-bearing portion of the dielectric is designed to ensure that any tensile load applied between the cap and the pin is

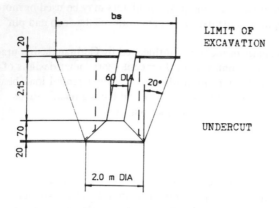

Concrete volume

0.3^2 . . 0.2   = 0.056m³

0.3^2 . . 2.15   = 0.607m³

$\dfrac{0.7}{3}$ . . $(1.0^2 + 0.3^2 + 1.0 . 0.3)$ = 1.018m³

1.0^2 . . 0.05   = 0.157m³

                                                        ‾‾‾‾‾‾‾‾‾‾
                                                        1.838m³

Concrete weight

1.838 . 2240 = 4117kg   = 4.117t

Earth volume

b_s = 2. 2.85 . tan 20° + 2.0 = 4.07m³

$\dfrac{2.85}{3}$. $(2.035^2 + 1.0^2 + 2.035 . 1.0)$ = 21.40m³

− (0.607 + 1.018)   = -1.625m³

                                                        ‾‾‾‾‾‾‾‾‾‾
                                                        19.775m³
                                                        ‾‾‾‾‾‾‾‾‾‾

Earth weight

19.775 x 1600   = 31.640t

Total weight resisting uplift   = 31.640t
                                                        + 4.117t

                                                        ‾‾‾‾‾‾‾‾‾‾
                                                        35.757t
                                                        ‾‾‾‾‾‾‾‾‾‾

Fig. 3.11   Undercut pyramid foundation

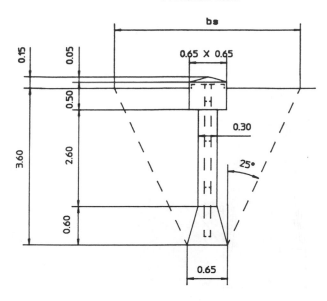

Volume of concrete

$$x\ 0.6\ (0.65^2 + 0.3^2 + 0.65\ x\ 0.3)\ /12 = 0.111$$
$$x\ 0.3^2\ x\ 2.60/4 = 0.183$$
$$0.65^2\ x\ 0.5 = 0.211$$
$$0.65^2\ x\ 0.05/3 = 0.007$$

$$\overline{0.512^3}$$

Volume of frustum

$$b_s = 2\ x\ 3.6\ \tan 25 + .65 = 3.632$$

$$\text{volume} = 3.6\ (3.632^2 + 0.65^2$$
$$+ 3.632\ x\ 0.65)\ /12 = 10.81m^3$$

Weight of concrete $0.512\ x\ 2240\ kg/m^3$ $= 1146kg$
Weight of ground $10.81\ x\ 1600\ kg/m^3$ $=17296kg$

Total uplift capacity $=18442kg$

Fig. 3.12   Concrete auger foundation

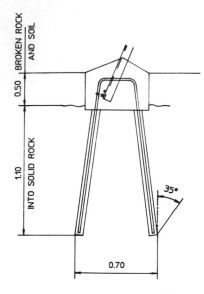

Yield point of anchor bars 2400kg/cm^2
Bond strength of bars = 4 x 1.1 x 10^2 x 10.49 x x 2   = 29.000kg
Volume of uplift frustum:
   Width of frustum at g.1 = 2 x 1.6 tan 35 + .7 = 2.66m   = 2.66m
   Volume of frustum 1.6 x 12/12 (2.66^2 + .7^2 + 2.66 x .7)   = 3.94m^3
   Weight of rock = 2300kg x 3.94   = 9.079kg
Uplift strength of re-bars = 4 x 2400 x .2^2/4   = 30.138kg

Fig. 3.13   Rock anchor foundation

transferred from the load bearing cone of the pin to the re-entrant bell mouth of the cap, through the cement so that the dielectric is loaded largely in compression. The compressive strength of porcelain and toughened glass being about 10 times the tensile strength.

In the event of the dielectric being destroyed, perhaps due to rifle fire, resulting in the skirt being completely broken away, the mechanical security of the broken insulator is assured by the design of the metal fittings.

Generally, a correctly designed cap and pin insulator will, in clean non-polluted areas, provide a satisfactory service life of at least 25 years. It is worth noting that there are many cap and pin insulator units which are still in service after 40 years in non-polluted areas. In regions with high levels of marine or industrial/chemical pollution however, the galvanised coating of the metal fittings is often quickly degraded and, if the pollution is severe, the "tell-tale" sign of rust can become visible after five or six years in service. There are instances where insulator pins have been so severely attacked by corrosion that the original pin diameter of 16 mm has reduced to 7 mm, resulting in a severe loss of mechanical strength.

In recent years, this type of attack has been largely eliminated by means of zinc collars alloyed to the galvanizing coating of the insulator pin. The collar is arranged to intercept any leakage currents which pass over the insulator surface and terminate at the pin near to the cement surface. By this means it is possible to protect the "pin" from the corrosive effects of leakage currents for ten or twelve years even in severely polluted areas.

3.6.4 Dielectric materials

3.6.4.1 Porcelain

The material used to manufacture electrical porcelain is usually a combination of ball clay, china clay, quartz, alumina and various flux materials, for example feldspar. The constituents are crushed and mixed together with water to form a suspension. The clay suspension, known as slip, is then pumped to a de-watering press where the water content is reduced to about 20% by volume, resulting in the formation of a plastic clay cake. The air is then removed by means of a vacuum process in a pug mill from which the clay is extruded in a plastic state. The soft clay body can then be pressed or turned to form the insulator shape, this is then dried to enable the shaped body to be handled without distortion.

The clay form is then dipped into a glaze (a suspension in water of metallic earth elements and clays), and subsequently fired in a gas or oil fired kiln where the temperature is slowly raised to about 1200 °C and then carefully cooled to room temperature. During firing, the clay materials are converted to form a glassy matrix with a dielectric strength of about 200 kV/cm with a dielectric constant of approximately 5. The porcelain matrix has a coefficient of thermal expansion of about $4 \times 10^{-1} K^{-1}$ and a volume resistivity at 20 °C of 10 ohm.

After firing, metal fittings are fixed by means of a cement mortar. When the cement mortar is cured, the insulator is subjected to a routine mechanical test load of 40 or 50% of the guaranteed minimum failing load (MFL) for 10 s. Finally, the complete insulator is subjected to a flashover test for 5 minutes at a voltage equal to the dry flashover voltage of the insulator.

3.6.4.2 Toughened glass

Glass insulators are most usually made from soda-lime-silica glass. The crushed raw material being melted in a furnace called a "tank" in which the temperature of the raw material is raised to about 1250 °C. The molten glass slowly flows through the tank and is fed in "tear" shaped blobs of constant weight into steel moulds where the viscous glass is moulded to shape. If at this stage in the manufacturing process the glass is allowed to cool slowly an annealed glass will form. This type of glass, like window glass, can easily be cracked. However, if the hot, shaped insulator shed is cooled rapidly by blowing cold compressed air onto the surface of the glass, the material becomes toughened.

This process arises because the outer surface of the glass chills and becomes solid but the inside of the glass remains molten and cools more slowly. The result is that the outer layer of the glass goes into compression and the inner layer goes into tension. The toughening process provides a mechanically strong material with a dielectric strength of some 1350 kV/cm, a dielectric constant of 7.0, resistivity at 20°C of 1012 ohm and a coefficient of expansion of 8.5 x 10^{-6} K^{-1} similar to that of the ferrous fitting which is approximately 11.5 x 10^{-6} K^{-1}.

Assembly with metal end fittings uses similar methods to those employed with porcelain insulators although high alumina cement is usually employed for the assembly. Toughened glass, however, has a unique property which is that if the shed is in one piece the disc must be electrically sound. There is no need therefore to carry out routine electrical tests on toughened glass insulators, only routine mechanical tests are required.

The unique feature of toughened glass is used to advantage in service since, if the glass shed of an insulator can be "seen", the electrical properties of the disc are good. Toughened glass insulators can therefore be examined from a distance and seen to be electrically sound. Toughened glass insulators are more resistant to the melting effects of high power fault currents than porcelain insulators. Generally, for equivalent mechanical ratings, toughened glass units are lighter than porcelain. However, toughened glass insulators are much more susceptible to surface damage resulting from the effects of leakage currents which develop in service in polluted conditions. Damage by leakage currents results in melting of the glass surface and the formation of small molten channels of glass which, when these become sufficiently deep, cause the glass to shatter.

3.6.5 Long rod insulators

Long rod insulators are generally manufactured from high alumina porcelain, but some are marketed which are formed from boro-silicate annealed glass. A recent survey showed that 86% of all insulators covered by the survey were of the cap and pin type and 14% were of the porcelain long rod type.

With long rod insulators the tensile service load is supported entirely by the core of the porcelain. Because of the brittle nature of the porcelain, considerable effort is spent on quality control during the manufacture of long rods. The quality control procedures employ ultra-sonic crack detection equipment, both before and after routine mechanical tests during which the units are loaded to about 80% of their guaranteed minimum failing load. By means of the stringent quality control, faults are largely eliminated from these insulators. However, when accidental damage is sustained by long rod insulators the results, due to the brittle cracking of the core, are usually catastrophic.

3.6.6 Composite insulators

Composite insulators have been under development for more than 30 years, and to date they are still only in service in relatively small quantities. As stated earlier, the mechanical load of composite insulators is supported by means of a resin bonded glass fibre core. Typically, a 50 mm diameter core will have a minimum failing load of about 500kN. To protect the core from attack by leakage currents and to provide an extended creepage path, the cores are surrounded by weather sheds. These may be cast as one piece onto the core or as separate weather sheds which, after bonding, are effectively vulcanised to form a homogeneous insulating unit.

It is in the study of weather shed materials that most development effort has been concentrated. Early insulators utilised polysulphide rubber. Later, ethylene propylene dimethyl monomer (EPDM) and other monomers and polymers were used. Most of these suffered serious damage by attack from the ultraviolet content of sunlight. The attack caused the surface to lose the shiny finish produced during manufacture and become rough, which allowed pollution materials to build up. The rough surface results in the increase in the magnitude of leakage currents and when sufficient energy is available in the leakage current, the surface melts. The early materials were generally organic compounds and rapidly the surface became carbonized as a result of the leakage currents.

Considerable improvements have been made recently by the use of silicone rubber compounds. These compounds, when formed into weather sheds, are hydrophobic. As a result they cause any rain water to form into discrete droplets so that continuous wetted areas do not form. Accordingly, leakage currents are kept to low levels and good pollution performances are achieved with silicone rubber, certainly for the first 8 to 10 years of service life. It is found that the particles of pollution deposited on the weather shed also become impregnated with silicone exuded by the silicone compound. Hence the pollution itself becomes hydrophobic together with the surface of the weather shed. The hydrophobicity can be lost by the action of intense surface discharge activity. When the discharge activity stops, the surface and pollution particles recover their hydrophobic properties over a period of about 24 h. It is for this reason that prudent line designers would specify creepage path lengths equivalent to those used presently for the toughened glass or porcelain insulation. By this action, the level of discharge activity under heavy rain and pollution conditions is minimised.

Composite insulators are lighter in weight than the equivalent glass or porcelain insulators but more expensive. At present however, it is necessary to estimate the likely strength reduction which will occur with composite insulators in service using the methods described in [7, 8] since service life experience is limited to about 12 years.

3.6.7 Insulator sets

3.6.7.1 General description

The line conductor is supported from the tower by means of an insulator set, the name given to the string of insulators, associated clamps, fittings and frequently arcing horns or grading rings. The number of insulator units in the string is determined by the clearance requirements demanded by the lightning flashover performance and with systems above about 220kV, the required switching impulse performance requirements. The numbers and types of insulator are also fixed by the creepage path requirements necessary to cater for the likely pollution levels. Finally, the mechanical strength of the insulator set is established by the everyday load and overload which the conductors will apply to the insulator set.

Insulator sets are of two basic types. The first, nominated a suspension set, so called because the insulator set is suspended from the tower cross arm and the conductor is suspended from the clamp at the lower end of the set. The simple suspension set is free to swing from the cross arm transversely or longitudinally to the line route. The second basic type of insulator is called the tension set since the full line conductor tension is supported axially by this set.

The shape of insulator unit selected depends largely on the type of pollution to which the line will be exposed. With normal relatively low levels of pollution requiring Class I creepage nominated in BS 137, i.e. 20mm/kV of system highest voltage, the normal or plane-shedded insulator is employed. This type of insulator has comparatively small ribs on the under surface.

If the line is to be constructed in an area where significant amounts of salt spray or heavy fog is likely to occur then an anti-fog insulator is usually selected. This type of insulator has deep "anti-fog" ribs to enhance the creepage path.

In desert situations, an open profile or aerodynamic insulator is often employed. This type of insulator is basically a flat disc, generally with no ribs on the underside. To achieve the appropriate amount of creepage the diameter of these units is greater than that of standard insulators. The open profile insulator operates successfully in the desert because the wind, which usually supports dust and salts (frequently sodium chloride), passes by the insulator with little disturbance. In dusty situations, eddies would form around the ribs of conventional insulators and the change in velocity of the air stream would cause the deposition of polluting particles onto the insulator ribs. The rate of disposition can be quite high and this can result in the insulator becoming heavily polluted which results in the increased probability of flashover.

3.6.7.2 The effects of lightning

When lighting strikes an overhead line, the voltage at the stricken point rapidly rises, and surges in the form of travelling waves progress rapidly along the conductor away from the stricken point to a region where the clearance to earth is sufficiently low to allow the air between conductor and earth to become totally ionised allowing a flashover to develop. Usually, the location with minimum clearance is at an insulator set. To ensure that a preferential path between conductor and earth exists, arc horns are fitted to the insulator sets. The arc gaps are arranged to be to the side of suspension sets and above tension sets. The length of the arc gap is less than the dry arcing distance of the insulator string. Thus, when as a result of a lightning induced surge an arc gap becomes ionised, a sparkover to earth will result. The spark will be fed by the power frequency energy of the line and the result will be a power frequency arc which will "run" safely between the arc-horn tips away from the insulator units until the line protection trips.

The energy of a power-follow-through arc is considerable and it is necessary to design the arc-horns and associated metalwork to have sufficient cross-sectional area to withstand the heating effects of the arc.

3.6.7.3 The effects of switching surges

During switching operations surges can be generated, and in spite of special arrangements for suppression, surges of 1.7 times the system voltage can occur. The surges can cause flashover and the preferred paths for these will be across arc gaps. Power-follow-through arcs will generally develop and arc gaps are therefore required to provide the same protection as is afforded to the insulator set during lightning impulse flashovers.

3.6.7.4 Voltage distribution across an insulator string

In service, the transmission voltage shared across each unit of an insulator string is not uniform. Unless special precautions are taken, up to 15 % of the line voltage will be developed across the line-end insulator and the voltage across each of the remainder of the insulators will progressively decrease towards the earth end of the string. The voltage is shared in relationship to the self-capacitance of each insulator unit and the stay capacitance of each insulator to earth.

The voltage sharing can however be significantly improved by specially designing the arc-horns to serve the purpose voltage grading rings. With well-designed grading rings the line-end unit may be found to support only 8 or 9 % of the phase voltage, resulting in significant improvements in the RIV performance of the insulator set.

3.7 Electrical discharges

3.7.1 Radio interference

Transmission lines sometimes produce radio interference often heard as the crackling, buzzing noise on a car radio when a car passes beneath a power distribution or transmission line. Radio interference (RI) is essentially a random radio signal radiated from some source. Depending on the relative strengths of the broadcast signal and the interfering noise, the signal to noise ratio (SNR), (or perhaps more accurately defined as the "protection ratio") varies from being a minor annoyance to a level which results in the total blotting out of a radio station. RI can have serious consequences if essential services involving air or marine radio navigation equipment are affected. Equally, power line carrier (PLC) equipment which generally operates at frequencies of 50 kHz to a few 100 kHz must not be affected by RI.

The main source of RI radiated by overhead lines are corona discharges formed on the conductor, hardware or insulators. Corona discharges form when the electric field intensity exceeds the breakdown strength of air and localised ionisation of the air occurs. There are a number of factors which affect the voltage at which corona forms, including the air density, the air humidity, the amount of photo-ionisation and to a limited degree the material of the electrode.

Corona will form when an electrode is charged to a sufficiently high voltage from a power frequency voltage source. The mechanism of corona production is different for each half cycle of voltage. The negative half cycle tends to produce short duration pulses of current up to perhaps 1 mA in magnitude lasting for about 0.03 μs. The radio noise generated by a negatively charged electrode tends to be of a low level compared to the noise generated by a positively charged electrode.

With an electrode charged with a low positive voltage, but which is sufficiently high to cause the formation of corona, a low level of steady corona forms giving very little radio noise. When the voltage is raised the radio interference voltage (RIV) becomes much greater and is usually about 100 times as high as the noise produced during the negative half cycle. Positive current pulses have a much longer duration than the negative pulses, positive pulses lasting up to 0.5 s. This corona produces a frequency spectrum shown below:

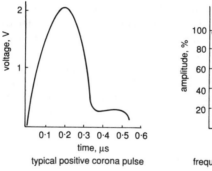
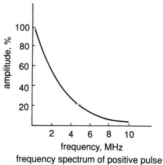

typical positive corona pulse   frequency spectrum of positive pulse

The value of E, the surface gradient at which corona forms on a cylindrical conductor is determined from the relationship established by Peek:

$$E = 30 \, \rho \, m \left(1 + \frac{0.0301}{\sqrt{r}} \right)$$

where ρ = air density
m = roughness or stranding factor
r = radius of the conductor (cm)
E = surface gradient (kVpk/cm)

Air density can be seen to affect the corona starting voltage. The higher the density, the higher the corona starting voltage in the range of normal atmospheric pressures which we consider.

Water vapour is electro-negative, that is, it will absorb free electrons and thus tends to inhibit the formation of corona pulses. Therefore, a high humidity atmosphere tends to provide an improved RIV performance for insulator sets. However, when water condenses to form droplets, or when drops resulting from rain attach to the conductor, corona forms on these water drops because the surface voltage gradient is increased by the water droplet and RIV is generated and power loss due to corona increases.

Although corona is the main source of RIV generated by overhead line, RIV can often be generated by small sparks developing across very small gaps. These gaps may be caused by poor electrical contacts between, for example, metal fittings of an insulator as shown below.

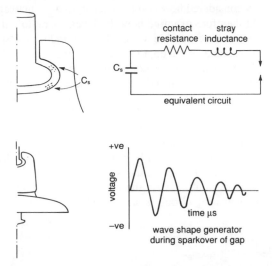

This type of breakdown generates very high frequency noise in the region 10 to 200 or 300 MHz. Contact-generated radio noise is much more prevalent in dry weather because during rainfall the small spark gaps caused by the poor contacts are "shorted out" by the conducting electrolytes formed from the rain water and dissolved salts.

Contact gap generated noise can usually be avoided by ensuring that there is sufficient mechanical load on the insulator sets to break down the insulating oxides or carbonates which form at the contact surfaces of the metallic ball and socket fittings. It is also possible to eliminate the noise source by coating the contact surface with conducting grease which inhibits the formation of sparks across the otherwise poor contact area.

The insulators of an insulator set can also give rise to radio interference. Generally, the noise produced by cap and pin insulators is caused because the surface voltage gradient exceeds the breakdown strength of the air surrounding the insulator. Proper design of the insulator can improve the RIV performance of the insulator by careful improvement in the shape of the electrodes (the caps and pins) and careful quality control of the level and surface finish of the cement used to fix the metal parts to the dielectric weather shed.

The surface resistance of an insulator has a very significant effect on the RIV generated by an insulator. Glass adsorbs moisture from the atmosphere and adsorbed moisture collected in the surface molecules of the insulator may well improve the RIV performance of an insulator by tending to improve the voltage grading around the insulator. It is therefore essential before commencing any RIV tests to allow an insulator to stabilise to the humidity of the atmosphere in which the test will be carried out. A period of about 24 h should be allowed for proper stabilisation. In addition, during testing the insulator units should not be exposed to the humidity of the exhaled breath of the test operator.

We have briefly considered how corona discharges and small sparks cause radio noise. We should therefore examine how the levels of radio noise are measured. RI is heard as a buzzing crackle or click in a radio loudspeaker. At low distribution voltages, and even up to system voltages of 66 kV, with normal conductor sizes, the conductor operating stress is very low, about 6 or 7 kV rms/cm. As a result, corona and RIV does not occur. The only sources of RIV at these low voltages are the insulators, bindings and stirrups or extraneous debris, for example short pieces of wire which might be thrown onto the conductor.

As 132 kV was replaced by 275 kV in the 1950s, people began to consider more seriously the problem of RIV, although in the United Kingdom at that time only a corona test was specified for 275 kV insulator tests. Nevertheless, the corona discharge was recognised as a potential source of power loss as well as being likely to give rise to difficulties with the operation of PLC and producing RI.

A typical RIV measuring circuit is shown below:

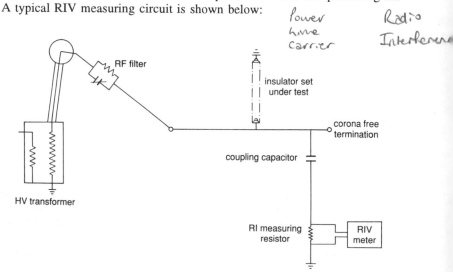

[handwritten annotations: "Power line Carrier", "Radio Interference", "RIV = Radio Interference Voltage?"]

The RIV measuring circuits and equipment currently used to determine the RIV performance of line hardware and insulator sets are defined by a number of National and International standards [9, 10, 11, 12].

As radio and TV signal wavelengths have been progressively reduced and with the much more frequent use of frequency modulated (FM) signals, problems caused by RIV are tending to lessen. In addition, the signal strength measured at the receiver resulting from the more general use of local radio stations is generally higher. Radio interference, in spite of increases in system voltages, is therefore becoming less of a problem. Nevertheless, any corona generated by the line represents a power loss and because energy costs continue to rise, the corona loss must be kept to economically low levels. Consideration must also be given to the safe operation of PLC equipment.

During the design of an overhead transmission line, the acceptable RIV level which can be generated by the line is established taking into account the signal strengths of the local radio transmission signals, the density of population living near to the proposed line route. A typical calculation of the RIV levels likely to be generated by a 500 kV transmission line is included as Appendix 3.4.

3.7.2 Corona loss

Consideration is also given to the likely corona loss from a transmission line during the design of the line. Generally, if the line is designed to have a good RIV performance, the corona loss from the line will be relatively low in dry weather. Under rainfall or in the period following rain when water droplets are still present on the conductors, losses of some tens of MW occur with 500 kV systems due to corona. A typical calculation of corona loss from a transmission line is given in Appendix 3.5 employing the method established by V.V. Burgsdorf [13].

3.8 Line construction

3.8.1 General

At the inception of an overhead line project, as part of the feasibility study, the location of the load centres both present and projected, together with present and future generation sources, will be established. The required transfer capacity of the line will then be known and the type of construction fixed. The economic study of the line design will have established the optimum span length and tower heights to provide proper ground clearance from the line conductor to satisfy the local ground clearance requirements. In most parts of the world the electricity supply authority or utility will have a schedule of mandatory ground clearances for the range of transmission voltages in use.

3.8.2 Route selection

Initially, the available maps covering the line route are obtained and a straight line between the line terminals drawn on the overall planning map. A convenient scale for use for a line 200 km or more in length is 1 : 250 000. The major topographical features of the regions crossed by the route should then be considered, for example areas with major access problems or regions with known areas of bad or poor quality ground should not be crossed if excessive foundation prices are to be avoided. Areas of outstanding natural beauty should also be considered. It may be necessary to avoid crossing areas with particular types of vegetation.

Other features crossed by the route must be considered, for example airports, major highways, pipelines, the lines of sight between microwave aerials and river crossings. The proximity of sea coasts or major isolated sources of industrial pollution must also be considered.

It will be necessary at this stage of the route selection to discuss the proposed route with town and country planners and other major planning authorities, for example highway planners, post and telephone, civil and military aviation authorities to ensure that the line would not cause obstruction or require to be re-routed as other developments take place along the route.

Bearing in mind the various points mentioned above, a preliminary line route can be established and transferred to 1 : 50 000 scale maps if these are available. If the maps are reasonably up to date it may be possible to establish the route with sufficient accuracy to hand over to the overhead line contractor at the start of the contract. However, in many parts of the world where line of sight access is limited, perhaps because of dense vegetation or where towns and villages are developing rapidly, it is sometimes useful after the proposed route has been marked on the 1 : 50 000 scale maps to carry out an aerial photographic exercise. A photomosaic strip map, about 1 km wide, of the proposed line route is produced. Careful examination of the photomosaics will allow existing maps to be updated with the minimum of disturbance to the local landowners.

A complete aerial survey of the line route can also be made, locating angle points and other features of special interest by means of satellite signals and Global Positioning Systems (GPS) to produce the necessary accuracy particularly in areas of dense vegetation. The aerial survey is followed by a ground survey which allows close inspection of soil and geology of the area crossed.

3.8.3 Basic span

The economic span was referred to in section 3.2.4.2. It is determined by calculating the effect the span length has on the installed cost of the line. If a short span is selected then short towers can be employed and foundation loads are minimised. There are, however, more towers per km and more insulator sets and hardware. If a long span is considered then increased tower lengths will be necessary to provide proper ground clearance to cater for the increased conductor sag. The number of insulator sets per km will be reduced but the working load of the insulator sets will be increased and may become critical. Further, conductor working tension must be limited to about 20% of conductor failing load to avoid problems with vibration.

The economic span can therefore be established by calculation by allocating costs to the various components. The optimum span established is sometimes referred to as the basic span. The length of the basic span fixes the distance between suspension tower centres in level ground to provide the correct clearance to ground. Because lines are frequently constructed in undulating ground it is necessary to cater for uneven ground. Similarly, it may be necessary to site towers in particular locations to avoid rivers or roads etc. Accordingly, for reasons of economy, it becomes necessary to provide towers with a range of increased heights. Usually, this is done by providing extended heights in steps of 3 m. In addition, for areas where short spans will be used, towers with "minus" heights are provided. Further consideration should be given to towers to be constructed on sloping sites and towers with a range of individual leg extensions are designed.

3.8.4 Line route profile

Having established the range of tower heights to provide the correct ground clearance for the various ground levels and span lengths it is next necessary to establish a convenient method of drawing a profile or cross section of the line. In section 3.3.4 we stated that the conductors in a section will be strung at virtually the same tension throughout the section. The towers throughout a section will usually be located at irregular intervals. It is found convenient to use a template cut from transparent plastic material to the shape of the conductor catenary at maximum operating temperature and with allowance for creep. This sag template, on which the heights of standard and minus and plus height towers are inscribed, can then be overlaid on the profile drawing of the ground surface. The profile of the centreline of the route is prepared by the survey team. The sag template can then be positioned on the profile to provide the correct ground clearance in each span and the tower heights can also be marked on the profile. In ground subject to side slopes, profiles left and right of centre line should also be marked so that ground clearances are achieved; this is especially necessary for horizontal configuration towers. It should be noted that for accuracy of plotting it is usually necessary to provide 3 sag templates, the first for the basic span, one for spans 10% less than the basic span and the third for spans 10% greater in length than the basic span.

During the survey, a record will be taken of the types of ground crossed and this information will be backed-up by soil investigations. Foundation designs will be based on the prevailing soil conditions. In preparation for construction, the range of types of foundation are tested by installing sample foundations in the various types of ground along the route. The test foundations are usually stressed in uplift to progressively higher loads relaxing the uplift load to zero after each increment. By this means, it is possible to demonstrate that the foundation will withstand the ultimate load applied by the tower and that any movement of the foundation is "elastic" and that plastic movement does not occur.

The foregoing preparation of profiles can be seen to be a time-consuming labour-intensive operation. For a number of years now, overhead line construction companies have adopted computer profiling programmes designed to operate interactively and thus optimise the location and types of tower in all line sections. The programmes print out a continuous profile drawing using the digitised levels and chainage down-loaded from the survey team's computerised records.

3.9 European standards and their impact

Many European countries have in the past produced their own national standards for insulators and fittings required for overhead line construction. In the early days there were considerable differences, particularly in dimensions — and in some instances test procedures — between these various standards.

In the early 1950s, under the auspices of the International Electrotechnical Commission (IEC), considerable international standardisation was achieved in details of dimensions and testing of insulators. The IEC, with its central office in Switzerland, has always been fully supported by European manufacturers and utilities. As the IEC Standards for dimensions and testing procedures of insulators have been agreed and published, European countries have generally adopted the text of these documents and translated them into their own language as the national standard. Thus, there is already a very large degree of consensus regarding the requirements for overhead line insulators.

In a few countries, notably Germany and Austria, there are additional standards in the DIN or VDE series which provide for standardisation in areas not covered by IEC Publications, e.g. caps for long rod insulators. These still employ the standardised IEC couplings.

In direct contrast to some areas of standardisation, there has been little attention given to European standards drawn up by the Comité Européen de Normalisation Electro Technique (CENELEC) for overhead line insulators, or indeed for insulators generally. The requirement for European Norms has now been recognised. In principle the existing IEC Standards will be adopted without amendment as the basis for European Standards. In a very few instances special national conditions will need to be recognised as deviations from a completely unified practice throughout Europe. An example of this does occur on overhead line construction where one of the couplings for 28 mm diameter pins was introduced in the UK many years before IEC 120 was extended to include this size. For excellent technical reasons the UK Standard differs from the IEC Standard, and will therefore have to be included in the European Standard as a special national condition.

In future, there will be very close co-operation between IEC and the European authority CENELEC so that new or revised standards will be submitted to dual voting procedures. With a long history of co-operation between the UK and its European partners, and on a worldwide basis, it is not envisaged that there will be any significant problems in reaching agreement on common IEC and CENELEC, European Standards, at least so far as insulators and their testing are concerned.

The situation on overhead line fittings in relation to European Standards is potentially more complex than for insulators. In the absence of any IEC Standards, other than those for coupling dimensions, each country has produced its own range of standard designs and testing procedures. Although the design requirements are in many cases identical, the details between one country — or manufacturer — and another often differ. In the UK BS3288: Parts 1 and 2, covering fittings, have been applicable to testing and design for many years.

At the present time, both IEC and CENELEC have Technical Committees and Working Groups responsible for preparing overhead line standards. It is hoped that the documents dealing with fittings will be identical, but achievement of this aim may involve considerable compromise by the various countries involved to reach agreement. The UK is actively participating in both organisations in order to protect its own interests where possible and to achieve a smooth transition from individual national standards to a common European Standard.

3.10 References

1   IEC 652 (1979) "Loading tests on overhead line towers"

2   IEC 1089 (1991) "Round wire concentric lay overhead electrical conductors"

3   BS3288 "Insulator and conductor fittings for overhead power lines".
        Part 1: 1973 (1979) "Performance and general requirements"
        Part 2: 1990 "Specification for a range of fittings"

4   BRADBURY, J., KUSKA, G.F. and TARR, D. J.: "Sag and tension calculations in mountainous terrain". *IEE Proc. C*, 1982, **129**, (5)

5   " The thermal behaviour of overhead conductors" *Electra*, October 1992, **144** (CIGRE Working Group 22.12)

6   VANNER, M. J.: "Foundation uplift resistance: the effects of foundation type and of seasonal changes in ground conditions". *IEE Proc. C*, 1982, **129**, (6)

7   IEC Draft Document 36 (Secretariat) 65. " Tests on composite insulators for AC overhead lines with a nominal voltage greater than 1000 V"

8   ANSI American National Standards Institute C29-11 1989. "Tests for composite suspension insulators for overhead transmission lines"

9   CISPR 16 (1987). "Specification for radio interference measuring apparatus and measurement method"

10   IEC 437 (1973). "Radio interference tests on high voltage insulators"

11   BS5049: 1987. "Methods for measurement of radio interference characteristics of overhead power lines and high voltage equipment"

12   BS5602: 1978 (1990). "Code of practice for abatement of radio interference from overhead power lines"

13   BURGSDORF *et al.*: CIGRE 413 (1960). "Corona investigations on extra high voltage overhead lines"

Appendix 3.1: Conductor sag and tension calculations

Referring to the diagram in section 3.3 the true length of the conductor from C to B may be calculated from:

$$\frac{S}{2} = C \sinh \frac{L}{2c}$$

Hence the complete conductor length S from A to B is obtained by substituting

$$\text{for } c = \frac{T}{W}$$

$$S = \frac{2T}{w} \sinh \frac{WL}{2T}$$

and the conductor sag is:

$$D = Y - C$$

$$\text{but } D = \left(\cosh \frac{LW}{2T} - 1 \right)$$

and cosh x can be expanded:

$$\cosh x = 1 + \frac{x^2}{2!} + \frac{x^2}{4!} + \text{etc.}$$

The equation for sag, D, can also be expanded as follows:

$$D = \frac{T}{W}.1 + \frac{LW^2}{2T}\frac{1}{2!} + \frac{LW^4}{2T}\frac{1}{4!} + \text{etc.} - 1$$

$$D = \frac{WL^2}{8T} + \frac{W^3L^4}{384T^3} + \text{etc.}$$

Typically, a 400 m span and a tension of 2500 kg with a conductor of mass 1.74 kg/m will have a sag of:

$$D = \frac{1.74 \times 400^2}{8 \times 2500} + \frac{1.74^3 \times 400^4}{384 \times 2500^3} = 13.942 \text{ m}$$

The second term of the calculation is only 0.15 % of the first term and, therefore, to a first approximation it is usual to ignore it. Accordingly, the equation:

$$D = \frac{WL^2}{8T}$$

is used to calculate sags and tensions for level spans when the sag is less than about 10 % of the span. This equation describes the shape of a parabola.

For use in the case of supports on sloping ground, i.e. with supports at A and B in the following diagram, the parabolic equation for sag and tension is adapted as follows:

$$X = \frac{L_2^2 W - 2TL}{2L_2 W}$$

where, as shown in the diagram below, X is the distance from the lower support, B.

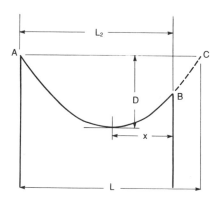

It should be noted that the parabola follows the same curve from A to B as for a parabola with level supports fixed at A and C. The above equation is sufficiently accurate for determining sags and tensions of slack spans from the terminal towers to the substation gantry.

In mountainous regions, however, where there are steep sided slopes, often resulting in uplift being applied to cross arms, a more rigorous analysis is required [7].

Appendix 3.2: Change of state equation

An equation for the change in conductor tension with temperature change is:

$$T_2^3 + T_2^2 \left[\frac{EAW_1^2 \, L^2}{24 \, T_1^2} + EA\alpha \left(t_2 - t_1 \right) - T_1 \right] = \frac{EAW_2^2 \, L^2}{24}$$

where

T_1	=	initial conductor tension at t_1 (kg)
T_2	=	final conductor tension at t_2 (kg)
E	=	modulus of elasticity (H Bar)
A	=	conductor area (mm^2)
W_1	=	initial conductor mass (kg/m)
W_2	=	final conductor mass (kg/m)
L	=	span (m)
α	=	coefficient of linear expansion ($^\circ$C^{-1})
t_1	=	initial temperature ($^\circ$C)
t_2	=	new temperature ($^\circ$C)

Example: We wish to calculate the conductor tension at 75 °C and know the tension at 25 °C (ambient temperature) that is at 20% nominal breaking load.

Using a "canary" ACSR conductor:

$E = 6960 \, \text{kg/mm}^2$, $A = 515.2 \, \text{mm}^2$, $W = 1.726 \, \text{kg/m}$, $T_1 = 2826 \, \text{kg}$, $\alpha = 23 \times 10^{-6} \, ^\circ\text{C}^{-1}$, $L = 400 \, \text{m}$, $t_1 = 25 \, ^\circ\text{C}$, $t_2 = 75 \, ^\circ\text{C}$, and substituting in the formula with zero wind:

$$T_2^3 + T_2^2 \left[\frac{6960 \times 515.2 \times 1.726^2 \times 400^2}{24 \times 2826^2} + 6960 \times 23 \times 10^{-6} \times 515.2(75 - 20) - 2826 \right]$$

$$= \left(\frac{6960 \times 515.2 \times 1.726^2 \times 400^2}{24} \right)$$

$$T_2^3 + 10\,627 T_2^2 = 7.1216 \times 10^{10}$$

Hence, $T_{75} = 2378$ kg

If we consider wind loading acting on the conductor then the calculation is carried out as shown below:

$$T_2^3 + T_2^2 \left[\frac{EAW_1^2 \, L^2}{24 \, T_1^2} + EA\alpha \left(t_2 - t_1 \right) - T_1 \right] = \frac{EAW_2^2 \, L^2}{24}$$

in which W_2 = conductor resultant weight with wind (kg/m)

$$W_2 = \sqrt{W_1^2 + F^2}$$

where
F = wind force = 0.0047 x V^2 x d (kg/m)
V = wind speed (km/h)
d = overall diameter (m)

Appendix 3.3: Capacity of OHL conductor

To calculate the ampacity of overhead line conductor based on its thermal rating.

(1)   A simple method for calculation of current rating on a thermal basis is as follows:

$$I^2R + \alpha' S'd = H_c + \varepsilon s \left[(\theta + t + 273)^4 - (t + 273)^4 \right] \Pi d$$

where

I = current rating (A)

R = a.c. resistance ohms per cm at operating temperature and current, see (2)

α' = solar absorption coefficient (1.0 for aged conductor and 0.6 for new conductor)

S' = intensity of solar radiation (Wcm^{-2})

d = overall diameter of conductor (cm)

H_c = convection loss in (Wcm^{-1})

ε = emissivity of conductor (1.0 for aged conductor and 0.3 for new conductor)

s = Stefans's constant = 5.7 x 10^{-12} (Wcm^{-2})

t = ambient temperature (°C)

θ = conductor temperature rise above ambient (°C)

(2) Calculation of a.c. resistance is made using the following equation which relates conductor resistance at any given temperature:

$$\frac{R't_2}{R't_1} = \frac{M + t_2}{M + t_1}$$

in which

t_1 = temperature of conductor

t_2 = new temperature of conductor

$R't_1$ = d.c. resistance at t_1

$R't_2$ = d.c. resistance at t_2

M = constant for type of conductor material = 228.1 for ACSR

To calculate the a.c. resistance of a conductor knowing the d.c. resistance, the following relationship for skin effect is used:

$$R_{t2} = K \times R'_{t2}$$

where R_{t2} = a.c. resistance at t_2.

The value of K may be read from the following curve. The derivation of the values of K is based on the method described by V.T. Morgan (1965, *IEE Proceedings*, **112**, pp. 325-334).

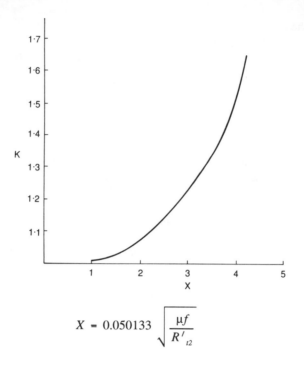

$$X = 0.050133 \sqrt{\frac{\mu f}{R'_{t2}}}$$

and where
R'_{t2} = d.c. resistance at t_2 (Ω/km)
f = power supply frequency (Hz)
μ = permeability = 1 for nonmagnetic materials.

(3)　　Calculation of convection loss (H_c). The experimental evidence upon which values of H_c are based is found in Electrical Research Association (ERA) reports. The values of H_c are significantly different for stranded as compared with smooth conductors but substantially independent of the number of strands from 3 to 37.

　　　　The values of H_c for the various types of conductor and wind conditions are derived from:

Natural convection (no wind):

$$H_c = \left(12.8 \times 10^{-4}d\right)^{0.699} \theta^{1.233}$$

for stranded conductor

$$H_c = \left(19.5 \times 10^{-4}d\right)^{0.561} \theta^{1.187}$$

for smooth conductor.

Forced convection (with wind):

$$H_c = 13.8 \times 10^{-4} \left(V \times d\right)^{0.448} \theta$$

for stranded conductor

$$H_c = 15.95 \times 10^{-4} \left(V \times d\right)^{0.462} \theta$$

for smooth conductor

and V = wind velocity (cms^{-1})

(4)   Example of calculation of thermal rating, based on ACSR conductor code name "Canary".

α' = 1.0 for aged conductor
S' = 0.135 (Wcm^{-2}) for the island of Java
d = 2.95 cm
ε = 1.0
s = 5.7 x 10^{-12} (Wcm^{-2})
θ = 50 °C
t = 25 °C
V = 50 cms^{-1}
R'_{t1} = 0.06351 Ω/km at t_1 = 20 °C
t_2 = 75 °C

$$\frac{R'_{75}}{0.06351} = \frac{228.1 + 75}{228.1 + 20}$$

∴ $R'_{75} = 0.07759\ \Omega$ per km at 75 °C

and a.c. resistor is derived from:

$$X = 0.050133 \sqrt{\frac{1 \times 50}{0.07759}} = 1.27264$$

and from the curve, $1.27264 \equiv k$ of 1.01.

$R'_{75} = 1.01 \times 0.07759$ ∴ $R_{75} = 0.0783659 \, \Omega\text{km}^{-1}$

using the values of R, V given below:

$R = 0.0783659 \times 10^{-5} \, \Omega\text{cm}^{-1}$
$V = 50 \, \text{cms}^{-1}$

$$H_c = 13.8 \times 10^{-4} \left(50 \times 2.95\right)^{0.448} \times 50$$

$$\therefore H_c = 0.64635$$

and substituting the values in the heat balance equation:

$$I^2R + \alpha' S'd = H_c + 1.0 \times 5.7 \times 10^{-12} \left[\left(50+25+273\right)^4 - \left(25+273\right)^4 \right] \Pi \times 2.95$$

$$I^2R + \alpha' S'd = H_c + 0.35816$$

$$\therefore I = 879 \, \text{A}$$

Appendix 3.4: Calculation of RIV generated by a typical transmission line

Calculation of surface voltage gradient using the following equation:

$$E = \frac{V}{\sqrt{3}} \cdot \frac{\beta}{r \ln \left(\frac{a}{Re} \cdot \frac{2h}{\sqrt{4h^2 + a^2}} \right)}$$

in which

$$\beta = \frac{1 + (n - 1)\frac{r}{R}}{n}$$

$$r_e = R \, n \sqrt{\frac{n.r}{R}}$$

$$R = \frac{S}{2. \sin\frac{\pi}{n}}$$

where

E	=	conductor surface voltage gradient (kV/cm^{-1})
V	=	Rated voltage (kV)
β	=	Factor for multiple conductor (= 1 for tube)
r	=	Radius of conductor (cm)
R	=	Outside radius of bundle (cm)
r_e	=	Equivalent radius of bundle conductor (cm)
s	=	Distance between component conductor centres (cm)
a	=	Phase spacing (cm)
h	=	Height of conductor above ground (cm)
n	=	Number of component conductors in bundle

The stress on a quadruple bundle of conductors codenamed "Dove" set at 45 cm centres:

"DOVE" ϕ = 2.355 cm
        r = 1.1775 cm

$$a = 12.0\,\text{m} = 1200\,\text{cm}$$
$$h = 1530\,\text{cm}$$
$$s = 45.0\,\text{cm}$$
$$V = 500\,\text{kV}$$

(i)

$$R = \frac{45.0}{2\,\sin\dfrac{\pi}{4}} = 31.8 \text{ cm}$$

(ii)

$$\beta = \frac{1 + (4-1)\,\dfrac{1.775}{31.8}}{4} = 0.2777$$

(iii)

$$r_e = 31.8 \times 4\sqrt{\frac{4 \times 1.1775}{31.8}} = 19.72 \text{ cm}$$

$$E = \frac{500}{\sqrt{3}} \cdot \frac{0.2777}{1.1775\,\ln\left(\dfrac{1200}{19.72} \cdot \dfrac{2 \times 1530}{\sqrt{4 \times 1530^2 + 1200^2}}\right)}$$

$$\therefore E = 17.79\,\text{kV/cm}$$

We then calculate the mean stress for the conductor bundle under consideration:

$$\text{mean stress} = 17.79 \times \frac{1}{1 + 2.12 \times \dfrac{2.355}{45.0}} = 16.01 \text{ kV/cm}$$

RI at 60 m from line at 500 kV using the comparison formula:

$$E - E_0 = 3.9 \left(g - g_0\right) + 40 \log_{10} \frac{d}{d_0} + 10 \log_{10} \frac{n}{n_0}$$

Use E_0 = the noise generated by the "Apple Grove" 750 kV overhead line, reported in IEEE paper No. 69, TP 688 PNR,

where

E, E_0	=	RI noise level dB > $1\,\mu$V/m
g, g_0	=	mean voltage gradient $1 > \mu$V/cm
d, d_0	=	conductor diameter
n, n_0	=	number of conductors in bundle.

$$E - 36 = 3.9 \left(16.01 - 15.62\right) + 40 \log_{10} \frac{23.55}{35.1} + 10 \log_{10} \frac{4}{4}$$

$$= 30.58 \text{ dB} > 1\,\mu\text{V } 1 \text{ m at } 60\,\text{m, say } 31\,\text{dB}$$

Ratio of noise at 60 m from the line to noise at 20 m from the line is established from the relationship:

$$30 \log_{10} \frac{60}{20} = 14.31 \text{ dB} > 1\ \mu\text{V/m}$$

Therefore the noise level at 20 m from the line will be:

$$31 + 14.3 = \underline{45.3} \text{ dB} > 1\,\mu\text{V/m}$$

Appendix 3.5: Calculation of corona loss dissipated by a typical transmission line

Using the method developed by V.V. Burgsdorf *et al.* [13].

Transmission line parameters

Conductor radius	1.288 cm
Number of conductors in bundle	4

Bundle centres	45.0 cm
Conductor configuration	horizontal
Phase spacing	12 m
System voltage	500 kV RMS = 707 kV pk
Line altitude	500 m

Average duration of weather types:-

(a) fair weather	7300 hrs
(b) wet weather	1460 hrs

(1)   Calculate E_0 from:

$$E_0 = 30.3 \times m \left(1 + \frac{0.3}{\sqrt{r}} \right)$$

where m = surface factor = 0.82 for stranded conductor and unity for polished cylindrical surfaces and r = conductor radius (cm)

$$E_0 = 30.3 \times 0.82 \left(1 + \frac{0.3}{\sqrt{1.288}} \right)$$

$$\therefore E_0 = 31.41 \, kV$$

(2)   Determine value of E_{pk} using Maxwell's coefficients:

$$E_{pk} = \frac{1.66 \times 707}{52.9} = 22.18 \, kV_{pk}/cm$$

(3)   Determine E_{AV} centre phase from:

$$E_{AV} = E_{pk} \times \frac{1}{\left(1 + \frac{2.12 \, d}{A} \right)}$$

where

E_{pk} = maximum conductor surface stress per phase (kV/cm)
d = subconductor diameter (cm)
A = bundle spacing (cm)

$$E_{AV} = 22.18 \times \cfrac{1}{\left(1 + \left[2.12 \times \cfrac{2.576}{45}\right]\right)}$$

$$= \underline{19.77\,\text{kV}_{pk}/\text{cm}}$$

(4)   Determine E_{AV} outerphase:

$$E_{AV} = 0.943 \times 19.77$$

$$= \underline{18.64\,\text{kV}_{pk}/\text{cm}}$$

(5)   Correction due to air density at 500 m
Correction factor = 0.96

E_{AV} (corrected) centre phase = $19.77/0.96$ = $20.59\,\text{kV}_{pk}/\text{cm}$

and

E_{AV} (corrected) outer phases = $18.64/0.96$ = $19.41\,\text{kV}_{pk}/\text{cm}$

(6)   Determine values of E_{AV}/E_0

for centre phase = $20.59/31.41 = 0.655$

for outer phases = $19.41/31.41 = 0.617$

(7)   Determine from Burgsdorf's curves the values of

$$\frac{P_k}{n^2\,r^2}$$

for fair weather and wet weather:

	centre phase	outer phases
fair weather	0.026	0.019
wet weather	0.545	0.36

(8)   Calculate average annual line losses from:

$$P_k = \frac{n^2 \, r^2}{8760} \, \Sigma \, fi \, \frac{E_{AV}}{E_0} \, Ti$$

where

P_k = average annual phase losses due to corona (kW/km)
n = number of conductors/bundle
r = subconductor radius (cm)

Dry weather loss, power loss, centre phase

4^2 x 1.288^2 x 0.026   = 0.6901 kW/km

power loss, outer phases

2 x 4^2 x 1.288^2 x 0.0199   = $\underline{1.00863}$
                              1.6987 kW/km

Wet weather loss, centre phase

4^2 x 1.288^2 x 0.545   = 14.465 kW/km

outer phases

2 x 4^2 x 1.288^2 x 0.36   = $\underline{19.11}$
                            33.576 kW/km

with say 1000 h rain per annum and 7760 h fair weather

annual loss = 7760 x 1.6987   = 13 181
            + 1000 x 33.57   = $\underline{33\,576}$
                              46 757.9 kW h/km

Chapter 4

High voltage cables

M. Simmons

4.1 Introduction

The first rule to understand when considering the design of high voltage cables is that many of the design parameters are set by the system into which the cable will be incorporated. The designer must take a humble stance of respecting, for example, the circuit current rating and impulse withstand voltage of any overhead line to which the cable may be connected. Apart from such technical aspects there is the impact on cable design of the economics of manufacture, in-service running costs and the ever-increasing influence of environmental concerns.

The expectation of an insulated high voltage cable is that it will be an entirely passive element in the power transmission system, non-perturbing to associated equipment and providing unremarked service for many (at least four) decades. This life expectation has arisen because previous generations of cables have already provided such service and, understandably, the customers for cables see no reason to accept a lesser performance. Table 4.1 [1], shows that the incidence of faults of the British 275 and 400 kV transmission network due to insulated cables is a small proportion of the overall number of system faults.

The design of cables is to a large extent regulated by a number of industry, national and international standards and guides. These may be related to voltage level, technology of production or customer preference. Table 4.2 shows how the technology of manufacture, and in particular the primary insulation material, relates to the various broad voltage categories and some of the standards reigning in these areas. The document which guides the calculation of power rating of a cable by a systematic and agreed method is IEC 287.

The brevity of this account of high voltage cables necessarily means that much detail of the technology, science and philosophy must be omitted, but it is hoped that such a broad approach will be useful to practising engineers.

Table 4.1   *Failure rates in the various components of insulated HV cables*

Component	Index of failures	Cable type	
		Oil-filled 220-500 kV (Av. age 20 yrs)	XLPE 60-220 kV (Av. age 7 yrs)
Cable	Failures/100 circuit.km.yr	0.05	0.05 - 0.07
Straight joints	Failures/100 units. yr	0.08	0.01
Stop joints	Failures/100 units. yr	0.16	-
Terminations	Failures/100 units. yr	0.07	0.04
Ancillary equipment	Failures/100 units. yr	0.32	-

4.2 The components of an insulated power cable system

The cable itself is only one of the components in any system. It may take several forms depending on technical considerations and customer preference. For example, in some cases all three phases are contained in the same sheath providing minimised installation space and economy in the laying operation with lower magnetic interference fields as a bonus. The sacrifices to be made are reduced power rating for equal conductor sizes, less flexibility in installation practice due to the largeness and weight of the cable and greater complexity of the accessories.

No cable ever operates without being included in a system which at the very least will have two terminations per phase. These terminations may be into air, either for outdoor or indoor locations, or directly into other equipment with which they must be compatable, e.g. as oil immersed for generator transformers or gas immersed for buss-bars or switch-gear. More generally, a cable system will also contain a variety of joints: to allow lengths greater than can be transported (about 500 m for the largest cables) to be made-up in the field (straight joints), to join differing technology cable (transition joints), to provide a hydraulic barrier (stop joints in oil-filled cables) or to branch a circuit (branch joints). Some cables require ancillary equipment such as that needed to maintain the hydrostatic oil pressure in self-contained oil-filled cables. Fig. 4.1 shows a typical variety of cable accessories.

The few faults that arise in insulated cable systems tend to be associated with these accessories, very few derive from failure of the cable itself. Table 4.1 illustrates this [1].

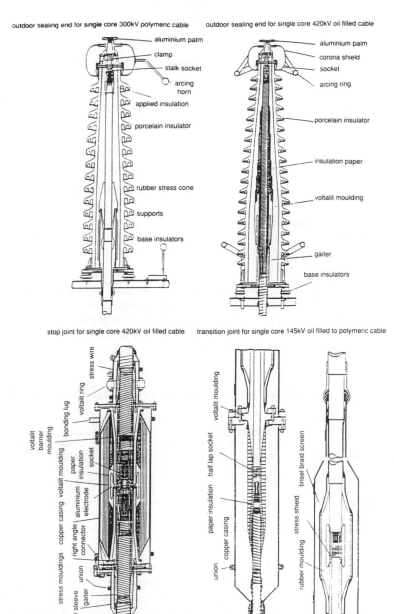

outdoor sealing end for single core 300kV polymeric cable

- aluminium palm
- clamp
- stalk socket
- arcing horn
- applied insulation
- porcelain insulator
- rubber stress cone
- supports
- base insulators

outdoor sealing end for single core 420kV oil filled cable

- aluminium palm
- corona shield
- socket
- arcing ring
- porcelain insulator
- insulation paper
- voltalit moulding
- gaiter
- base insulators

stop joint for single core 420kV oil filled cable

- stress wire
- voltalit ring
- bonding lug
- voltalit barrier moulding
- paper insulation
- socket
- voltalit moulding
- copper casing
- aluminium electrode
- right angle connector
- stress mouldings
- union
- gaiter
- rubber sleeve

transition joint for single core 145kV oil filled to polymeric cable

- voltalit moulding
- half lap socket
- paper insulation
- copper casing
- union
- tinsel braid screen
- stress shield
- rubber moulding

Fig. 4.1 *Typical cable accessories. Termination for polymeric and oil-filled cables, a stop-joint for oil-filled cable and a transition joint from oil-filled to polymeric cable*

Table 4.2 Insulation and technology of HV cables

Medium voltage, i.e. up to 33 kV	Super tension, i.e. 33 to 400 kV+
Lapped paper, dried and impregnated with viscous or non-draining compound BS 6480, ESI 09-12, IEC 55	Lapped paper, dried and impregnated with low viscosity oil. Eng. Recom. C47, NGTS 3:5.1, IEC 141-1
Extruded PVC	Extruded vulcanised (EP) rubbers
Extruded vulcanised (EP) rubbers. BS 6622, IEC 502	Extruded cross-linked polyethylene TPS 2/12
Extruded cross-linked polyethylene BS 6622, IEC 502	

To grasp what these failure rates would mean in practice it is informative to interpret them in terms of a model three phase HV circuit. If we propose a three kilometer circuit with two sets of terminations, one set of stop joints and ancillary equipment in the case of oil-filled cables, and five sets of joints in all, we arrive at the failure rate estimates shown in Table 4.3.

Table 4.3   *Relative failures in three circuit kilometers of insulated HV cable*

	Oil-filled system	XLPE system
System failure rate/yr	0.033	0.0084 - 0.0102
Proportion due to cable/%	14	54 - 62
Proportion due to accessories/%	86	38 - 46

These statistics are more indicative than exact, due particularly to the limited service time with XLPE cables. However, they support the experience of those involved in insulated cable technology that the cable itself, taking into account its length, is the least problematic item in any system. Section 4.4.2 will deal specifically with those components of a system that perhaps provide the greatest reliability challenge; namely the accessories.

4.3 Design features

4.3.1 Conductor

Despite adventurous attempts to make power cable conductors from metals such as sodium (provoked by the high and variable price of copper at times and early difficulties of joining aluminium) and the future promise of high-temperature superconductors or intrinsically conducting polymers, only copper or aluminium are used as conductor material in present day practice. However, there is a great deal of variety in the form of the conductors made from these two metals. The choice of which metal, the cross-sectional area and the form and construction of the conductor is integrated with the overall cable design. This is related closely to the current rating and the jointing methods desirable or available and, in some cases, the handling characteristics. At the medium voltage level a solid circular or shaped rod (segments to permit close spacing of several conductors within a circular section) may be used in either metal up to $150\,mm^2$ but, for reasons of flexibility, most conductors are made up from strands compacted in one of several ways to reduce the free space and hence overall size (such a saving in size can have a significant influence in the costs of all the materials external to the conductor).

Apart from merely compacting a bundle of initially round wires there is the possibility of forming larger shaped strands or segments which fit together to make up a very area efficient conductor (known sometimes as Conci after the shell-like shape of the individual segments). With very large conductors, e.g. $2000\,mm^2$ and above, consideration is given to the skin depth for conduction at 50 Hz and its effect on a.c. resistance. The resulting conductor is made up from a large number of individually insulated wires formed into segmental shaped bundles which are also insulated from each other. This design is known as a Milliken conductor. Various conductor forms are illustrated in Fig. 4.2.

In choosing the metal the other major consideration, aside from those already mentioned of cost and joining method, is electrical conductivity. This is of course tied closely to the rating calculations as described in the IEC publication 287. The ratio of conductivities is approximately 0.64:1, aluminium to copper. Also interesting from design viewpoints are the ratios of specific conductivities relative to cost and weight these being 2:1 and 3:1 copper to aluminium respectively. The apparent disadvantage of copper is balanced in higher voltage cables by the smaller cross-sections and consequently lesser quantities of insulating and protection materials required to finish the cable.

uncompacted
circular

compacted
oval

compacted
circular

conci

Milliken

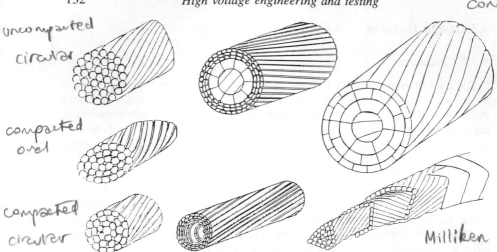

Fig. 4.2　　*Illustrations of various conductors: uncompacted circular, compacted oval and circular, flattened strips on steel spiral and on segments, wholly segmented (Conci) and Milliken*

4.3.2 Insulation system

General requirements of an insulation system may be established irrespective of the materials used.

- Electric strength. This is NOT the value often quoted in a manufacturer's literature determined by a BS 2918 test but a set of values related to the nature of voltage, e.g. waveform, 50 Hz or lightning impulse, and expressed as a set of statistical parameters (most usefully of a Weibull distribution) related to physical dimensions giving the probability of failure at a particular electric stress.
- Dielectric properties. The product of permittivity and loss tangent of the insulation determines the dissipation of energy as heat. Its value therefore bears on the rating of the cable, especially at the highest voltages as it has a "V squared" relationship, e.g. at 400 kV the dielectric losses may represent 20% of overall losses. The permittivity governs, with dimensions, the capacitance of the cable which is relevant to system design and economics. Permittivity is largely invariable once a dielectric material has been selected but may be deliberately chosen to control electric stress distribution in a.c. service. The loss tangent provides a dimension-free measure of the perfection (or lack of perfection) of the insulation which may be interpreted diagnostically. Electrical resistivity and its relationship with stress, time and temperature becomes important under d.c. service conditions as it then controls the stress distribution within the insulation. This may be as equally relevant in a.c. systems as d.c. as commissioning tests most often demand the application of high d.c. voltages.

- Ageing characteristics. Accepting that all organic materials degrade with time at a rate which is enhanced at higher temperatures it is necessary to know that none of the critical electrical or mechanical properties of an insulating material will become inadequate during the required service life of the system at the temperature of operation. These critical properties will be different with differing materials but, in general, it is possible to define the rate of degradation and its relationship with temperature using the parameters of an Arrhenius expression and a rate law to predict the end of safe service operation.

- Mechanical properties. The greatest demands on mechanical properties occur during the manufacturing and installation when materials are subjected to intense mechanical stresses the tolerance of which depends on rupture strength, elongation, elastic moduli, tear strength etc. However, during service there are situations where vibrations, thermally-induced movements or circuit changes may challenge the mechanical integrity of the insulation in its aged condition.

- Thermal conductivity. The ability of the insulation to transmit heat is one of the several influences on the temperature arrived at by the innermost parts of a cable and hence the degradation rate and therefore the cable system rating. The thermal resistance of the insulation may be the major contributor to the overall thermal resistance.

4.3.2.1 Paper-based (laminated construction)

The construction of an insulating wall (such as the dielectric of a high voltage cable) by building up its thickness with a large number of layers of a thin sheet material like paper has proven very successful. This is due to its great design and manufacturing flexibility, the availability of a high electrical quality material (cellulose paper) and the statistical merit of a multiple layer material where a defect in one layer only marginally weakens the whole either electrically or mechanically.

For paper to function as insulation at high voltage, and therefore high electrical stress, it is necessary to remove the water it naturally absorbs when exposed to ambient, normally moist, air. The paper used for electrical insulation is made (by the Kraft process) from a cellulose wood pulp chosen for long fibres and freedom from contaminants to be, when dry, physically strong and of low electrical losses. The moisture content of paper is approximately proportional to the relative humidity to which it is exposed up to a maximum of about 20 % by weight. Typically paper will be received from the suppliers with about 5 % but for even medium voltage applications this must be reduced to < 0.2 %. For the highest voltages a moisture level in the region of 0.05 % is sought and must be maintained. Fig. 4.3 illustrates a typical relationship between the critical electrical parameter of loss tangent and moisture content for an oil impregnated paper.

The purpose of the impregnant is two-fold: it radically reduces the reabsorption speed of water where the insulation is exposed to moist air as during installation operations (the diffusion coefficients of water through and along oil impregnated paper layers are 2×10^{-13} and 4×10^{-9} m^2/s respectively compared to 3×10^{-5} m^2/s

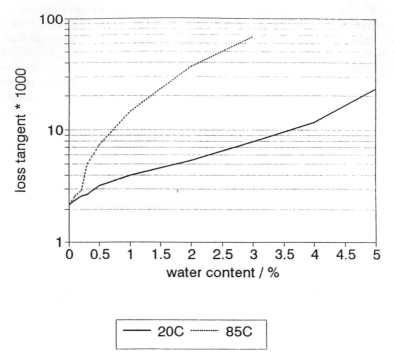

*Fig. 4.3   Relationship between loss tangent and moisture content of oil-
impregnated paper*

through stationary air); and it radically increases its electric strength (under a.c.
stresses the short term electric strength of oil impregnated paper is about 50
kV/mm compared with only 1 kV/mm for air).

The nature of the impregnant depends on the cable type as indicated in Table
4.2. Viscous impregnants in current use are based mostly in polybutenes, chosen
for their temperature viscosity characteristic, their chemical stability and
compatability, tendency to emit gas when subjected to small electrical discharges
(thus increasing the local pressure and surpressing the discharges) and their
generally good dielectric properties. There still exist, however, many cables
impregnated with viscous materials from the previous generations of impregnants
based in mixtures of mineral oils with various waxes, resins and polymers. Low
viscosity impregnants in current use are also largely synthetic oils and belong to
the family of alkyl benzenes. These offered a selection of advantages over the
previously used mineral oil blends. They exhibit good chemical stability, even in
the presence of oxygen, excellent dielectric properties with the low permittivity
characteristic of a non-polar organic liquid, may have a low viscosity and vapour
pressure but, above all, their attractive feature is a very high tendency to absorb
hydrogen under electrically active conditions. This means that should ever a void

appear in an impregnated dielectric (which implies that electrical activity must occur under normal a.c. service stresses) then it will be self-correcting as the product of the activity, hydrogen, will be absorbed chemically into the oil. Under d.c. stresses the electrical activity in any void is greatly reduced so that viscous impregnants are more admissable at high stress.

The set of properties of an insulating paper (or other laminate material), including its dimensions, depend on the exact application in which it is to be used. The thickness of each individual layer of a laminated dielectric is inversely related to its electric strength (as it is the impregnant in the interturn butt gaps which fails first) so that the most highly stressed regions adjacent to the conductor call up the thinnest layer. Electric strength is also related to the impermeability of a cellulosic paper which, in turn, is coupled to its density. The higher the density of paper the higher and more disadvantageous will also be the permittivity and loss tangent. Attempts to overcome this conflict have included the manufacture of paper where a high impermeability to retain electric strength has been contributed by one layer of a multilayer paper while a second layer of low density has ensured an overall low dielectric loss. For the purposes of design, an a.c. stress of up to 15 kV/mm (or 100 kV lightning impulse) is typical for low viscosity oil impregnated paper. Typical values of dielectric properties are loss tangent of 0.0025 and permittivity of 3.5. Vital to the ability to lap on an adequate thickness of a laminate such as paper are its mechanical characteristics. The elastic moduli in three dimensions and the coefficient of friction govern the relationship between the tension of a laminate tape during the lapping operation and the ability of the completed structure to bend without breaking or creasing (which would lower the electric strength). This last requirement has impeded the simple replacement of cellulosic paper (typical Young's modulus 10 GPa) with polymers (typical Young's modulus below 2 GPa) that have inherently lower dielectric losses coupled with high electric strength. The compromise that has some promise is a laminate that combines some of the electrical properties of a polymer such as polypropylene with the mechanical characteristics of cellulosic paper in a sandwich construction (PPLP). The advantage in current rating achieved by adoption of this compromise for a large 400 kV directly buried cable varies from 14 to 26 % depending on installation conditions and ambient temperature [2].

Essential to the successful performance of a laminated dielectric are the conducting screens which complete the insulation system. These must remain intimate with the insulating layers under all conditions of operation and therefore have closely matched physical properties. This is achieved by making a laminar material, generally of cellulose paper heavily loaded with carbon to have a volume resistivity in the region of 10 ohm m, sometimes laminated to insulating paper ("Duplex") and othertimes laminated to aluminium foil. Tapes made from these materials may then be lapped on to the cable, both on the conductor prior to the insulation and over the insulation before further processes. The choice of screen and its adequate placing can have significant effects on the electric strength of the cable and on the dielectric losses.

4.3.2.2 *Polymeric*

Extrusion of a polymer to form, in a single operation, the insulating wall of a cable, screen layers included, offers the prospect of high production speed with the merits of polymeric materials that can be demonstrated on a small scale. Polymers can be selected to have dielectric losses less than 10% of cellulosic paper or 20% of PPLP, "intrinsic" electric strength four times as high as oil impregnated paper and thermal conductance 30% greater. The disadvantage is that a single defect can have a diastrous effect on the integrity of the whole cable due to the homogeneity of the dielectric. This fact is reflected in the parameters of the statistical distributions of electrical breakdown in polymeric insulation systems compared with laminated systems; whereas a lapped paper cable may have an electrical break-down level predictable within a few percent, that of a polymeric cable will only be predictable within some tens of percent with the same confidence. This inherent characteristic of extruded polymeric cables has forced the use of statistical methods employing the Wiebull distribution into the field of cable design [3, 4]. Using this distribution, the failure probability of an insulation system is described with reference to a particular size specimen of cable, for example based on a 20kV 70mm² design, with a characteristic parameter for time to failure (t_o) qualified by a scatter parameter (a), a characteristic parameter for electric stress at breakdown (E_0) qualified by a scatter parameter (b_p) and a further scatter parameter (b_t) which governs the effect of physical dimensions. Determining the values of this set of parameters following a change in materials or manufacturing technology is a major task consuming much time, manpower and materials. However, the expected gains in performance and eventually in overall costs and reliability have led to manufacture of 400kV cable at design stresses of up to 15kV/mm using the "super clean" cross-linkable polyethylene (XLPE) now available and thus matching oil impregnated paper cables stresses.

As indicated in Table 4.2 there is more than one candidate polymer at some voltages. At 150kV and below it is possible to use compounds of ethylene-propylene (EP) rubber without disadvantage of its ten times higher dielectric losses than polyethylene. The benefit of water tree resistance (see section 4.5.3) allows designs in which the cable core is unprotected from water and lesser rigidity eases installation practice and so offsets the small cost disadvantage of the insulation system. An alternative to EP rubbers to overcome the susceptibility of XLPE to water tree deterioration is tree retardant polyethylene (TRXLPE) with a cost level in between XLPE and EP rubber.

The conducting screen materials which are extruded simultaneously with the insulation are compounds of carbon with polymers chosen to accept a high carbon loading, be bondable to the insulation, be compatible with the extrusion conditions and be removable when required without damaging the dielectric. This last demand is most important in medium voltage cables where the operations of jointing and terminating may have severe time, tooling and skill constraints. For these cases an "easy strip" screen is called for, presenting some problems with certain insulation types, e.g. TRXLPE.

4.3.3 Containment

The hermetic containment of impregnated paper cables is obviously essential to prevent loss of the impregnant and to maintain the paper in a state of dryness. The only materials of containment that are totally impervious to water are inorganic substances such as metals and ceramics. This leads to a necessary set of properties for self-contained cables including being non-magnetic to prevent excess generation of heat, flexible in the case of sheaths to allow spooling of the cable and being sealable by site-available methods to permit termination and jointing. In practice these requirements are met by lead alloy and aluminium sheaths (some very special situations dictate welded stainless steel) and by porcelain insulators at terminations. The choice, to some extent, is made on customer preference but factors which are considered include the laying conditions and permitted weight of the finished cable, the internal oil pressure, external corrosion vulnerability and available jointing skills.

Lead alloys for sheaths are specified in BS 801. Particular alloys are indicated for various designs, e.g. unarmoured cables alloy E, unarmoured anti-fatigue alloy B and anti-creep (but reinforced with bronze tape) alloy 1/2 C. The relatively ductile lead sheaths require reinforcement by means of a non-magnetic tape (e.g. bronze) to counter the internal forces from oil pressure. Aluminium, however, which is generally corrugated, needs no reinforcing but has a predisposition to suffer from corrosion; lead alloys with armouring are not immune so it is common to find anti-corrosion protection layers on both (see section 4.3.4).

Containment of polymeric cables is necessary at all but the lower end of the high voltage cable range to prevent deterioration by water treeing. There is evidence and service experience to demonstrate that EP rubber cables tolerate the absence of a hermetic protection more ably than XLPE, thus eliminating the cost and complication of this component of a cable, and there are prospects that TRXLPE will be able to be used equally unprotected. This possibility has allowed the installation of submarine cables of tens of kilometres of circuit with no sheath, merely with armouring and soft sacrificial coverings. Otherwise, sheathing of polymeric cables and anti-corrosion coverings are closely similar to those of self-contained oil-filled cables. One special feature is the water blocking device incorporated under the sheath of polymeric cables to avoid longitudinal movement of water should the sheath be breached at any point. This generally takes the form of a spiral lapped tape loaded with a material which expands rapidly and enormously when in contact with water but the material may alternatively be disposed as a powder filling of the space under the sheath.

In an alternative philosophy of cable containment the processed cables are installed into pre-laid rigid steel pipes which are then filled with the impregnant. Such a procedure demands that the impregnant be more viscous than that used in self-contained designs and incurs higher initial costs but allows the possibility of replacing the insulated core to extend the life of the installation or, with an improved core, uprating the circuit. It therefore has most appeal in highly developed and densely populated cities.

One new technology which attempts to bring the nearly zero permeability of metals together with the low cost and weight of polymeric sheaths is the metal laminate. With this concept a foil of a metal such as lead is supported by a robust polymeric layer so that any path for moisture migration through an overlap in the foil is extremely long. Countering the advantages cited above, this form of sheath is likely to be less tolerant of physical and electrical abuse (i.e. short circuits) than the thick metal types.

Maintenance of hermetic containment and accessories presents its own problems. At joints the seals are effected by solder, welding and mechanical (with O-ring) means and various corrosion and mechanical protections applied. This may become somewhat complex if hydraulic pipes and earth bonding leads have to brought out which perhaps explains the high proportion of failures attributed to ancillary equipment in oil-filled cables (Table 4.3). At terminations the further problem is the interface with the ambient conditions which, as likely as not, will be moist and polluting on a surface along which a significant electric field will be acting. The inevitable resultant electrical activity can even be agressive to ceramic materials (e.g. porcelain) with inadequate design of this surface (i.e. the length of creepage path and shape of the "sheds"). For lower voltage cables with either viscous compound impregnated paper or polymeric insulation, joints and terminations may be contained by polymeric components as described in section 4.4.2 but with the acceptance of lesser hermeticity and resistance to surface electrical activity.

4.3.4 Protection

Armouring layers consisting of galvanised steel wires (BS 1441 and 1442) are a necessary protection against external forces in all except the most benign installation conditions for lead sheathed cables and polymeric insulation not using metallic sheaths. The variants might be designed to deter penetration by sharp objects or to resist crushing forces. Aluminium wires provide a lesser protection but which is adequate for some situations.

An effective means of corrosion protection, both for the metal of the sheath and armour, is the application of bitumen layers under extruded polymeric over-sheaths. The polymer is chosen for toughness required to resist installation conditions and, in some cases, compatability with the installation environment (e.g. presence of chemicals in the soil or voracious insect species). Commonly used polymers are various grades of polyethylene (lower density for flexibility and insensitivity to cracking, higher density for abrasion resistance) and PVCs (formulated for the expected temperature conditions, e.g. tropical, temperate or arctic, and for easy jointing; see BS6746). All of these polymers must be protected against UV degradation for those locations where the cable may be exposed to prolonged sunlight by the addition of carbon or, when the sheath cannot be black, with a chemical UV degradation inhibitor. For protection against tropical insect attack specific insecticides are also added, previously gamma BHC but in this environmentally conscious time more likely to be a synthetic pyrethroid.

4.3.5 Thermal and mechanical environment

Except in tunnel installations, few insulated cables are laid in air although ducted (air filled ducts) cables may be said to be not "directly buried". Whether they be in air, in ducts or directly buried, the thermal resistance of their environment is a highly significant part of the overall thermal resistance between the conductors and the outside world and thus vitally important to the current rating of the cable (see IEC 287). In directly buried cables the material which fills the trench in which the cables are laid must have a predictable, stable thermal resistivity as low as possible. Conservative values are generally used for design purposes, e.g. 1.2 °C metre/Watt for a dried but granulametrically specified sand, whereas in its most normal moist state its resistivity will be 0.8 °C metre/Watt. Unfortunately moisture tends to migrate away from warm objects such as cables so that where a guaranteed lower value is required, options such as weak-mix concrete (also providing mechanical support and protection) or wax-mixed sand are available. Artificial irrigation has been used to maintain the cable trench filling moist despite the consequent non-passivity of the installation. To avoid the thermal disadvantage of air-filled ducts but retain their advantage of post-installation recabling the duct may be filled with a semi-solid/semi-liquid mix of a clay slurry (bentonite) and cement.

There was, in the 1960-70 period, a great deal of interest in forced cooling of cable systems to enhance the heat removal and thus increase their rating. Systems were devised and constructed in which, in the simplest form, heat was removed from the soil adjacent to the cables by water pipes with flowing water, thus defining the ambient thermal conditions. In opportune locations some cables have been laid in troughs immersed in water where the water flow has been controlled by a series of weirs. Considerably more complex are designs in which water which is in direct contact with the sheath or a fluid such as the insulating oil of an impregnated cable is circulated and cooled. But, to gain rating, these systems sacrifice simplicity and passivity and must therefore have increased probability of non-availability for service.

Tunnel (and bridge) installations have at least three complicating features: the determination of the heat flow path and hence external thermal resistance, the non-stationary nature of a cable held in cleats and perhaps subject to recurrent vibrations and exceptional corrosion conditions. Mathematical models have been successfully used to determine the rating of such cables incorporating convective modes of heat flow but also having to recognise that in deep tunnels there may be a source of heat (e.g. hot rocks) other than the cable. Thermal expansion and contraction of a cleat suspended cable will flex it and challenge the integrity of the insulation system and sheathing especially after the insulation may have thermally aged or the metal changed from its original crystal structure. The corrosive nature of a submarine tunnel may demand titanium alloys to survive for the life expectancy of a cable installation.

Both thermal and mechanical design features may become critical at accessories as, at these points of a system, the heat paths from the conductor to the environment are more lengthy and less well defined and the mechanical forces built up along a constrained length of cable may be concentrated. The lower electrical design stresses of joints, for example, increase the insulation thickness some fivefold. This may be compensated for to an extent by the use of materials with higher thermal conductivity than the insulation of the cable (e.g. impregnated paper has a thermal resistivity of about 5 °C metre/Watt but silica loaded epoxy castings may have a thermal resistivity of less than 2 °C metre/Watt). More important is the longitudinal alleviation due to the high thermal conductivity of the conductor (attempts have been made to augment this further by the inclusion of heat-pipes into accessories). Tolerance of the thermally derived mechanical forces must generally be designed into the accessory by ensuring the component parts are strong enough (e.g. the epoxy castings of a stop joint) or movements are absorbed (e.g. by flexible connections in terminations). These forces may turn from compressive to equally high tensile forces after thermal cycling.

4.4 Manufacturing processes and materials

4.4.1 Cables

4.4.1.1 Conductors
The essential feature of low resistance of a conductor may be readily sabotaged by inadequate manufacturing processes. Of the seven categories of copper described in BS 6926, one particular grade is chosen (C 100). The electrolytic refining method produces an extremely pure material but trace quantities of certain substances can have a significant effect on such characteristics as annealability and conductivity. Copper "cathodes" (ingots) are turned into a few sizes of circular section rod for further drawing down to the particular sizes and, in some cases (see section 4.3.1), shapes. This drawing process of pulling the rods through very hard and well finished dies in the presence of very special lubricants plastically deforms the copper and inevitably causes changes in state of anneal. This may have to be remedied between and after the various stages of drawing until their desired size and shape is achieved. In the case of conductor wires destined for low viscosity oil impregnated cables it is essential to remove all traces of the drawing lubricant to prevent subsequent contamination of the oil and possible disastrous effects on the dielectric properties of the insulation. Aluminium rod is drawn and shaped in a similar manner to copper but, due to its different set of physical properties, requires somewhat different treatment in areas such as choice of lubricant and speed of production.

The building of a conductor in a laying-up machine may bring together from a very few to several hundred wires spiralled generally in alternate directions (lays) to cancel twisting forces. Passage through a die or sequence of dies gives the final form to the conductor before, in many cases, it is bound with tapes to maintain its integrity until it may be passed to the next manufacturing stage of insulation.

4.4.1.2 Laminated insulation

The lapping operation consists of spirally winding any number of layers of tape from 20 to 200. These layers must be placed one on the other so that the butt gaps between adjacent turns in any one layer do not coincide with those in the next layers above or below as this would cause an electrical weakness — they are neither so tight that the tapes are unable to slip to allow the cable to bend without creasing or breaking the tapes nor so loose that the tapes are able to shift with thermo-mechanical cycling to cause alignment of butt gaps, in either case threatening the electric strength of the cable. The machinery which accomplishes this is a coordinated line of up to 20 individual machines through which passes the conductor, each of which may place, typically, a 12 layer thickness of insulation. The critical setting of these machines is the tension maintained on the tapes as they are applied as this controls the initial tightness (interfacial pressure). Interacting with the tension of the tape is its elastic modulus; low values of modulus demand low tensions which are more difficult to set and control. The low modulus of all synthetic polymer laminate materials compared with cellulosic papers has been one of the difficulties only recently overcome by sophisticated tensioning devices, as mentioned in section 4.3.2.1. It is necessary to locate these lapping machines in a controlled humidity environment as cellulosic paper varies in thickness and, to a lesser extent, in length with humidity so that the tightness of the lapping would radically change with the subsequent drying operations. Depending on the cable, relative humidities between 1 and 50 % are chosen.

The lapped cable core is generally wound on to a transportable bobbin for further processing unless it is of exceptional length (for example submarine links). In this case it will be wound directly into a coil in the processing chamber as illustrated in Fig 4.4.

The removal of residual moisture from the insulation is accomplished by heat and vacuum over periods which may be only a day for the lower voltage thinner insulation wall cables but much more for supertension cables. In the case of submarine cables the quantities of water removed are measured in tons, all of which must be extracted from a compact thickness of perhaps 20 mm of paper surrounded by hundreds of turns of cable core. The heat applied must be carefully controlled so as not to prematurely age the cellulose paper and the vacuum cycle deigned to optimise the transfers of heat and water vapour. In most operations the conductor is heated by passing very high electric currents through it and the conductor temperature monitored by measuring its change in electric resistance. The end point of drying may be, in some cases, judged by measuring the dielectric properties of the insulation.

The viscous impregnant of the lower voltage and submarine cables is treated by vacuum and heat to reduce its gas and water content to a reasonably low level before introducing it into the processing tank to saturate the cable core. However, in the case of the low viscosity oils used for supertension cables, an extremely

Fig. 4.4   Operations in a submarine cable factory; coiling the cable into the vacuum drying and impregnation tank

thorough degassing procedure is employed to achieve a residual equilibrium gas pressure (RGP) of less than 1 mb (oil received from suppliers will have an RGP of 1000 mb, i.e. be saturated with air). An alternative processing method, used when an unusual (possibly experimental, sensitive to handling or expensive) impregnant is called for, is vacuum sheathing. The cable is first dried in a large tank under vacuum and then sheathed whilst maintaining this vacuum. The impregnant is then introduced into the sheath to saturate the insulation and fill the sheath space.

Following the normal mass impregnation process with viscous impregnants, the cable core may be exposed to air (or better, dry nitrogen) whilst the metal sheathing process is carried out. The low viscosity oil impregnated cables, on the other hand, are maintained under degassed oil at all times during this next manufacturing procedure. The current practice is to extrude directly on to the cable core either lead from a pot of the molten alloy or aluminium from high purity billets. Previous technologies have made the sheath separately as a tube (particularly aluminium) into which the cable has been drawn or, in the special case of stainless steel, seam welded the sheath over the core. Drying of a core which is already sheathed is a lengthy and inefficient process so that pre-processing sheathing is generally reserved for short experimental cables.

The final stages of manufacture require the placement of reinforcing, armouring and corrosion protection sheathing which demands one or two more passages of the cable through appropriate machinery. In the case of the low viscosity oil impregnated cables these processes will be carried out without losing the hydraulic pressure of the oil and without admitting any inward leaks of air that would degrade the electrical integrity of the oil/paper system.

4.4.1.3 Extruded insulation

The formation of the extruded insulation system of inner conducting screen, dielectric and outer conducting screen is almost always carried out in a single pass through a tandem arrangement of two or three extruders. The dimensions of these layers is determined by a series of dies in the extrusion head which are carefully designed to respect the rheological characteristics of the polymer melt. This melt must be at a temperature high enough to allow thermo-plastic forming to take place but low enough not to activate yet the organic peroxide cross-linking agent. It is extremely important to consider every stage of the delivery of the polymer from the supplier to the entry point of the extruder. Nowhere, especially in that polymer destined for the mostly highly stressed insulations, is it permissible to allow contamination to enter as, with extruded rather than laminated dielectrics, we can no longer tolerate any defects. For this reason such humdrum details as the packaging and granule transfer methods can assume high importance in the assurance of quality of the cable core, even with filtering of the polymer melt and monitoring the melt for the presence of foreign particles.

As the core leaves the extruder the polymer is extremely soft and incapable of supporting the conductor or even maintaining the form given by the dies through which it has passed. It will remain this weak until it has been cross-linked (cured/vulcanised) and cooled to a temperature well below the crystalline melting point of the polymer (in the case of unfilled polymers such as polyethylene, filled EP rubbers are more robust). Three solutions have been applied to minimise distortion:

— extrude vertically — this demands the expensive and inflexible arrangement of mounting the extruders on a high tower;
— extrude into a fluid of sufficient density to buoy up the cable core — such fluids tend to be difficult to handle (e.g. molten salts) and matching the average density of a range of cable sizes is needed;
— extrude nearly horizontally onto a conductor tensioned into a catenary curve — this overcomes the conductor weight problem and the weight of the insulation system can be buoyantly compensated for by a relatively innocuous heat transfer fluid such as silicone oil.

To cross-link the polymer, the peroxide compounded into it must be raised to its activation temperature for a given time by passing the core through a tube filled with some heat transfer fluid held at a pressure sufficient to suppress bubble formation. Without pressuring to about 1 MPa, bubbles would form in the polymer due to the large amounts of by-products (typically methane gas and several other volatile substances). The most commonly used heat transfer fluids are:

—   steam — very efficient heat transfer but leaves the insulation system saturated with water, provides no buoyancy and possibly induces microvoids;
—   nitrogen — inert but no buoyancy and poor heat transfer;
—   silicone oil — efficient heat transfer medium, good buoyancy for thick insulations but requires complex handling equipment.

Following cross-linking, the cable cores are heat-conditioned to adjust the levels of cross-link byproducts. Otherwise the long term effusion of methane, for example, could lead to hazardous installations. The finishing of the core proceeds with lapping operations to emplace copper screens and water stopping tapes etc. before applying outer protection. In the case of medium voltage cables made with water resistant dielectrics such as EP rubbers, this may be merely a mechanical protection or low permeability polymer sheath but for supertension XLPE cables it is likely to be an extruded metal sheath (as with laminated and impregnated insulations).

4.4.2 Accessories

The conflict inherent in accessory design is that they must be finally assembled in the field where conditions are less controlled than the factory environment in which the cable was manufactured but that they are by far the most vulnerable part of the system. To some extent this may be compensated for by reduction of electrical stresses and limitation of mechanical forces and thermal resistances. The manufacturing philosophy is to prefabricate as much of the accessory as possible to minimise site work and to ensure, as far as is possible, that the final assembly consists of easy to perform non-critical operations.

4.4.2.1 Medium voltage

Until recently (in cable terms), a medium voltage cable joint would have been made of materials of similar nature to those used in manufacture of the cables being joined, i.e. insulated by layers of paper, dried and impregnated with a viscous oil. The factory dried and and impregnated paper would perhaps be creped to accommodate the changing profiles of the ferrules joining the conductors. This hand constructed joint would then be sealed in a cast-iron or equivalent metal case sealed to the metal cable sheath. The terminations would be similarly constructed using porcelain insulators to provide the hermetic protection. Clearly the prefabrication level of this type of accessory was low and consequently the skill level of the personnel doing the assembly was correspondingly high.

The advent of polymeric cables and the drive to de-skill cable installation practice has led to several philosophies of accessory design which spilled over into paper cable use. The target has been to provide kits containing the fewest number of components, covering the largest range of cable sizes which may be assembled with the minimum skill and tool use in the shortest time at the lowest cost. The path followed has been to exploit the elastic or thermo-elastic character of polymers together with their ready formability and, in some designs, to use synthetic resins which may be poured into a shell surrounding the accessory, there to undergo a cross-linking chemical reaction to result in a form stable material. Because of size differences brought about by conductor cross-section, voltage levels and stress variations, the range-taking capability of an accessory presents particular challenges. The economics of production and procurement of stock (as ready-to-assemble kits) depends on how many sizes are required to cover the range of possible situations. This number is reduced to a minimum by a combination of design and material selection. For example, in an elastic accessory such as a joint, the tubular insulation wall may consist of a 6 mm thickness of rubber for an 11 kV component, the property of elastic modulus of this must be chosen to allow it to be stretched by up to 100 % to accommodate the size range as the jointer threads it over the cables. This operation must not require forces outside those available from a jointer, e.g. 100 kN sustained thrust. In addition to the elastic modulus several other mechanical properties of the rubber need to be defined such as sensitivity to tear when under strain during assembly or in service. In some designs elastic components are pre-stretched onto carriers for assembly but clearly must retain their elastic nature even after long periods of storage in such a stretched state. Both silicone and EP rubbers have been widely applied in this area.

The thermo-elastic behaviour of polymers has produced a range of designs in which the ability of a cross-linked polymer to be "frozen" in the expanded state for location in the accessory. It is then unfrozen by applying heat to the component to shrink it down to the size required. Again there is a limit to the shrink ratio and the need for a controlled and carefully applied heat source to unfreeze the component. The classic material is polyethylene cross-linked by one of several means prior to expansion.

Cold cross-linkable synthetic resins have, in some designs, been used as primary insulation and in others as secondary to rubbers. The balance of desirable properties has included the safe handleability of active chemicals under site conditions, their sensitivity to moisture (in some cases), their inherent expense, tendency to degrade in service (certain families of resins) and limited shelf life in storage. Despite these considerations a number of designs exist using acrylic, epoxy and polyurethane resin systems.

Termination insulators have a specific requirement: they must be manufactured from polymeric compositions that tolerate the exposure to weather and pollution and the surface electrical activity that inevitably occurs. Porcelain is a very robust (electrically) material and the development of competitive polymer compositions has been the objective of several decades of work. Current materials based on silicone and EP rubbers which may be moulded by conventional polymer technology perform well in laboratory trials and are gaining successful service experience.

An example of the harnessing of polymer technology with electrical design is the use of voltage variable resistance components in both joints and terminations to relieve electrical stresses at the ends of the conducting cable screens. Particular constituents of the polymeric compounds (e.g. silicon carbide) impart this characteristic of decreasing electrical resistivity with increasing electrical stress. These compounds may be moulded to form the stress relieving tubes as with the termination insulators.

It is appropriate to mention the additional considerations necessary when applying polymeric accessories to impregnated paper cables. Firstly, there is the incompatibility of many polymers with the impregnants leading to swelling or cracking. Secondly, the finite permeability of all polymers to water does not fit well with the hydrophilic nature of paper even when impregnated. Very special selection and application of water and oil barrier materials (e.g. polyvinylidene chloride and varnished nylon fabric, respectively) resolves both these difficulties although their presence as separate items (e.g. as tapes) provided in termination or jointing kits is a move away from the sought after simplicity.

4.4.2.2 Supertension

At the higher electrical stresses employed at supertension voltages the benefits of factory manufacture on accessory performance are even greater than at medium voltage. The main insulation systems for impregnated paper cables are also of paper dried and impregnated in factory conditions and hermetically sealed for storage and transport. These "tube sets" must then be briefly exposed to ambient conditions during assembly operations (which may, in some cases, happen in an air-conditioned enclosure despite being on-site) but the reprocessing is much shorter and easier. In the case of hydraulic stop-joints, the greater part of the joint, its centre compartment, remains unexposed following factory assembly but the two cables it joins still need the building by hand of a paper tube cone to complete the assembly. In addition to impregnated paper, highly filled synthetic resins (principally epoxy filled with silica) provide the hydraulic but electrically insulating barriers and offer improved thermal performance in straight joints and terminations (section 4.3.5). Mouldings in these filled resins often include metallic electrodes to geometrically control the electric fields. An alternative method of electric stress control is by the inclusion of conducting foils between layers of paper to form a series of concentric capacitors. Although effective, this method is very demanding of manufacturing skills and time.

Terminations still use porcelain insulators despite the disadvantageous weight, cost and their impact intolerance. These are part-assembled prior to transport to site where, apart from the sweating of cable sheaths to end bells, connections are made by bolt-up means with O-rings seals (employing oil resistant rubbers).

Vital to the integrity of oil-filled cables is the system of pre-pressurised tanks which maintain the hydraulic pressure of the oil. In these tanks are a number of totally sealed but flexible metal cells which contain an inert gas at an appropriate pressure. The oil surrounding these cells enters and exits from the tank as required by the thermal expansion and contraction of the oil and other items in the cable system, compressing or relieving the cells as it does so.

The design and construction philosophy of accessories for polymeric supertension cables is evolving as these compete with impregnated laminated cables at increasing voltage levels. Approaches adopted have included hand made insulation from tapes lapped and then heat consolidated and cross-linked, on-site injection moulding to reproduce the solid insulation of the cables and factory made mouldings. The first of these methods has been successfully applied despite the high skill level required of the jointers, the need to maintain factory cleanliness standards and the necessity of controlling a critical time and temperature sensitive chemical process in site situations. Injection moulding of sizeable quantities of (generally cross-linkable) polymer, as in the second method, adds the requirement that a large and complex machine be brought to site. This is not always feasible. The third approach allows the major and critical components to be fabricated, inspected and mostly tested in the controlled environment of a factory. It does, however, rely on the careful matching of dimensions of the accessory with the cable — not always a trivial matter — to avoid electrical weaknesses at the interfaces. This problem is alleviated by the use of elastic materials for the mouldings which, despite the great wall thicknesses at these voltage levels, accommodate some dimensional variation.

4.4.3 Environmental issues

4.4.3.1 In manufacture
Conversion of raw materials into any product involves the consumption of energy, the use of packaging for transport and the generation of by-products. Cable manufacture itself is, relative to many other industries, benign in that it uses materials which are already in an advanced state of conversion and merely require treatments and forming as described in previous sections rather than transformation from mineral or natural sources. However, this is not an area that can be ignored by cable manufacturers. In forming conductors from the as-delivered rod, lubricants are used which consist of oil emulsions. These become contaminated with metal fines and metal compounds and therefore must not be disposed of carelessly but must be treated as toxic waste. As metioned earlier, the residual lubricant must be removed to ensure good electrical characteristics of oil-filled cables. Until recently this process involved scrubbing with chlorinated solvents, which have also been used in the cleaning of many other parts of cable systems such as oil pressure tanks. Since the implication of these solvents in destruction of the ozone layer, alternative environmentally friendly materials and processes have been substituted, for example employing water-based detergents, water/hydrocarbon solvent mixtures and various hydrocarbon mixtures.

The sheathing of cables with lead demands that care is taken to prevent discharge of lead-bearing fumes into the environment. As a process registered with the HSE it is necessary to monitor such emissions to ensure that hazardous levels are never approached. Monitoring is an essential part of the production operation in any areas where the potential for causing a deterioration in the air, soil or water systems in and around the production facility are registered.

4.4.3.2 In service

Buried in the ground, mostly, and largely passive in operation, cables offer little threat to the environment in service. Some concern has been expressed over lead in sheaths and the possible contamination of surface waters. This is despite the corrosion protection plastic oversheaths, the low solubility of lead alloys and the presence of naturally occurring lead ores in many regions. Concern over the effect of cable oil spillages on flora and fauna have been addressed by demonstration that they are very readily bio-degraded. The currently used straight chain alkylbenzene was shown to achieve an 80% ratio of biological to chemical oxygen demand in a 20 day test (test to OECD Guideline 301D).

Although not material related, there are two further environmental concerns:

*   Aesthetically, buried cables provide no detriment to the environment compared to the alternative of overhead transmission lines. This aspect, although claimed to be an expensive luxury by some, is an important gain wherever power has to cross a beautiful landscape.
*   The lesser contribution of buried cables to electromagnetic fields can also be claimed as an advantage compared to overhead lines. This is especially true with certain installation configurations.

4.4.3.3 In disposal

Due to their longevity, disposal of cables falls to a subsequent generation. Recovery of valuable metals generally ensures care in retrieval. The less valuable materials are incidentally disposed of, generally by controlled incineration, whilst recovering the metals. As with the in-service spillage of cable oils, their intrinsic bio-degradability prevents long term damage should any be released into the soil.

4.5 Testing

4.5.1 Routine

To deliver a product such as a cable with the integrity that promises a service life of four decades demands that, as a routine matter, the quality of raw materials, the semi-finished products and the finished items be closely monitored. Due to the high proportion of the cost of a cable contributed by the raw materials it is a wise policy to determine their adequacy at an early manufacturing stage. This may be accomplished by measuring critical characteristics of the materials as they enter the cable makers domain or, preferably, prior to dispatch from the suppliers. However, such testing costs time and money. It is customary, therefore, to avoid duplication of testing by sharing the tasks with the supplier, ensuring by audit that the supplier possesses satisfactory quality control schemes. It is the target of the cable manufacturer to be able to trace every material that forms a part of the product back through the processes of manufacture to its origins at the supplier. This is no mean task as the list of raw materials used by a major manufacturer of a wide range of cables runs to over a thousand items.

The laboratories carrying out routine testing are likely to possess a sub-set of the capabilities of the laboratories that would make the first selection of materials and be instrumental in establishing the agreed specification for each material. These capabilities would include electrical, mechanical and chemical testing methods with special areas such as polymer rheology and fire performance testing.

Semi-finished products such as cable core prior to laying up or accessory components prior to further assembly are often more inspectable than at any later stage. It is certainly less costly to reject sub-standard items as early as possible before more work and further materials are invested in them. Apart from dimensional checks, these tests may include electrical quality of oil from an impregnated cable or accessory, the completeness of cure of polymeric insulation or the freedom from defects of a resin casting determined by partial discharge measurements or ultra-sonic testing.

Routine testing of the finished product is often regulated by the standards to which it is made and will include examination of small samples as well as tests on the ready-to-dispatch cable. These latter involve high voltage tests appropriate to the nature of the cable with specific parameters to be met such as loss tangent, capacitance, conductor resistance and, in the case of polymeric cables, partial discharge levels. This last area has been pressed in recent times to reach higher sensitivities in longer cable lengths. This has been achieved by employing sophisticated means of extracting the very small radio frequency signals evidencing the partial discharges from the electrically noisy environment of a heavy industrial plant.

4.5.2 Type Testing

The various standards documents indicated in Table 4.2 prescribe performance levels that must be demonstrated for each new design offered or significant change in materials or technology of production. Some of the principal requirements are shown in Table 4.4. These are typical of existing specifications but the individual documents must be examined in each case to determine the exact testing details. These type tests have evolved to provide a high level of confidence that the cable system, including of course the various types of joints and the terminations, will perform well in normal and predictable excursions from normal service. In practice it is the thermal, mechanical and electrical stresses that are experienced by the system during the excursions from normal service that regulate the cable and accessory design. For example, the current rating of a buried cable will presume the highest likely thermal resistance of the soil with the highest temperature, the strength of cleating of a non-buried cable (e.g. in a tunnel) relates to maximum short circuit currents and the insulation thickness will be such that the electric stress during the pulse from a lightning strike will be less than the breakdown strength of that insulation. Some parts of type test protocols have a bearing on the longer-term performance under normal working conditions. Rapid thermal cycling, for example, is expected to accelerate any changes in dielectric properties of an oil-filled cable that might otherwise only show up after some time in service. As another example, more relevant to polymeric insulation, is the application of very high a.c. electric stresses for a period of time which may be relatable to life expectation at service stress using statistical transforms [4]. However, the trend in type testing philosophy is towards longer tests carried out in conditions closer to those which the system will meet in its normal working life — this is in recognition that nothing is more comforting than proven service experience.

Table 4.4   Typical type test requirements

System voltage (kV)	A.C. test voltage (kV)	Lightning impulse withstand voltage (kV)	Some of the other tests
11	18	95	Bend tests: 3 reverse cycles
33	54	194	Load cycles: 20 x 6-8 h load + 16-18h cooling with 1.33-1.5 U_0
66	114	342	Loss tangent and capacitance at temperatures of 95 °C and 2 x U_0
132	228	640	Partial discharge (polymeric only) at 1.5 U_0
275	375	1050	Switching impulse (oil-filled)
400	500	1425	d.c. voltage (polymeric)

4.5.3 Special testing

There is no limit to the special testing that may be required to satisfy particular conditions of employment of a cable system but two quite different areas of testing are worthy of individual mention; water tree resistance (see section 4.3.2.2) and fire performance.

The mechanism of deterioration of polymeric insulation by the growth of large numbers of separate tree structured channels in the presence of water and quite modest a.c. electric fields has been known since the mid 1960s. Eventually a few of these interfere with the local electrical field, enhancing it to a level where discharge activity commences and electrical breakdown channels propagate rapidly to provoke a complete dielectric failure. Low density polyethylene is particularly prone to this phenomenon and a large quantity of initially low-cost cable, installed principally in the USA in the 1970s, failed prematurely causing disaster to both manufacturers and customers alike. Since then the use of cross-linked polyethylene and EP rubber compounds has considerably reduced the hazard and, more recently, special treeing resistant grades of polyethylene (TRXLPE) have become available. The claim is that these improved materials permit cables to be made and installed in moist locations without a protecting water barrier and therefore more cheaply than metal sheathed cables. To test this claim a variety of test methods of differing complexity and running time have been proposed. The common aim is to discriminate between designs and materials that promise long trouble free performance and those that do not. The conditions of test are intended to accelerate deterioration without departing from the mechanism that would occur in service. Table 4.5 illustrates some of the principal features of the dominant test protocols at this time in the UK.

The feature to note is the period of two years demanded in some cases which means that a manufacturer has to wait that time at least before the product is acceptable to the customer. Should it fail along the way s/he must start again with a new product. This means that a manufacturer cannot afford to gamble on passing such a test; it is necessary to run a programme of testing which includes in parallel all possible candidates. The resources, both in staff and equipment, required to meet this commitment are substantial and costly. It is to be hoped that the emerging understanding of the fundamental mechanisms of this form of deterioration will enable more certain selection of candidate materials before embarking on such long testing protocols.

Fire performance of a cable system implies a number of differing criteria. There is the ability of some cables to be resistant to the propagation of a fire to which they are exposed and of others (but generally lower voltage cables) to maintain some working capacity in a fire situation. There is the requirement that the smoke emitted from cables in certain locations should not exceed a certain density and not contain significant amounts of either corrosive species or toxicologically hazardous substances. Although not all cables, and as yet few supertension cables, need to meet such requirements (e.g. a buried cable is unlikely to be caught up in a fire), this is a field of cable design growing in importance. Table 4.6 shows some of the common test regimes in specifying cables and Fig 4.5 illustrates a laboratory dedicated to such work. Once again these special tests represent a considerable investment by the manufacturer.

4.6 Diagnostics

Despite the low incidence of failure in cable systems demonstrated in Table 4.3 and subjected to extensive type testing as illustrated by Table 4.4, a number of diagnostic methods have been developed. These are appropriate to different types of cables and permit some reasonable approach to quantifying a deterioration, proposing its cause, finding its location (if it should be discrete), determining if it is progressive or stable and estimating its effect on system life. The following

Table 4.5   *Some testing protocols for water tree resistance of polymeric cables*

Origin of test	Conditions of ageing and test criterion	Time to complete
AEIC CS5	6 kV/mm, tap water conductor & screen, temperatures 90°C and 45°C, 8/16 h cycling, test: a.c. steps 7 kV/5 mins	2880 h
CIGRE 21-11	5 kV/mm, 0.2 g NaCl/litre conductor & screen, temperatures 40°C and 40°C constant, test: a.c. steps 1 kV/mm/min	3000 h
Eastern REC	$2.5 \times U_0$ (16 kV for 11 kV), tap water at screen only, temperatures 70°C and 70°C constant, test: a.c. $8 \times U_0$ (50 kV for 11 kV) 6 h	3000 h
Midland REC	U_0, tap water at screen only, temperatures 70°C and 70°C constant, test: a.c. $8 \times U_0$ 20 h after first 7000 h, further 7000 h survival at U_0	7000 h + 7000 h
London REC	As Midland but final $8 \times U_0$ 20 h at end of second 7000 h	7000 h + 7000 h
Unipede	$25 \times U_0$, 0.3 g NaCl/litre conductor and screen, temperature 30°C and 30°C constant, test: a.c. steps U_0/5 mins, 100% > 14 kV/mm, 70% > 18 kV/mm, 40% > 22 kV/mm.	17500 h

are examples of diagnostic methods — there are many others in current use.

4.6.1 Impregnated paper insulation

As mentioned in section 4.3.2.1, ageing of cellulosic paper is an inescapable phenomenon and is accelerated by higher temperatures. As ageing proceeds, various properties of the paper change and the chemical degradation of the cellulose liberates a variety of chemical species into the impregnant [5]. Access to the paper insulation is generally only possible when carrying out works such as

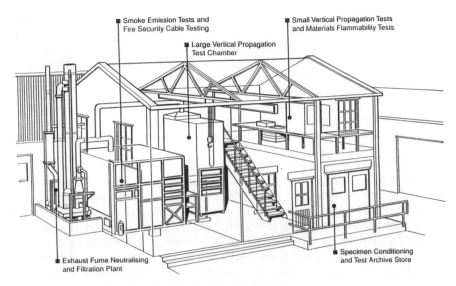

Smoke Emission Tests and
Fire Security Cable Testing

Small Vertical Propagation Tests
and Materials Flammability Tests

Large Vertical Propagation
Test Chamber

Exhaust Fume Neutralising
and Filtration Plant

Specimen Conditioning
and Test Archive Store

Fig. 4.5   *A fire test laboratory for cable testing*

Table 4.6   *Some testing methods for fire performance of cables and accessories*

Characteristic	Standard	Conditions
Propagation of fire	BS 4066/1	Single cable, mounted vertically, ignited by a 1 kW gas flame.
	BS 4066/3	Bunch of cables with volume of non-metallic material up to 7 litres/metre, mounted vertically, ignited by 20 kW gas flame.
Smoke emission	BS 6724	One metre of cable, mounted horizontally, ignited by 1 litre of burning alcohol.
Acid gas	BS 6425	One gram of material from non-metallic parts of cable, pyrolysed in a tube furnace.
Toxicity	NES 713	Four grams (approximately) of material from non-metallic parts of the cable, ignited by a gas burner in a sealed chamber.

diversion of a route or reinstatement of damage. On these occasions, small specimens of paper retrieved from the cable can tell a great deal about the life history of the system and, by comparison with laboratory and field studies, allow prediction of future life under a particular working regime. Fig. 4.6 illustrates the deterioration of some readily measured paper characteristics; one of these at least, burst strength, may be measured under site conditions. Setting a lower limit to such a property based on the service demanded of each cable system is not an exact science but reasonable end-of-life criteria may be proposed. Ten percent of original physical properties has been assumed by some authorities [6] provided the cable is not disturbed.

In low viscosity oil-filled cable it is possible to withdraw small quantities of the oil from locations at the accessories without entering and disturbing the insulation system proper. Extraction and analysis of the dissolved gases provides a complex but informative view of any deterioration mechanisms in progress in the cable. The main complicating factor in interpretation is the geometry of the system as the source of production of the dissolved gases may be kilometres or only a few metres from the sampling point. Although similar analyses are performed for other impregnated paper insulated equipment such as transformers, its application to cables has an additional demand, that of sampling and transporting the oil under high vacuum conditions to permit the fullest understanding of the appearance of any gas into a hermetically sealed system. The qualitative interpretation of analyses is close to that applied to other equipment: hydrogen with methane indicates partial discharge activity, acetylene principally points to arcing under oil, carbon monoxide dioxide signals thermal ageing of cellulosic paper [7]. Special to cable systems can be hydrogen-diffused from aluminium sheaths, carbon dioxide leaked from pressure elements in oil pressure tanks and the interpretation of the oxygen to nitrogen ratio in terms of residence time of air leaked into a system. Particular interest is focused on the build-up of carbon monoxide as this is a relatively low solubility gas which, if sufficient were to be produced by thermal ageing of the paper, would lead to saturation of the oil, bubble formation and then rapid cable failure.

4.6.2 Polymeric insulation

Retrieval of specimens of polymeric cable insulation is sometimes possible, as with impregnated paper cables, when work is being carried out on a system. When there has been exposure to water (and even when not) a microscopic examination of carefully prepared slices may reveal water trees or micro-voids and any progression of these to electrical trees. The slices have to be only a few tens of micro-metres thick, cut with a well-honed microtome blade and then stained to show up water trees when viewed with an optical microscope. Some automation of these observations is possible with image processing to reduce the subjectivity of the determination of water tree number and size but interpretations are still a matter of debate. Such examinations adjacent to a failure induced during developmental testing can reveal the less strong regions of a polymeric insulation

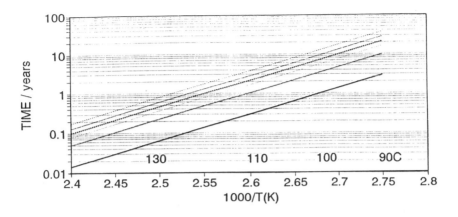

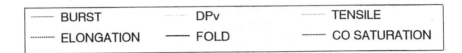

Fig. 4.6   *The ageing of paper insulation. The relationship between temperature and time to reach critical values of properties of the insulation*

system. For example, it may become apparent at which screen the failure initiated with implications on the perfection of the extruder dies, or which path through an accessory the failure ran indicating the relative importance electric field distribution and materials interfaces. One technique widely applied to unfilled XLPE as used in supertension cables is to prepare a substantial piece of cable, say 50 mm long, with smoothed (microtomed) end faces. On heating to above the crystalline melting point of the polyethylene (generally to about 150 °C), the insulation becomes glass-like and transparent allowing a visual inspection of significant volumes of material and areas of surfaces — very much more than can be observed by means of slicing.

Ageing of polymeric insulation other than by treeing may be estimated by deterioration of physical properties of specimens cut from cable samples and by a variety of chemical analytical techniques. These may, for example, determine the depletion of anti-oxidant or generation of species formed by the oxidation of the insulation. A technique (differential scanning calorimetry) that is used to reveal the thermal life of cross-linked polymers measures the heat flow into a microscopic specimen of insulation as its temperature is raised at a predetermined rate. The temperatures at which these heat flows occur and the quantity of heat flowing may be interpreted as changes in the micro-crystalline structure of the polymer, this structure having been established by the time and temperature history of the cable. Thus it is often possible to say that a cable has been loaded at a certain level or, more interestingly, has suffered an overload reaching temperatures in excess of the design limits.

4.7 Case studies

Although this section has been divided into "Typical" and "Specialist" it is one of the features of the high voltage insulated cable field that almost every product is special, particularly so as the voltage level increases. However, there are, as already indicated, general design philosophies and these may be illustrated by reference to an individual cable.

4.7.1 Typical

For distribution of power in urban and suburban areas at medium voltage (principally 11 kV), cable might well be supplied on a contract covering a period of years for a few particular conductor sizes, constructions and accessory sets. In this case the customer was intending to adopt polymeric insulation to realise the economic benefits of lower installation skill and time requirements. Clearly these cost savings needed to exceed any price disadvantage compared with the previously used impregnated paper system. For both cable price and complexity of installation reasons it was necessary to adopt a design without a metallic sheath, therefore demanding a polymeric insulation system with proven resistance to water treeing. An EP rubber composition was chosen which had been developed to perform well in the Cigre water tree resistance test (Table 4.5, section 4.5.3) and was of a family of materials with some 20 years satisfactory service experience. The type test included the protocol indicated as Eastern REC in Table 4.5, otherwise the cables were to be in compliance with BS 6622. The conductor was chosen to be solid circular (the circuit was to be laid as separate cores rather than three core to simplify installation techniques) aluminium. The outer conducting screen was matched to the insulation system to be cold strippable (to minimise jointer skills and tooling). The earth screen was to be formed by copper wires laid over the outer conducting screen. As the final cover, a sheath of extruded medium density polyethylene was chosen to provide an abrasion resistance and water impermeability superior to low density polyethylene without the sensitivity to cracking of high density polyethylene. This sheath needed to be coloured red for identification purposes with careful choice of a pigment and anti-oxidant combination to avoid deterioration with exposure to solar radiation.

For transmission of power at supertension system voltages a cable will be described that involved a single 275 kV circuit passing through a road tunnel under a wide river, the Mersey, and which was replacing an old technology gas compression cable. Of major interest to the customer were safety due to the presence of the public in the tunnel and reliability, this latter being one of the motivating reasons for recabling. The overall circuit length was 14 km, of which 2.7 km was in the tunnel, the two sub-stations at either end of the route being 4 and 6 km distant inland. The system was rated at 748 MVA summer load (831 MVA winter) and to meet this, copper conductor sections of. 1300 mm^2 were required in the tunnel and 2000 mm^2 in the land section (this difference due to the better cooling brought about by forced air circulation in the tunnel), and the circuit was fully cross-bonded. The insulation system was chosen to be oil-impregnated paper at a design service electrical stress of 15 kV/mm (lightning impulse withstand of 99 kV/mm). The containment was of corrugated aluminium with a medium density polyethylene protection. The installation in the tunnel was in concrete troughs filled with cement bound sand to provide the highest level of security due to the nature of the location, allied to good and stable heat transfer capability. The land sections were directly buried but also in cement bound sand to ensure stable and optimum heat transfer to the soil. The circuit was arranged to exclude the more complex stop joints (which require oil feeds) from the tunnel section and, where straight joints were needed, these were also enclosed in cement bound sand.

4.7.2 Special

Although long submarine power links are generally thought of in terms of d.c. connections there are situations where an a.c. system is more appropriate as when the power flow is relatively modest into a discrete area and without other interconnections. Such a case is the single cable circuit linking the mainland grid to the Isles of Scilly. The route distance is 56 km passing through extremely busy sea lanes and avoiding the very rough sea bed lying directly between St Mary's and Lands End and the important fishing grounds. The cable was to be a three core construction with conductor chosen to be 70 mm^2 compacted circular copper. The insulation system of EP rubber designed for a service stress at the conductor screen of 3.1 kV/mm with a highly bonded conducting outer screen covered with a double layer of interlocked tinned copper tapes. A "wet design" was adopted benefiting from the service-proven water resistance of the dielectric and providing a light weight (without metal sheath), flexible and significantly less costly to lay cable. The three cores were laid-up with polypropylene string fillers to provide a good overall circularity and two layers of galvanised steel wire with polypropylene string bedding as protective armouring. A final serving of UV and oxidation resistant polypropylene string finished the cable.

A very different mainland to island link is the cable supplying Sardinia from Italy where the maximum laying depth of 500 m and transmission voltage of 200 kV d.c. with a submarine route of 105 km demanded very special solutions. Tolerance of the pressure at these depths (some 5 MPa) was achieved by an oval cable cross-section (circular cables suffer pinching of metal sheaths during thermal cycling) based on an oval copper conductor of 420 mm^2. The forming operations produced a very hard conductor; this hardness was retained so that some 70% of the tension on the cable during laying at these depths was supported by the conductor itself. The insulation system of viscous compound impregnated paper was dimensioned to give a maximum calculated stress of 25 kV/mm on the conductor at uniform (non-loaded) temperature. At working temperature profiles this stress is relieved on the conductor as the resistivity of the warmer insulation falls, thus throwing more stress onto the outer layers of cooler insulation. On cooling, following loading, very high (up to double) stresses can occur under d.c. conditions threatening the integrity of the insulation; otherwise, a higher design stress would have been permissible. The containment chosen was a rather hard lead alloy (E) sheath which has a good fatigue characteristic and is able to withstand concentrated mechanical forces with an oversheath of low density polyethylene to act as a distributor or absorber of torsion and tension stresses. The consequent danger of overvoltage development between sheath and reinforcement during voltage transients was overcome by making electrical bonds at every third kilometre of cable. Above the polyethylene oversheath, a reinforcement of galvanised steel tapes was placed in two 50% staggered layers. The armouring of individually textile wrapped galvanised steel wires was laid on a jute bedding allowing them freedom to expand radially and avoid transmitting dangerous mechanical stresses to the conductor. A very important consideration in mechanical design was the minimisation of twisting tendency and maximisation of rigidity (within the bounds determined by handling operations) to avoid kinking on release of tension during laying.

4.8 References

1   FARNETI, F. *et al.*: "Reliability of underground and submarine high voltage cables". Cigre Montreal Symposium, 1991, Paper S 38-91

2   ENDERSBY, T.M. *et al.*: "The application of polypropylene paper laminate insulated oil-filled cable to EHV and UHV transmission". Cigre, 1992, Paper 21 307

3   OCCHINI, E.: "A statistical approach to the discussion of the dielectric strength in electric cables". *Trans. IEEE*, 1971, Paper No. TP 157-PWR

4   "Users' guide for Weibull statistics applied to cables reliability studies". Cigre Study Committee 21, Working Group 21-09, 1983

5   SIMMONS, M.A. *et al.*: "Thermal ageing of cellulose paper insulation". IEE Conference Paper A3-6/15/76, 1976

6   GAZZANA-PRIAROGGIA, *et al.*: "The influence of ageing on the characteristics of oil-filled cable dielectric". *IEE Proc. A*, 1960, **108**, pp.1-13

7   BS 5574:1978: "Guide for the sampling of gases and of oil from oil-filled electrical equipment and for the analysis of free and dissolved gases".

Chapter 5

System commercial considerations

S. Armstrong

5.1 Introduction

This chapter examines the development of the electricity system in the United Kingdom following privatisation in 1990. It looks at the development of a commercial market and the impact of the market on the strategies of the leading players. The chapter also looks at the development of markets throughout the world and focuses on three different countries where the social and political environments determine differing priorities for the electricity industries.

5.2 The UK electricity industry

5.2.1 The old structure

As discussed in Chapter 1, the previous structure of the nationalised industry in England and Wales was dominated by the Central Electricity Generating Board (CEGB). The CEGB was responsible for the generation of electricity in bulk and the transmission of the electricity through the national grid to the regional distribution boards, each of which enjoyed a local franchise.

This structure was characterised by a "cost plus" philosophy, with centrally planned investment, an engineering-led approach and a virtual barrier to new entrants. Almost all consumers had no option but to purchase from their local Area Board, and although the Energy Act 1983 attempted to stimulate competition in generation the overall structure of the industry effectively prevented this.

5.2.2 The structure following the Electricity Act 1989

The Electricity Act 1989 privatised the industry and set out to encourage competition where appropriate and regulate where the scope for competition was limited. The industry may be sub-divided into four areas:

• Generation — the production of electricity;

- Transmission — the bulk transfer of electricity;
- Distribution — the delivery of electricity over local networks;
- Supply — the purchase and sale of electricity to customers./

The transmission and distribution franchises are both natural monopolies and therefore regulated. The generation and supply business are open to competition.

The twelve Regional Electricity Companies (RECs) are the direct successors of the previous Area Boards and supply to a franchise market in their regions, although customers with a demand of at least 1 MW may seek their supplies from outside the franchise market. This threshold reduces to 100 kW in 1994 and will disappear in 1998.

In Scotland vertical integration has been maintained in the new structure with the creation of Scottish Power and Scottish Hydro-Electric but, as in England and Wales, nuclear generation was assigned to a separate company, Scottish Nuclear, which remains state-owned.

The most important change in the industry has undoubtedly been the development of competition. The new system is competitive at both ends of the market, in both electricity generation and supply. The system is also more transparent: the costs and revenues resulting from generation, transmission, distribution and supply are gradually being unbundled.

Choice of supplier has been introduced for large customers and will soon be extended to others. Smaller customers have also benefited from the increased choice of generators open to their regional electricity company. One of the major changes brought about by this more competitive environment is the increase in cost consciousness within the industry, which has led to an abandonment of the technology-led approach.

5.2.3 Generation   CEGB

The old Central Electricity Generating Board was broken up and divided into four new companies. The National Grid Company was established to own and run the transmission system in England and Wales and enable the bulk transmission of electricity between generators and suppliers, and two fossil-fuel based generating companies were created: National Power, which inherited 30,000 MW of CEGB capacity, and PowerGen, with 18,000 MW. The nuclear capacity of the CEGB was allocated to Nuclear Electric, which was retained in Government ownership. However, the CEGB successor companies are not the only players in the wholesale market. The Scottish electricity companies, Electricité de France, and a growing number of independent power operators are all now generating electricity for sale in England and Wales. All this has led to an increasingly competitive market.

There are several key features of the system. Firstly, power is traded through an open commodity market, the Pool, which will be described later. Secondly, the generators no longer have any obligation to supply and they have to compete for their market share. As generation is an increasingly competitive business, there is no price regulation in the wholesale electricity market.

5.2.4 Transmission

The National Grid Company (NGC) as the network operator has a focal point within the industry. The NGC owns and runs the main high voltage supergrid system which facilitates the transport of electrical energy from the producers to the customers who require electricity. It consists of 7000 kilometres of high voltage transmission lines and over 200 sub-stations. This grid system is operated and controlled through the national and area control centres under the direction of national control. The primary role of the National Grid Company is to develop and maintain an efficient, co-ordinated and economic transmission system, and secondly, to facilitate competition in both generation and supply.

5.2.5 Distribution and supply

The main activities of the twelve regional electricity companies (RECs) in England and Wales fall into four distinct but closely related areas: the distribution and supply of electricity to over 22 million customers, the retailing of electrical appliances, and electrical installation contracting. The first two activities, distribution and supply, account for about 90% of turnover.

Since privatisation, the regional electricity companies have diversified to varying extents and moved into generation. All but one of the companies have now become involved in independent generation schemes.

5.3 Commercial arrangements

5.3.1 The electricity pool

the electricity Pool.

All the major generating companies are required to sell all the electricity they produce into an open commodity market known as "the Pool". The Pool is a simple name for what is, in effect, a rather complex trading mechanism. This accommodates two special features associated with the sale of electricity from many power stations to a number of different customers over a fully interconnected system:

- The amount of power required by customers at any time must be matched by that generated at the power stations; and
- It is physically impossible to distinguish between electricity generated at one station and that generated at another.

Each generating unit has to declare a day in advance its availability to the market, together with the price at which it is prepared to generate, for each and every half hour of the day. The units are then called to generate by the National Grid Company in ascending order of price. The most expensive unit used sets the system marginal energy price (SMP) which all others receive for that half hour. In addition, there is a separate pricing mechanism for capacity made available to the Pool. In this case, a half hourly capacity value is derived from a calculation of the balance of surplus demand for each half hour period.

5.3.1.1 Derivation of pool prices

(1)   NGC derives an operating regime for all the plant on the basis of the forecast demand and declared availabilities of plant but assuming that the transmission constraints did not exist. The regime is known as the "unconstrained schedule" and is distinct from the schedule used to despatch plant.

(2)   Each generating unit's offer prices are converted into a "Generator Price", which is the average price (per kWh) of providing power at the unit's maximum declared available output, excluding the start-up charge.

(3)   In "Table A" periods, SMP is defined as the highest Generator Price of plant required to operate in each half hour according to the unconstrained schedule, as long as the plant does not receive a marginal plant adjustment. In "Table B" periods, SMP will be the highest incremental price of a plant that is not labelled inflexible.

(4)   The capacity element is LOLP(VOLL-SMP) where LOLP is the "Loss of Load Probability". This is the probability of capacity being inadequate to supply demand in the particular half hour because of a sudden unexpected increase in demand or a sudden failure of plant such as a generating station. It will be calculated by NGC. VOLL is the "Value of Lost Load". It is a measure of the price that pool customers are willing to pay to avoid a loss of supply. It will be set at a level to ensure that the quality of supply will be maintained and is set by the Director General of Electricity Supply.

(5)   The price paid every half hour to generators for each kWh produced, the "pool input price", PIP, will be:

           SMP + capacity element

       Generators are also be paid for reserve, marginal plant operation, and for any ancillary services. In addition, they receive payments to recompense them for transmission constraints and forecasting errors and for having the plant available to operate. Generators are penalised if they do not follow NGC's instructions.

(6)   The costs associated with (5) and the transmission constraints are spread over the units of electricity purchased through the pool during "Table A" periods. This results in an 'uplift' being added to PIP to arrive at 'pool output price' (POP). The difference between PIP and POP covers the costs of:

- reserve
- availability of plant
- forecasting errors
- transmission constraints
- ancillary services
- marginal plant adjustments

5.3.2 *The transmission system*

An important element of the new commodity market is the requirement for the network operator to be independent of generation and supply in order to operate a level playing field.

Access to the grid is unrestricted. However, there are several important aspects associated with open access which need to be mentioned. There is a Grid Code which defines the specific technical access requirements. All generators and distributors seeking connection must meet the appropriate standards to ensure that technical difficulties are not caused for others connected to the system. The Grid Code covers items such as planning requirements, connection conditions, demand forecasting, scheduling and despatch, operational liaison, safety co-ordination, and testing requirements, together with the registration of appropriate data. The terms for access are non-discriminatory, as the transmission charges are identical for new and existing customers.

In order to facilitate access to the grid there are licence obligations which require the National Grid Company to respond within three months to a request for connection to the grid system. During this period the design, planning, legal and economic aspects associated with such proposed connections are investigated. Following this period, the applicant has a further three months in which to accept the offer of connection. One further important aspect associated with access to the grid is the NGC's Seven Year Statement. This is a statement based on current information that gives a view of the future in order that potential customers can determine investment decisions and opportunities.

The connection and use of system charges cover four elements — entry, exit, system service and infrastructure charges. These are paid by the suppliers and the generators which connect with and use the grid system. The charges identify differentials between 14 zones in England and Wales, and these differentials indicate the costs associated with the connection of generation in relation to the location of demand within England and Wales. These terms are published to ensure awareness of the costs associated with the use of the grid system and are non-discriminatory. Grid charges are regulated on the basis of an RPI - x formula.

The transmission licence contains strict standards covering system security and quality of supply. Integrity has been maintained in the period since privatisation when the electricity market was first formed, even during extremely adverse conditions. During December 1990, the grid system was subjected to over 800 faults in the space of 36 hours caused by very severe weather. The integrity of the system was maintained and the implications of the instructions to generation plant were covered through the mechanisms for payment within the electricity market place.

Another interesting operational impact occurred during the last football World Cup semi-finals, when a 2800 MW increase in demand (equivalent to 28 million electric lights) was met in the space of 3 to 4 minutes This occurred when people in England and Wales knew the result of the televised semi-final, when England lost to West Germany. In unison, millions decided to turn on lights and kettles, thus increasing significantly the demand for electrical energy, which was met through instructions to power stations within England and Wales. Again, the associated payments were made through the mechanisms of the marketplace.

The structure of the new system has ensured that system integrity and quality of supply have been maintained, and the changes have enabled the NGC to focus more clearly on the costs associated with the work the company undertakes to maintain the supply of electricity. The system has benefited from the unbundling of costs which has taken place. The segregation of transmission from generation and distribution has allowed attention to be focused on the costs associated with the transport of electrical energy.

NGC has a number of objectives in setting its structure of charges. These derive primarily from the Electricity Act 1989 and the Transmission Licence. In particular NGC is required not to discriminate between users or classes of users and to facilitate competition in supply and generation.

In addition to these licence obligations NGC has a number of other objectives:

- clarity of principles and transparency of methodology;
- to inform existing users and potential new entrants with accurate and stable cost messages;
- to charge on the basis of services provided and on the basis of incremental rather than average costs, and so promote optimal use of and investment in the transmission system; and
- to be implementable within practical cost parameters and timescales.

The requirement of NGC not to discriminate in its charges is of great importance, and has a number of implications in the light of the types of services provided by the transmission system and the requirement to offer use of system terms to all who apply. In particular, in order to achieve non-discrimination it is necessary to distinguish between the types of services provided to different users. The type and amount of services provided to a particular user will vary according to both the location of a particular user and to the type of use (i.e. whether a generator or supplier).

At a more theoretical level, NGC takes the view that, given its objectives of charging, it is appropriate to charge on a basis of marginal costs. In other words, each user on the system should pay charges which reflect the cost of an increment of usage of the system rather than the average cost of the total use made of the system. As certain incremental costs will vary according to the location of the user on the system and the type of user involved, the approach assesses the charges appropriate to an increment at the particular location for a particular type of user.

Transparency of charges enables the customer to identify where costs are being incurred and promote a downward pressure on prices. It should be noted that roughly 73% of costs are associated with the production of electricity, 7% are due to transmission and 20% due to distribution.

The open access to the transmission system has stimulated a significant increase in new generation projects. Up to April 1992, electricity generators had signed connection agreements with the National Grid Company for 21600 MW of new generating capacity. This is approximately 40% of the maximum demand for electricity in England and Wales.

While not all of these new generation projects will come to fruition, approximately 10000 MW looks extremely likely. CCGTs account for all but 2000 MW of the new plant, and over 11000 MW is planned by companies other than the two large generators. Consequently, the trend is very much towards increasing competition in generation. In addition, competition in supply has started. 1800 customers of 1 MW or more now take their supply from sources other than their own local distribution company. This is approximately 40% of those companies which could contract for their supply of electricity with any supplier in the electricity market.

5.3.3 Distribution and supply

The Regional Electricity Companies such as East Midlands Electricity (EME) operate two businesses associated with electricity; the distribution business which operates and maintains the distribution networks, and the supply business which purchases electricity in bulk through the wholesale market and sells it to customers.

To ensure competition in the supply business, regional electricity companies are required to provide open access to their distribution networks on a non-discriminatory basis. As a result, not all consumers to whom a regional electricity company distributes electricity are necessarily supply customers of that company.

The distribution business operates and maintains the assets which carry power from grid supply points to individual customers within each authorised area. This involves a network of overhead lines, underground cables, switches and transformers operating at voltages from 132 kV down to 11 kV.

The distribution business is in effect a monopoly, so the charges made by a regional electricity company for the use of its distribution network are subject to regulatory price control. The formula currently in use is RPI - x, where RPI is the retail price index and x is a number which varies between the companies. This formula applies to the tariff charged to users for transporting electricity across the network, the "use of system" charge.

The price formula requires that the average use of system charge per unit distributed should not increase year on year by more than RPI + x, after adjusting for an incentive to reduce electrical losses. The initial values for x were agreed by the Government and the electricity companies, and range from 0 to 2.5

according to the company. In fact, increases in use of system charges have generally been kept well below the permitted limit.

The business of supplying electricity is restricted, with only a few exceptions, to companies holding electricity supply licences. These licences are of two types. Firstly, each of the regional electricity companies holds a licence giving its rights and obligations relating to supplies to customers within its authorised area. These licences are called Public Electricity Supply licences. The other type of licence is called a Second Tier licence. Regional electricity companies holding such a licence can supply customers outside their region, and the licence also allows generators to supply customers direct.

Since the public electricity supply licence covers supplies to a monopoly market, price controls have been incorporated. These price controls operate in a similar manner to the distribution business control, which as mentioned above, is related to the Retail Price Index and an x factor to ensure prices fall over time. One essential difference has been introduced into the supply price regulation. Only the supply business' added costs and margins are limited by this formula. This means that all the uncontrollable costs (electricity purchase, transmission and distribution costs) can be passed through. This is because these charges are either regulated elsewhere or are subject to competitive pressures.

Competition is also being introduced progressively in the retail market. Already, customers with a demand of over 1 MW can shop around and buy their electricity from either their local regional electricity company, another of the RECs, a generator, or even buy from the Pool itself. They are no longer captive to their local supplier.

In 1994 the threshold between the franchise market and the free market is lowered from 1 MW to 100 kW, and in 1998 the franchise is to be abolished altogether. All customers will then be able to shop around for supplies in a totally free market. How far this is realistic will ultimately depend on the continuing development of remote metering and control systems.

5.3.4 Contracts for difference

The volatility of the mechanism for calculating pool price has meant that the buyers and sellers to the pool are exposed to tremendous risk. In the longer term payments should be predictable based on the cost of new generating capacity but the experience to date suggest that the mechanism coupled with the duopolistic power of the generators show that the pool price is extremely difficult to predict.

To counteract this volatility a market is hedging contracts, "Contracts for Differences" or CFDs have been developed. These contracts are not directly related to the flow of electricity but are financial instruments between the generators selling to and the buyers buying from the pool. The effect of the CFDs is to provide each party with a stable price for a given amount of MW or MWh (each REC and Generator has developed a portfolio of contracts related to their supply requirements or generation output). The range of contracts developed includes one way, two way, firm, non-firm, restricted hours etc. The list is continuing to develop as parties change their requirements and needs.

5.3.5 The non-fossil fuel obligation

In addition to purchases from the pool and arrangements under contracts for differences, RECs engage in a third form of buying which is also governed by condition 5 of their licences but is to be assessed separately. The Government introduced explicit incentives for non-fossil fuel generating capacity by obliging each REC to make arrangements so as to secure the availability of a specified amount of non-fossil fuelled power station capacity. The first non-fossil fuel obligation (NFFO) order pertained to nuclear power and expires on 31 March 1998. The initial amount of capacity the RECs were obliged to contract for was 8553 MW which corresponded to capacity owned by Nuclear Electric, BNFL and UKAEA. Subsequent to this the Secretary of State has issued orders pertaining to generating capacity from renewable sources. The two renewable NFFO orders issued to date correspond to approximately 550 MW of capacity and also expire on 31 March 1998.

The contracts associated with the NFFO orders are different from the other contracts for differences within the industry. RECs undertake to pay a premium price to the generator for all output up to a level which the generator is almost certain to exceed. Thereafter they pay the pool price. However, all the excess payment above the pool price is recovered through the fossil fuel levy. The REC itself pays only the pool price and, since this is what it pays on all its uncontracted purchases of electricity, is commercially unaffected by the contract except, like all suppliers, in so far as the levy is increased by it. The fossil fuel levy is set as a percentage of the final price of electricity and was 10.5% in 1990/1991 and 11% in 1991/1992 and 1992/1993 and 10% in 1993/94.

Nuclear Electric receives most of the funds raised by the levy. The effect of this has been to provide the company with a payment from the levy amounting to £1.2 bn in 1990/91, £1.3 bn in 1991/2, and an estimated £1.1 bn in 1992/3 in addition to its marginal revenue from selling extra units of electricity.

Contracts signed under the Non-Fossil Fuel Obligation do not affect the cost of an individual REC's generation portfolio. Such contracts have been examined in detail, either by the Secretary of State in the case of nuclear generation or by OFFER in the case of renewables, at the times when the obligations were set.

One feature of the obligation may affect a REC's contract. Electricity generated by non-fossil fuel power stations which are not subject to these arrangements carries a "green certificate" and is levy exempt. At present this includes certain transfers over the interconnectors linking the electricity system in England and Wales with systems in Scotland and France. Such electricity can therefore command a premium price in the generation contract market.

5.3.6 Regulation

The licensing regime and price regulation described above are implemented by the Regulator, the Director General of Electricity Supply, who is appointed by the Government to enforce the various conditions and regulations in the Electricity Act and the licences. The Regulator's principal roles are to ensure that competition develops smoothly and effectively and, where competition is inappropriate, that adequate safeguards are in place to protect customers.

A series of codes of practice have been adopted by the regional electricity companies and approved by the Regulator to govern the way in which business is undertaken with franchise customers. These cover energy efficiency, ways of paying bills, services for the elderly and disabled customers and complaints procedures. To ensure that customer service is maintained and improved, a series of performance standards have been agreed with the Regulator. In addition, the regional electricity companies are required to ensure that quality of supply is maintained, and that levels of interruption due to faults do not increase.

5.4 Summary of the major changes to the UK electricity industry

5.4.1 The market

Since vesting in March 1990 there has been tremendous change in the Electricity Supply Industry. The generators and Regional Electricity Companies are now operating in a competitive market and although the market is only three years old there are many different strategies being followed by different companies.

With the exception of one company, all RECs are involved in generation projects within the UK. This has and will cause market pressure on the generators to reduce costs and be more competitive as the new schemes will displace many existing stations in the merit order.

The RECs have also diversified in varying degrees. Some, like EME, have declared an intention to develop non-regulated businesses, others have withdrawn from such sectors to concentrate on the distribution and supply businesses.

There has been considerable debate on the power of National Power and PowerGen and their control over the pool. The Director General is monitoring their behaviour closely.

The Director General has also looked closely at the RECs obligations to purchase economically and published two reports in December 1992 and February 1993 on Economic Purchasing. These reports plus the White Paper published in March 1993 on the future of the energy market will obviously determine future developments in the industry.

5.5 Other markets

It can be seen that there has been a major change in the way electricity is treated in the UK. This market approach has obviously brought some problems, primarily associated with the duopoly power of the two leading fossil fuel generators but the developments of the UK are now being seen as a role model for many other countries.

However, there are a number of other factors which must be addressed before the market economy can be introduced. To demonstrate the problems associated with the different worldwide markets, three different areas are identified where East Midlands Electricity plc is currently providing technical support. The experiences in these areas show that the theory of the market cannot always be the main determinant in structuring the electricity industry. The three case studies are described below.

5.5.1 Eskom - South Africa

Eskom supplies more than 90% of the electricity used in South Africa, which is as large as the combined areas of Western Germany, the Netherlands, Belgium, France and Italy. Although this is only 4% of the surface area of Africa, Eskom's production represents more than half of the electricity used on the entire African continent.

Electricity supply in South Africa is more than a hundred years old. It was one of the first countries to use electricity on a commercial basis. Initially, various generating authorities were formed and some of the mines and municipalities generated their own power. The need for a central generating authority soon became evident and in 1923 a public utility, the Electricity Supply Commission — today known as Eskom — was established.

South Africa is sparcely populated by European standards. Of its 38 million people, nearly half live in urban areas, which are far apart. For example, Durban is about 600 km and Cape Town more than 1500 km from Johannesburg. The vast distances between the metropolitan areas and the relatively low population density present unique problems for electricity supply. Consequently, Eskom operates one of the most sophisticated distribution networks in the world. Despite the distances, electricity can be distributed anywhere in South Africa and to neighbouring countries such as Botswana, Lesotho, Mozambique, Namibia, Swaziland and Zimbabwe. Eskom also imports power from Namibia when surplus capacity is available. It is also under contract to buy power from Mozambique's Cahora Bassa hydro-electric power station.

Eskom operates under two Acts of Parliament: the Electricity Act and the Eskom Act. It is not a government corporation, but an independent, self-financing undertaking. It has no shareholders and is funded entirely from debt and retained earnings.

Two bodies plan and direct Eskom's activities: the Electricity Council and the Management Board. The Electricity Council is a non-executive body appointed by the government. The Council determines policy, planning and the objectives of Eskom. It also appoints the Management Board. The Council consists of representatives of major electricity customer groups and independent experts appointed for their specialist knowledge in, for example, finance, industrial relations and technology. The Management Board is responsible for the day-to-day running of Eskom. It makes proposals to the Electricity Council and implements policy.

The organisation is divided into eight functional groups which are further divided into more than 50 business units. This ensures a high degree of decentralisation and closer contact with customers. Eskom employs some 50000 people who are engaged in hundreds of job categories, from unskilled labourers to highly qualified engineers and financial experts. They are employed all over South Africa and come from different cultural and language groups.

After recent investigating the prospects of a listing on the Johannesburg Stock Exchange, the South African Government has decided that it did not at this time regard Eskom as a suitable candidate for privatisation in the near future.

Eskom has legitimate right of supply to 191 geographic areas. It also supplies electricity in bulk to 467 other bodies. These include municipalities/local authorities, 4 independent states and 6 self-governing homelands. The distribution chain is extremely complicated because of this; overall there are 966 organisations dealing with reticulation of electricity in South Africa, causing considerable duplication of effort.

The Electricity Control Board (ECB) is appointed by the Government to oversee electricity reticulation. It is also responsible for issuing licences to local authorities and others who wish to reticulate electricity.

The political system in South Africa meant that local areas looked after their own distribution networks. As a result they usually employed consultants to design and build the system. There were no standards and little if any follow-up service on how to manage the system. Consequently the systems fell into states of disrepair. In addition, in many cases electricity accounts were not sent out regularly, causing problems for customers when large unexpected bills arrived.

Some local authorities use the profits from electricity to subsidise other inefficient services and are therefore reluctant to see any change to the present system. Customers have complained about the poor electricity service by staging rent boycotts. They neither paid for rent or electricity. These boycotts have battered and in some cases bankrupted these local authorities. This in turn meant these local authorities could not pay Eskom.

Faced with the prospect of large debts and loss of sales Eskom has negotiated agreements to carry out the reticulation in some areas. In the meantime civic associations, town councils and the Transvaal Provisional Administration continue negotiations concerning other areas.

At Alexandra Eskom manages the network for the Regional Services Council and is responsible for maintenance, billing and future expansion. Eskom still supplies electricity in bulk to the Regional Services Council. In Soweto, Eskom has taken over ownership and operation of the network in lieu of debt. Eskom's key strategy in managing these networks is the replacement of existing credit meters with new prepayment card meters or similar.

A national forum to discuss electrification is ongoing including representatives from civic authorities, the Development Bank of Southern Africa, Eskom, the ANC and the Government. The outcome of this forum should influence decisions on restructuring the industry and lead to a new Electricity Act within the next one to two years.

5.5.2 Andhra Pradesh State Electricity Board

When the state of Andhra Pradesh was formed in 1956, the installed power generating capacity was just 99 MWs. The Andhra Pradesh State Electricity Board (APSEB) was constituted in 1959 under the provisions of the Electricity Supply Act in order to give an impetus to the growth of the power sector. The fact that the state's demand rose from 213 MWs in 1960-1 to 4131 MWs in 1990-1, justifies the creation of APSEB. The monopolistic character of the Board has never been far away from social responsibilities and the Board has been sensitive to changing social needs.

Coming into being at a time when the state of Andhra Pradesh was predominantly agricultural in nature and was preparing to develop an industrial base, the APSEB took on the vital responsibility of developing the key infrastructural requirement — power. The demand for electricity is a process that is always on the rise in a developing economy and the APSEB has always been at the forefront of the regime's development.

During the year 1990-1, a record peak demand of 3465 MW was recorded. The energy generation in the State was increased by 21 % between 1989 and 1991. The state's installed capacity increased to 4893 MW with a significant addition of 351 MW, out of which 210 MW was thermal, 66 MW was gas, 30 MW was hydro and 45 MW was by way of increased share of power from Central Sector Projects.

During the period ASPEB completed 198 CKMs of 220 kV line, and 243 CKMs of 132 kV line, two 220 kV sub-stations and six 132 kV sub-stations. For the first time in the country an experimental HVDC link between Lower Sileru in Andhra Pradesh and Barsoor in Madhya Pradesh with earth as return path was commissioned on 20 August 1990. 100 MW of power can flow through the line which has been executed as an R & D project funded by the Government of India.

Thus it can be seen that the Electricity Company is vital to the development of the country and the industrial base in this particular environment.

5.5.3 State Fuel and Power Corporation, Zanzibar and Pemba

Public supply of electricity commenced in Zanzibar in 1954 with the commissioning of Saateni Power Station. Electricity was distributed around the town at 11 kV by an underground ring system. The substations, brick built in "Arab" style to fit in with the local architecture, were fitted out with English

Electric switchgear, transformers and LV distribution boards. Supplies were also taken to the outskirts of town on underground spurs to the residential areas. Later, three 11 kV overhead lines were built from the town extending up to 15 km into the surrounding country to supply a few water pumping stations and large residences. In the main, the indigenous population occupying the central part of the town did not receive supply. After the revolution the government redeveloped the central area by building a number of East German designed blocks of flats and to supply these an inner 11 kV ring was installed with 5 British GRP clad unit substations with a total capacity of 3.5 MVA. After this there was virtually no major extension or alteration to the system apart from uprating the transformers in the original substations from 250 kVA to 10000 kVA. No reinforcement was done to either the LV or HV cable systems. As a result, today all cables are at their maximum load.

On the generation front, the load grew from 2.5 MVA in the fifties to 8 MVA by the late seventies. Additional generators were installed at Saateni but lack of maintenance meant that comparatively new sets were being destroyed quicker than new ones could be added. Load growth was therefore limited. By 1979 Saateni Power Station was too full to accommodate any further sets so it was decided to build a new power station to the north of the town at Mtoni and a number of Mirrlees Blackstone K5 and K9 units were ordered, delivered and then left in the open to deteriorate for six years whilst the power station building was completed. When they were eventually installed approximately 50% of their new price had to be spent on rehabilitating them. The same situation occurred on Pemba at the new Wesha Station being constructed to replace the old Tibirinzi Station.

In the meantime, the rise in world oil prices meant that diesel generation was becoming prohibitively expensive and therefore a submarine cable operating at 132 kV was installed from the Tanesco (Tanzanian Electric Supply Company) Grid at Dar es Salaam. At the time (1980) this was the longest a.c. cable in the world (37 km). The capacity of the link is 50 MVA and now supplies all the load of Zanzibar, the sets at Mtoni now having a standby role only.

Public supply started on Pemba with the commissioning of Tiberinzi Power Station near to the town of Chake Chake. Unlike Saateni, these were 415 V units of about 200 to 400 kVA capacity. Pemba differs from Zanzibar in that there are three main towns; Wete in the north, Chake Chake in the centre and Mkaoni in the south, each about 30 km apart. To utilise the output of Tiberinzi, two 11 kV lines built to BS 1320 employing 0.0224 sq.in. copper conductors were built north and south. Tibirinzi Power Station was replaced by Wesha at the same time as Mtoni was built and with exactly the same mistakes. The maximum load of Pemba today is about 1.8 MVA and further growth is limited by two factors; the cost of diesel and the capacity of the 11 kV lines north and south which are each limited to about 500 kVA due to voltage drop considerations. Because of the depth of the sea from Pemba to Tanga it was not feasible to lay an oil-filled cable at present and the cost of diesel generation makes it prohibitive to connect new consumers.

Whilst little has been done to the existing distribution systems of Zanzibar and Pemba since their installation, the Norwegian aid agency Norad has been financing the erection of an extensive 33 kV system over the island of Zanzibar to supply water pumps and to give supplies to local dispensaries. To date about 400 km of line has been commissioned on Zanzibar and work is in progress with another 160 km on Pemba. The problem for the local utility is that it does not have the money or resources to maintain the lines, nor the transport to go to read meters and, in the case of Pemba, the income is many times lower than the generation costs.

Table 5.1   Selected statistics for Zanzibar and Pemba

	Zanzibar	Pemba
Population	375 000	265 000
Customers	16 000	2000
Max. demand	12 MW	1.8 MW
Generation	Hydro (Tanesco)	Diesel
Load: domestic	53%	90%
commercial	14%	8%
industrial	27%	2%
S. light	8%	
Load factor	51%	
Losses	33%	

5.5.3.1 Operating conditions

The climate is hot, humid and dusty. Temperature range is 20 °C to 39 °C throughout the year with the winter being in the region 20 °C to 25 °C. Humidity varies from 92% in the morning to 64% in the afternoon. Average wind speed varies from 6 to 9 knots with rare excursions up to 21 knots. Over the last 100 years there have been three hurricanes that have caused extensive damage to buildings and trees.

The main problem for the distribution system is the high humidity which causes accelerated deterioration of switchgear. British designed outdoor equipment is standing up to the climate very well, particularly when installed inside. However, indoor designs fare very badly, particularly the styles favoured by European mainland utilities and manufacturers, none of which have lasted more than 15 years of operation on the island before failing. The design requirements for overhead lines is undemanding and this has resulted in contractors installing lines of very low mechanical strength even when using conductors of 150 sq. mm. The result of contact by falling coconut trees or road traffic accidents tends to be cascade collapse of the entire line. Attack by termites on poorly treated wood poles is also a problem. It has therefore become necessary to start rebuilding lines only 12 years old to an improved standard before total collapse takes place. However, the BS 1320 lines erected during the 1950s are still reasonably sound and require only the replacement of poles and switchgear.

5.5.3.2 Political considerations
The political history of electricity supply on the islands is as follows:

1954-7	Public Works Dept Electricity Section;
1957-64	Zanzibar Electricity Board;
1964-70	State Fuel & Power Corporation, commercially independent institution under Ministry of Communication & Works;
1970-8	State Fuel & Power Corp. Govt Department, all funds pooled into one account under the treasury;
1978 - present	State Fuel & Power Corp. State owned Parastatal under the Ministry of Water, Construction & Energy.

Ownership of the system is a major problem for SFPC. Although it technically owns all the system it does not have working capital to extend the system to provide new supplies. Therefore whenever a potential customer requests supply, SFPC advise on what hardware to obtain (transformers, switchgear, cables, poles etc.) and the customer then purchases it for SFPC to install. On many occasions customers do not consult SFPC first but decide on their own design and present SFPC with the materials and demand connection. Furthermore, the consumers then claim ownership of the equipment and refuse to allow SFPC to connect nearby customers. The result is a proliferation of different makes and types of equipment on the system.

5.5.4 Summary

Thus it can be seen that different countries have different problems that cannot always be resolved on purely economic grounds. Although the concept of the market is being adopted for much of the developed world, in other areas social and political considerations must be taken into account and in some cases dictate the strategy.

5.6 The way forward

5.6.1 United Kingdom

The first three years of privatisation have seen tremendous change in all the companies in the electricity industry. The generators have commenced a massive programme of closures, staff reductions and rationalisation. The RECs have, with the exception of one, entered the generation market with equity involvement in CCGTs. The RECs have also restructured, recognising the vastly different markets in which their distribution and supply, retail and contracting businesses operate. Many RECs have embarked on expansion programmes into other areas, for example East Midlands Electricity plc is now one of the largest security groups in the UK.

It is widely believed that, in the longer term, many RECs could well leave the electricity supply industry totally and concentrate their efforts on other businesses. The heavily regulated regime may mean that other business opportunities offer better returns for their investors. Indeed in the supply business there are different strategies being adopted by a number of companies who are looking to expand across the country. Some are keen to maintain their traditional area and others are not interested in playing the market at all.

There will be further changes to the regulation and rules of the industry and if the duopoly continue to abuse their power there could well be further change to the structure of the industry. The development of the UK energy policy, highlighted by the drop in demand for British coal from April 1993 will also be a dominant factor in any future movement. This, when linked to the regulatory reviews of distribution and supply businesses plus the reduction in the franchise market, will all ensure that the environment will continue to be dynamic.

5.6.2 Europe

The experience of the UK ESI has been an important catalyst in the development of proposals for a more liberal European electricity market. The liberalisation of energy markets is seen as an important component of the single European market. The European Commission released a directive in 1992 that suggested the preferred way forward would be through gradual evolution, avoidance of excessive regulation and subsidiarity. It is hoped this will avoid problems associated with national energy concerns and cultures.

The Commission's new proposals introduce three further broad elements of change. The first is creating a "transparent and non-discriminatory system" for member states to grant electricity-generation licences and transmission licences for both gas and electricity. Secondly, vertically integrated utilities must "unbundle" the management and accounts of their various activities (production, transmission, distribution) to ensure transparent commercial relationships between market players. Finally, network operators will have to open their transmission capacity to third parties, in a limited way at first. The draft directives contain the common rules necessary to achieve these second-stage objectives.

The directives seek to increase competition in the supply of electricity and gas to final users. In the electricity sector, this will involve encouraging independent generators by assuring them of access on fair terms to transmission and distribution networks. The directive implies, in effect, the goal of a common European transmission and distribution grid with open access for independent generators, electricity distributors, and the largest industrial customers.

Major concerns have been expressed by policy makers and the ESIs of Europe. Can a complex operation such as an electricity grid, where every demand for power must be instantaneously met by an equal supply, and where failure to do so can have direct consequences for all users of the grid, be left to the vagaries of the market, as in England and Wales? The success of the UK ESI restructure will take several years before the full impact on investment in electricity generation and transmission capacity is known. This has caused a cautious start to the Commission's proposals.

At first, the benefits of greater access to transmission grids will be limited to only a few consumers directly: member states will be required to allow only the very largest industrial users of electricity (those who consume 100 GWh a year or more) to have free access to contract directly with any or all generators for electricity. There are few users so large: steelworks and aluminium smelters are the natural candidates.

Only for deals between such large final consumers and generators will transmission companies be obliged to make capacity available on the grid (provided such capacity exists) in return for a "reasonable" charge. However, member states are free to widen the extent of third-party access if they choose. In the UK, the scope of third-party access is already much wider than the EC requirement — access to transmission and distribution grids is available to users demanding a capacity of 1 MW or more, which is typically equivalent to about 4-5 GWh of annual electricity usage.

The commission has allowed until 1 January 1996 for third party access to have an effect. However, without an impetus such as privatisation gave to the development of pooling in the UK, it is not obvious that the 3-year experimentation period proposed by the Commission will give time for electricity markets to develop beyond a few large contracts between independent generators and a handful of very large consumers.

There is therefore still much debate and work to be done on the open European market.

Chapter 6

Switchgear fundamentals

G.R. Jones

6.1 Introduction

The interruption of current in a network cannot meaningfully be considered in isolation from the operating voltage of the system and the nature of the system components and structure. The operating voltage itself affects the type of interrupter chosen for duty whilst the system components and structure (e.g. the extent to which the network is inductive etc.) will influence the detailed design of the interrupter unit because of the voltage transients produced during the current interruption process. It is for these reasons that a discussion of current interruption in electric power systems is appropriate in a book on high voltage technology.

The discussion is limited to SF_6 interrupters since this medium is the only serious contender for use in circuit breakers across the entire range of medium, high and extra high voltages. Other contenders — air, oil and vacuum — have an inferior voltage withstand capability (Fig. 6.1) although this is not an exclusive consideration with regard to current interruption and there are ranges in which vacuum, in particular, can offer advantages.

The fundamental principles of current interruption as governed by high voltage considerations are therefore described with respect to both system based effects and the characteristics of circuit breakers. Various types of SF_6 interrupters are considered and different factors which limit their performance are explained. Some possible trends in future developments are indicated.

6.2 Principles of current interruption in HV systems

All methods of interrupting current in high voltage systems rely upon introducing a non-conducting gap into a metallic conductor. To date, this has been achieved by mechanically separating two metallic contacts so that the gap so formed is either automatically filled by a liquid, a gas or even vacuum. In practice, such inherently insulating media may sustain a variety of different electrical discharges

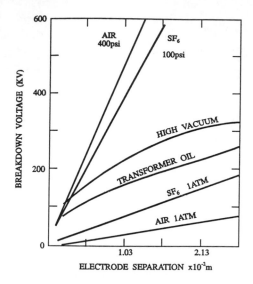

Fig. 6.1   Breakdown voltages of various interrupter media

which then prevent electrical isolation being achieved.

There are three major facets to such electrical discharges which then prevent electrical isolation being achieved. Firstly, as contacts are separated, an arc discharge is inevitably formed across the contact gap. The problem of current interruption then transforms into one of quenching the discharge against the capability of the high system voltage of sustaining a current flow through the discharge. Since this physical situation is governed by a competition between the electric power input due to the high voltage and the thermal losses from the electric arc, this phase of the current interruption process is known as the "thermal recovery phase" and is typically of a few µs duration.

The second facet of current interruption relates to the complete removal of the effects of arcing which only occurs many milliseconds after arc formation even under the most favourable conditions. The problem then is one of ensuring that the contact geometry and materials are capable of withstanding the highest voltage which can be generated by the system without electrical breakdown occurring in the interrupter.

The third facet bridges the gap between the thermal recovery phase and the breakdown withstand phase. The problem in this case is that the remnant effect of the arcing has cleared sufficiently to ensure thermal recovery but insufficiently to avoid a reduction in dielectric strength. This is known as the "dielectric recovery phase".

Based upon this understanding, circuit interruption technology is concerned on the one hand with the control and extinction of the various discharges which may occur, whilst on the other it relates to the connected system and the manner in which it produces post-current interruption voltage waveforms and magnitudes.

6.2.1   System-based effects

The basic premise which derives from the above considerations, as far as system effects are concerned, is that it is advantageous for the current to be reduced to zero in a controlled manner. The simplest and most common example is the symmetrical power frequency (50 Hz) current wave since the current reduces naturally to zero once every half cycle (Fig. 6.2) and at which point current interruption is sought. This represents the minimum natural rate of current decay (*di/dt*) so that for conventional power systems, which are inherently inductive, the induced voltage following current interruption is minimised. Consequently, the contact gap is less severely stressed transiently during both the thermal and dielectric recovery phases.

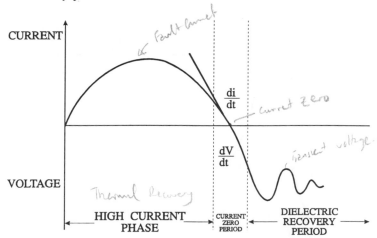

Fig. 6.2   *Voltage and current waveforms during the current interruption process*

The voltage transients of most interest in interrupting current in high voltage transmission systems are those produced by short circuit faults and short line faults. Short circuit faults occur close to the circuit breaker (Fig. 6.3). These produce the most onerous fault currents. In this case, the restrike voltage consists of a high frequency (ω_n) oscillation (governed by the system inductance (*L*) and capacitance (*C*)) superimposed upon an exponentially decaying component governed by the system resistance (*R*).

$$V_c \approx V_o \left[1 - e^{-R/2LT} \cos\omega_n t \right]$$

The maximum fault current is (e.g. Reference 3):

$$I_F \approx \frac{V_o}{\omega L} \sin \omega t$$

and the maximum possible voltage across the circuit breaker is twice the supply voltage. The system resistance R reduces this maximum voltage to lower values.

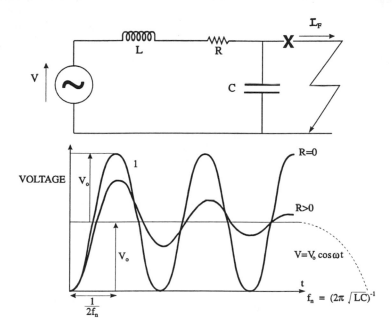

Fig. 6.3   Short line fault — onerous fault current

Short line faults occur on transmission lines a few km from the circuit breaker (Fig. 6.4) and constitute the most onerous transient recovery voltages. The voltage across the circuit breaker is the sum of the line side (V_L) voltage and the source side (V_S) voltage which occur at two different frequencies f_L and f_S, respectively (Fig. 6.4). Typically, at current zero $(dV/dt) \approx 10$ - $20 \text{kV}/\mu\text{s}$.

The situation is made more complicated in three phase systems because current zero occurs at different times in each phase implying that the fault is interrupted at different times leading to different voltage stresses across the interrupter units in each phase.

Apart from the symmetrical sinusoidal current waveform with its natural current zero, other current interruption situations exist (Fig. 6.5). For instance, the sinusoidal waveform may be superimposed upon a steady current to form an asymmetric wave with major and minor loops which cause different circuit breaker stresses. A related condition which occurs in generator faults corresponds to the

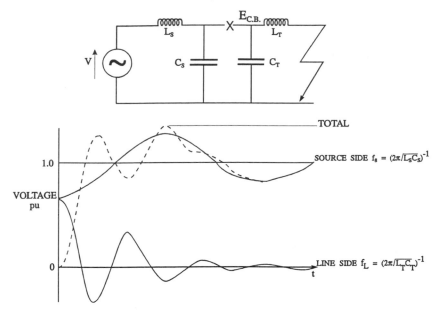

Fig. 6.4   Short line fault — onerous transient recovery voltages

power frequency wave superimposed upon an exponentially decaying component (Fig. 6.5) so that zero current crossing may be delayed for several half cycles. A further situation is the interruption of d.c. faults which is achieved by inducing an oscillatory current via arc instability in the circuit breaker and so forcing the current to pass through zero eventually (Fig. 6.5).

At the lower domestic voltages, current limitation can be conveniently induced leading to an earlier and slower approach to zero current than occurs naturally and with the additional benefit of reducing the energy absorption demands made of the interupter module (Fig. 6.5).

Finally, there are situations whereby high frequency (kHz - MHz) currents can be induced and these need to be tolerated by the circuit breaker. These occur, for instance, when switching on load inductors so that the line-side considerations on Fig. 6.4 are replaced by a lumped inductor load.

6.2.2 Circuit breaker characteristics

The basic characteristics of a circuit breaker relate respectively to the thermal and dielectric recovery phases.

The thermal recovery characteristic is in the form of a critical boundary separating fail and clear conditions on a rate of rise of recovery voltage (dV/dt) and rate of decay of current (di/dt) diagram (Fig. 6.6a).

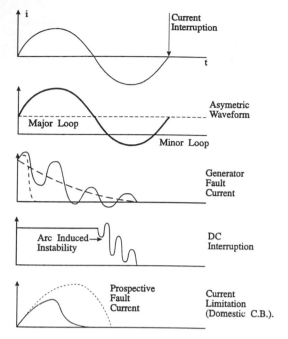

Fig. 6.5   Circuit breaker current waveforms

Typically, the boundary obeys the relationship:

$$\frac{dv}{dt} = \text{const.} \left(\frac{di}{dt}\right)^n$$

with $n = 1 \rightarrow 4.6$.

The thermal recovery performance may be improved by increasing the pressure of the circuit breaker gas, the nature of the gas or the geometry of the interrupter head (Fig. 6.6 b).

For the dielectric recovery regime the characteristic is represented by the critical boundary separating successful clearance and fail on a maximum restrike voltage (V_{MAX}) and rate of decay of current (di/dt) diagram (Fig. 6.7). The dielectric recovery performance may be improved by increasing the number of contact gaps (interrupter units) connected in series. By combining the thermal and dielectric recovery characteristics, the overall limiting curves for circuit breaker performance are obtained (Fig. 6.8).

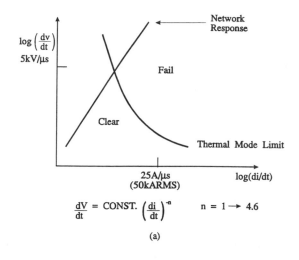

$$\frac{dV}{dt} = CONST. \left(\frac{di}{dt}\right)^{-n} \qquad n = 1 \rightarrow 4.6$$

(a)

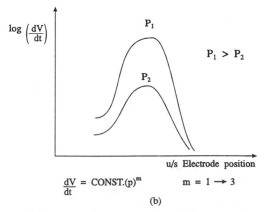

$$\frac{dV}{dt} = CONST.(p)^{m} \qquad m = 1 \rightarrow 3$$

(b)

Fig. 6.6   Thermal recovery characteristics
(a) Network response and interrupter characteristic
(b) Effect of pressure and geometry on performance

6.3 Arc control and extinction

The essence of good circuit breaker design and operation is to ensure proper arc control and efficient arc quenching to provide rapid voltage withstand capability after current interruption.   This, in turn, relies upon the removal of power dissipated in the arc discharge by Joule heating via the processes of convection, conduction and radiation (Fig. 6.9). They may be enhanced through the choice of arcing medium (e.g. sulphur hexafluoride) and arc confinement/movement method.

The properties of SF_6 with regard to high breakdown voltage, enthalpy removal capability and compressive effects have made it an almost universal choice

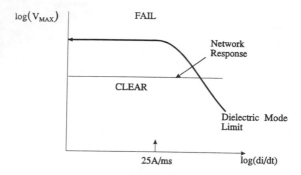

Fig. 6.7   *Dielectric recovery characteristic*

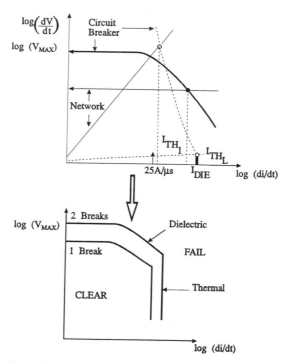

Fig. 6.8   *Overall circuit breaker performance — superposition of thermal and dielectric*

for extra high voltage applications. The influence of limited arc heating to form an enhancement of these qualities via controlled SF_6 dissociation (Fig. 6.10) should not be overlooked. The choice for arc control is between gas blast and electromagnetic-based methods.

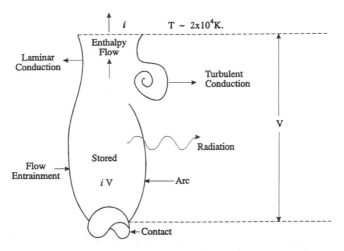

Fig. 6.9   Fundamental processes governing the control and quenching of an electric arc plasma

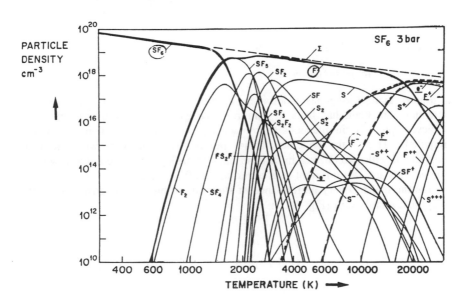

Fig. 6.10   Equilibrium concentration of species at various temperatures for sulphur hexafluoride (Ragaller, 1978)

6.3.1 Gas blast circuit breakers

In gas blast circuit breakers the contacts are separated along the axis of a gas flow guiding nozzle so that the arc is subjected to the convective effects of a co-axial gas flow (Fig. 6.11). The gas flow is produced by gas compression upstream of the nozzle.

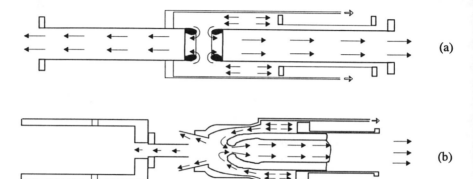

Fig. 6.11   *Gas blast interrupters*
   (a) duo blast
   (b) partial duo blast

Rather than storing gas indefinitely at high pressure until a fault demands attention, the trend has been to compress the gas transiently by piston action ("bicycle pump") simultaneous to contact separation when fault interruption is required. This is the principle of the "puffer circuit breaker" which may be configured with two nozzles in tandem, back to back, to provide a duoflow or, if the nozzles are of a different size, a partial duo flow unit.

The thermal recovery performance of such circuit breakers is governed by the empirical relationship (Fig. 6.12):

$$\left[\frac{dV}{dt}\right]_{CRIT} = ap^{1.5}\left(1+1.7\Delta p\right)A_o^{1.5}\left(1+\left(5 \times 10^{-5}\ \hat{\imath}\ \right)\left(\frac{di}{dt}\right)^2\right)^{-1} \quad (6.1)$$

Typical pressure time curves are shown on Fig 6.13.

The required gas compression may also be generated by relying upon arc induced gas heating to produce the pressurisation within a confined volume according to the gas law:

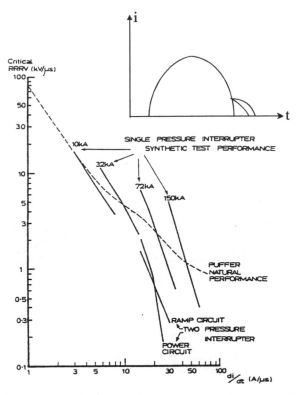

Fig. 6.12 *Effect of various parameters upon the thermal recovery characteristics of gas blast interrupters*

$$p = \left(\frac{R}{V}\right) T \tag{6.2}$$

(V = volume, T = gas temperature, R = gas constant)

The pressurisation produced at the end of a half cycle of fault current depends upon the parameter [1]:

$$Y = \frac{J_o \left(1 - \alpha\right) I \rho}{a \, \sigma \, \omega \, U_o \left(t\right)} \tag{6.3}$$

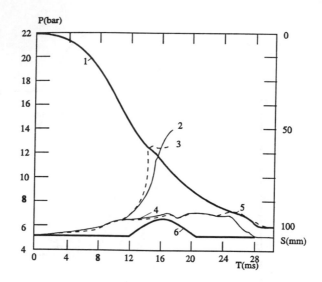

Fig. 6.13   *The pressure inside the puffer chamber:-*
1: The piston stroke curve;
2, 3: The calculated and measured chamber pressure for a peak of 18.6 kA;
4, 5: The calculated and measured chamber pressure in the absence of the arc;
6: Current waveform.

where

α	=	fraction of input power transferred by radiation
σ	=	electrical conductivity of arc plasma
ρ	=	effective length of the arc
$U_o(t)$	=	time varying volume of the arc chamber
$\hat{I}$	=	peak current
ω	=	angular frequency of current waveform

as shown in Fig. 6.14.

The thermal recovery performance is governed by equation 6.1 with the piston pressure term set to unity. The manner in which piston and arc induced pressures effect the thermal recovery performance of the circuit breaker is shown in Fig. 6.12.

6.3.2 Electromagnetic circuit breakers

In electromagnetic circuit breakers the arc is spun through the action of the Lorentz force produced by the fault current flowing through a *B* field producing coil. The arc may be spun azimuthally or helically (Fig. 6.15). By contrast with

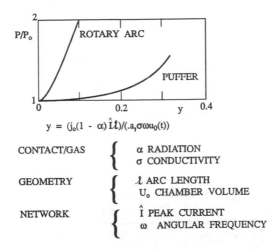

$$y = (j_0(1 - \alpha)\hat{I}\mathcal{l})/(.a_1\sigma\omega u_0(t))$$

CONTACT/GAS $\left\{ \begin{array}{l} \alpha \text{ RADIATION} \\ \sigma \text{ CONDUCTIVITY} \end{array} \right.$

GEOMETRY $\left\{ \begin{array}{l} \mathcal{l} \text{ ARC LENGTH} \\ U_0 \text{ CHAMBER VOLUME} \end{array} \right.$

NETWORK $\left\{ \begin{array}{l} \hat{I} \text{ PEAK CURRENT} \\ \omega \text{ ANGULAR FREQUENCY} \end{array} \right.$

Fig. 6.14 Arc-induced pressure elevation as a function of volume normalised current

the gas blast circuit breakers, arc controlling and quenching convection is generated by driving the arc through the surrounding flow rather than vice versa.

The thermal recovery performance is typically governed by [2] (Fig. 6.16)

$$\left[\frac{dV}{dt}\right]_{CRIT} = ap^{1.5}\left(\frac{di}{dt}\right)^n B^{1.9} f(\theta) \tag{6.4}$$

B = magnetic flux density.

The phase angle between i and B for a helical arc interrupter is governed by [2]

$$\tan\theta = \frac{\omega L_T \, \ell_T \, \Delta}{\zeta_T \, \pi \, d} \tag{6.5}$$

ω is the angular frequency, L_T, ζ_T are the self-inductance and resistivity of the material the annular contact, ℓ_T, Δ, d are the length, thickness and diameter of the cylindrical contact.

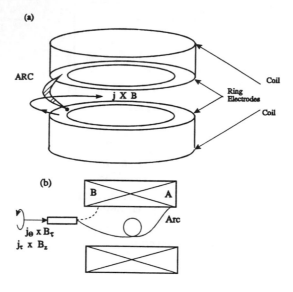

Fig. 6.15   *Electromagnetic interrupters*
(a) rotating arc, (b) helical arc

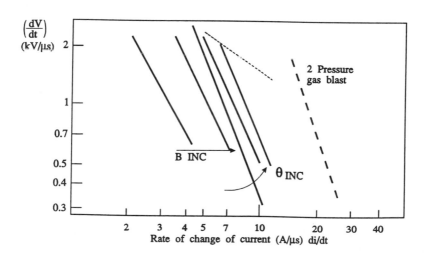

Fig. 6.16   *Effect of various parameters upon the thermal recovery*
characteristics of rotary arc interrupters
———— *Constant B* ----- *Particular unit*

The typical performance curves given in Fig. 6.16 are for constant B whereas in practice B increases with fault current level. The characteristic for a given interrupter therefore forms a locus across the constant B curves (much as the characteristic for the puffer breaker forms a locus across the constant pressure curves of Fig. 6.12). Note the similar gradients of the electromagnetic and gas blast characteristics.

6.3.3 Dielectric recovery

The dielectric recovery in SF_6 gas is a complex process governed by the dissociation chemistry of SF_6 which is only partially understood. Theoretical estimates of the time variation of concentration of dissociation by-products following arcing (and based upon unconfirmed equilibrium assumptions) show the complexity of the medium when subjected to the voltage stresses (Fig. 6.17a). The temperature of the dissociated SF_6 in the post arc column is deduced to recover via a number of steps which are governed by the thermal capacity and conduction properties of the dissociated by-products.

Estimates of the recovery of dielectric strength then yield characteristics of a step-like nature as shown in Fig. 6.17b. The evidence is that the precise nature of this curve is affected by the shape of the restrike voltage waveform because of the influence of the associated electric field upon the recombination chemistry of the ionic species.

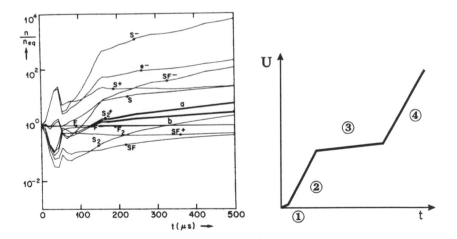

Fig. 6.17   *Dielectric recovery characteristics*
     (a) SF_6 dissociation products during dielectric recovery
     (b) Voltage withstand as a function of time

6.4 Other performance inhibiting factors

What has been described above are the fundamental factors which govern circuit breaker performance under almost ideal situations. For instance, the circuit breaker thermal recovery characteristics of Figs. 6.12 and 6.15 are, in principle, predictable from the physics shown in Fig. 6.9 using the laws of conservation of mass, momentum and energy — provided the boundary conditions and material properties are known.

However, in practice, complicating effects arise from a lack of knowledge of the material properties (e.g. enthalpy, radiation transport etc.) of the arcing medium which follows from the complex dissociation chemistry (Figs. 6.10 and 6.17a) and the entrainment of foreign species from, for example, the interrupter contacts and nozzle. The former leads to the formation of electrode plasma jets carrying contact material into the arc column which may be either in vapour form (copper) or particulate (tungsten) from the sintered copper tungsten electrodes utilised in such interrupters. The presence of particulate tungsten which has been detected spectroscopically can reduce the thermal recovery performance (Fig. 6.18) but its effect is minimised in duo flow circuit breakers with the flushing action of contraflows.

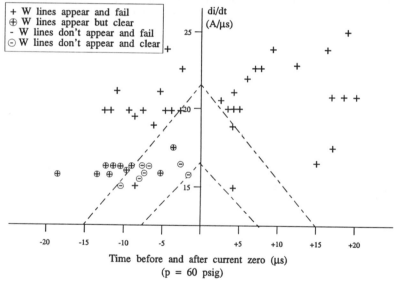

Fig. 6.18 Correlation between thermal recovery failure and the presence of tungsten in the arc plasma at current zero for various values current decay rate (di/dt)

Particulate material from contacts and from metallic ions reacting with dissociated SF_6 has a substantial effect in degrading the dielectric recovery of the interrupter (Fig. 6.19). There is evidence that it may be these particle related effects rather than the overheating of the SF_6 which may be the dominant factor in governing dielectric withstand (Fig. 6.19) at high peak fault currents.

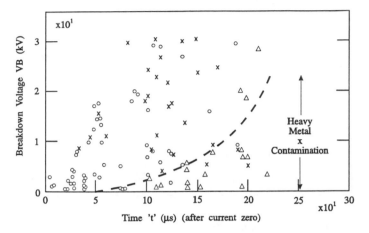

Fig. 6.19 Effect of particulate material on dielectric recovery

There is also concern about the occurrence of high frequency effects in interrupters and isolators and at least two such situations are now known to exist. The first relates to the formation of fast transients in Gas Insulated Substation systems. These are high frequency (kHz) electromagnetic waves which propagate within the SF_6 containing bus bar modules of such systems. They are only slowly attenuated because of the low loss nature (low R in Fig. 6.3) of the co-axial GIS which act essentially as waveguides. One of the sources of such transients is pre-arcing during the closing operation of an isolator or interrupter. Optical investigations of such switch operation have shown that events on the time scale of GHz may occur and that reflected waves within the GIS system interact to feed energy back into the arc which has insufficient time to be self-quenching (Fig. 6.20).

A second situation where high frequency arcing occurs is during on-load inductor or capacitor switching. In this case, arcing at frequencies in excess of kHz may occur during the closing or opening operations but at peak currents of only a few hundred amperes.

The effects are probably associated with network resonances close to the interrupter. In terms of the thermal recovery characteristics of Fig. 6.12 the implication is that although the peak current is low, the di/dt at current zero is high (due to the high frequency of the waveform) and typically lies off the (di/dt) axis of Fig. 6.12 at the high di/dt end.

Consequently, the voltage withstand and capability to interrupt is exceedingly low according to the characteristics of Fig. 6.12 so that the interrupter will tend to fail to interrupt such high frequency currents. Furthermore, the high di/dt at

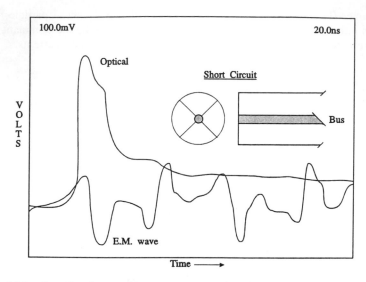

Fig. 6.20　Coupling between high frequency electromagnetic transients and an arc in an isolating switch

current zero coupled with the inductive nature of the load is inclined to produce high restrike voltages which may exceed the rating of the circuit breaker.

Evidence is now emerging that such high frequency phenomena within the interrupter unit may lead to unusual breakdowns which occur away from the main contact gap. This constitutes a serious situation since the potential for extinguishing the discharge by the normal operation of the interrupters no longer exists, leading to the possibility of the interrupter unit itself being damaged or destroyed.

Two basically different phenomena have been observed:

(1) The inductive coupling of high frequency currents onto peripheral parts of the circuit breaker assembly may produce such parasitic breakdowns (Fig. 6.21). The premature breakdown of one of two interrupter units connected in series leading to inductive parasitic breakdowns on the second gap is one such possibility;

(2) For particular points on travel of contacts with certain interrupter geometries, breakdown outside the interrupter nozzle has been shown to occur (Fig. 6.22).

Neither of these phenomena are, at present, well understood and both are the subject of continuing investigations.

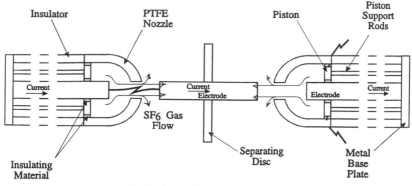

Idealised two break SF$_6$ puffer interrupter

Fig. 6.21 Parasitic breakdowns due to HF inductive coupling

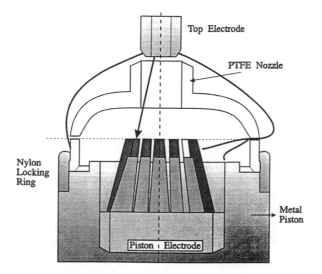

Fig. 6.22 Parasitic breakdowns due to E field distortions

6.5 Future trends

The advent of SF$_6$ circuit breakers has led to significant commercial benefits:

- At EHV, the interruption capability per interrupter unit has increased significantly (Fig. 6.23) and incorporation into Gas Insulated Systems is facilitated. The net effect is to reduce substation size with accompanying reduction in land costs;

- At distribution level, rotary arc circuit breakers have better immunity to current chopping than other types of circuit breakers.

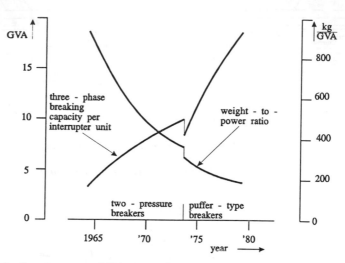

Fig. 6.23 Development of SF$_6$ circuit-breakers

However, these advantages have been gained only at the expense of other aspects. For instance, the puffer circuit breaker makes expensive demands upon operating energy (Fig. 6.24) because of the rapid piston action which is needed and the trend is to evolve designs which are less energy demanding.

The move towards various degrees of self-pressurisation by arc-induced gas heating is an example of such a trend. However, the least energy demanding operation is with the electromagnetic type of circuit breakers so that considerations of extending such principles to higher voltage levels is receiving some attention.

Problems with the production of high frequency transients are also gradually becoming apparent from extended service use. These appear to relate to both the low loss nature of co-axial bus bar systems of Gas Insulated Systems and also the rapidity of arc formation and quenching in SF$_6$. There is currently active investigation of these effects.

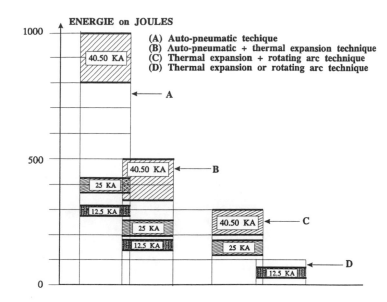

Fig. 6.24 *Evolution of opening energy as a function of the interrupting technique used and the interrupting capacity expressed in kA. The increasingly simple design and the utilisation of the arc energy result in a reduction in switching energy*

6.6 References

1   RYAN, H.M. and JONES, G. R.: "SF$_6$ switchgear" (Peter Peregrinus Ltd., 1989)

2   RAGALLER, K. (Ed.): "Current interruption in HV networks" (Plenum Press, 1978)

3   JONES, G. R.: "High pressure arcs in industrial devices" (Cambridge University Press, 1988)

4   FLURSCHEIM, C. H.: "Power circuit breaker theory and design" (Peter Peregrinus Ltd., 1982)

5   JONES, G. R., LAUGHTON, M. A. and SAY, M. G. (Eds.): "Electrical engineers reference book" 15th edition (Butterworth Heinemann, in press)

6   Greenwood, A.: "Vacuum switchgear" (Institution of Electrical Engineers, 1994)

Chapter 7

Switchgear design, development and service

S.M. Ghufran Ali

7.1 Introduction

This chapter describes the design, development and operation of switchgear. It also describes how the development of generation and transmission has influenced switchgear evolution (Appendix 7.1). Factors which have contributed to the simplicity of designs and increased reliability of SF_6 switchgear are addressed and the important features of various manufacturers designs in first, second and third generation interrupters and improvements in circuit breaker performance are highlighted. It also addresses issues associated with installation and on-site operations and monitoring.

7.1.1 SF_6 circuit breakers

A circuit breaker is a device which breaks or interrupts the flow of current in a circuit. It is used for controlling and protecting the distribution and transmission of electrical power. It is connected in series with the circuit it is expected to protect. It has to be capable of successfully:

- Interrupting (i) any level of current passing through its contacts from a few amperes to its full short-circuit currents both symmetrical and asymmetrical, at voltages specified in IEC56 and (ii) up to 25 % of full short circuit currents at twice the phase voltage;
- Closing up to full short-circuit making current (ie 2.5 x I_{sym}) at phase voltage and 25 % of full making currents at twice the phase voltage;
- Switching (making or breaking) inductive, capacitive (both line, cable or capacitor bank) and reactor currents without producing excessive overvoltages to avoid overstressing the dielectric withstand capabilities of a system;
- Performing opening and closing operations whenever required;
- Carrying the normal current assigned to it without overheating any joints or contacts.

This interrupting device becomes more complex as the short-circuit currents and voltages are increased and, at the same time, the fault clearance times are reduced to maintain maximum stability of the system.

A circuit-breaker has four main components: (1) interrupting medium (sulphur hexafluoride gas), (2) interrupter, (3) insulators and (4) mechanism.

7.1.2 Sulphur hexafluoride

The pure SF_6 is odourless and non-toxic but will not support life. Being extremely heavy (4.7 times denser than air) it tends to accumulate in low areas and may cause drowning. It is a gas with unique features which are particularly suited to switchgear applications. Its high dielectric withstand characteristic is due to its high electron attachment coefficient. The alternating-voltage withstand performance of SF_6 gas at 0.9 bar(g) is comparable with that of insulating oil. SF_6 has the added advantage that its arc voltage characteristic is low, hence the arc-energy removal requirements are low.

At temperatures above $1000\,°C$, SF_6 gas starts to fragment and at arc-core temperature of about $20000\,°C$, the process of dissociation accelerates producing a number of constituent gases including S_2F_{10} which is highly toxic (see Figs. 6.10 and 7.1). However, these recombine very quickly as the temperature starts to fall and the dielectric strength of the gap recovers to its original level in micro-seconds. This allows several interruptions in quick succession.

The solid arc products consist mainly of metal-fluorides and sulphides with elemental sulphur, carbon and metal oxides. These are acidic and must not be inhaled. Metallic fluorides formed during the arcing do not harm the switchgear components provided the moisture in the interrupting chamber is absorbed by the molecular sieve. The dielectric integrity of the equipment is not impaired by the presence of these fluorides and sulphides.

7.2 Interrupter development

The key component in any circuit breaker is the interrupter. Early EHV interrupters were two-pressure type and these were superseded by the single-pressure puffer type in the early 1970s.

7.2.1 Two-pressure system

In the late 1960s and early 1970s, SF_6 EHV interrupters were based on the well-established two-pressure air-blast technology modified to give a closed loop for the exhaust gases. SF_6 gas at high pressure (approx 15 bars) was released by the blast valve through a nozzle to a low pressure reservoir instead of being exhausted to atmosphere. The gas was recycled through filters then compressed and stored in the high pressure reservoir for subsequent operations. Heaters were

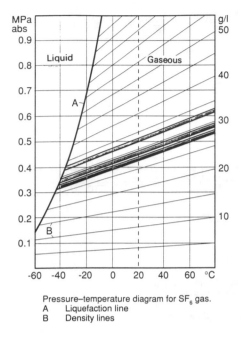

Pressure–temperature diagram for SF₆ gas.
A   Liquefaction line
B   Density lines

Fig. 7.1   *Pressure-temperature diagram for SF₆ gas - A is the liquefaction*
             line and B are density lines

used to avoid liquefaction of the high pressure gas at low temperatures.

The relative cost and complexity of design led to the development of inherently simpler and more reliable single-pressure puffer type interrupters.

7.2.2 Single-pressure puffer type interrupters

7.2.2.1 First generation interrupters

The principle of a single pressure puffer type interrupter is explained by the operation of universally known device — a "cycle pump" — where air is compressed by the relative movement of piston against a cylinder. In a puffer type interrupter, SF_6 gas in the chamber is compressed by the movement of the cylinder against the stationary piston. This high pressure gas is then directed across the arc in the downstream region through the converging/diverging nozzle to complete the arc-extinguishing process. The basic arrangement of the puffer type interrupters can be classified according to the flow of compressed SF_6 gas. These are generally known as mono, partial-duo and duo blast interrupters (Figs. 7.2 and 7.3).

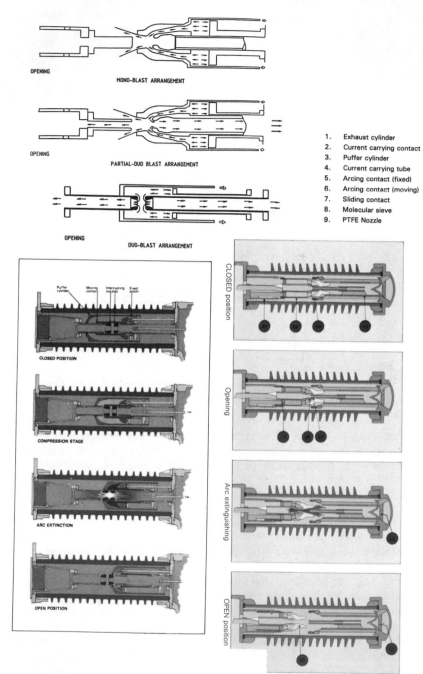

1. Exhaust cylinder
2. Current carrying contact
3. Puffer cylinder
4. Current carrying tube
5. Arcing contact (fixed)
6. Arcing contact (moving)
7. Sliding contact
8. Molecular sieve
9. PTFE Nozzle

Fig. 7.2   Principles of SF$_6$ puffer type interrupters

Circuit breaker interrupter

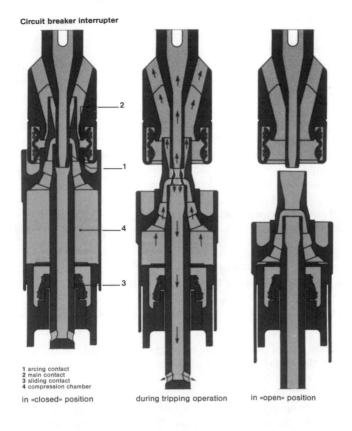

1 arcing contact
2 main contact
3 sliding contact
4 compression chamber

in «closed» position   during tripping operation   in «open» position

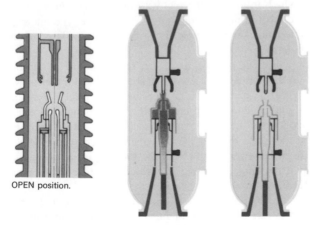

OPEN position.

Fig. 7.3   First and second generation interrupter circuit breakers

All types of interrupters are capable of high short-circuit current rating, but superior performance is generally achieved by either partial-duo or duo blast construction because most of the hot gases from the arc are directed away, giving the improved voltage recovery of the contact gap. Most of the puffer type interrupters have been developed for 50 to 63 kA ratings, and some even for 80 and 100 kA rating.

For a puffer type interrupter, retarding forces act on the piston surface as the contacts part. These forces are due to the total pressure-rise generated by compression and heating of SF_6 gas inside the interrupting chamber and are highest with maximum interrupting current and arc duration. Therefore, to provide consistent opening characteristics for all short circuit currents up to 100% rating, high energy mechanisms are required.

7.2.2.2 Second generation interrupters

Worldwide development in the second generation interrupter concentrated on:

—　　Rationalisation of designs;
—　　Improving the short circuit rating of interrupters;
—　　Better understanding of the interruption techniques;
—　　Improving the life of arcing contacts;
—　　Reducing the ablation rate of nozzles by using different nozzle filling materials.

Most of the present day SF_6 circuit breaker designs are virtually maintenance-free. This means that the arcing contacts and nozzles on the interrupters have been designed for long service life.

Most arcing contacts are fitted with copper-tungsten alloy tips. The erosion rate of these tips depends upon the grain size of tungsten, the copper to tungsten ratio, cintering process and production techniques. The choice of the copper-tungsten alloy is therefore essential for both the erosion rate of the tips and the emission of copper vapour which influences the recovery rate of the contact gap.

The nozzle is the most important component of a puffer type interrupter. The interruption characteristic of an interrupter is governed by nozzle-geometry, shape, size and nozzle material.

In the western world, at present there are only nine EHV circuit breaker manufacturers — ABB, AEG, GEC-Alsthom, Hitachi, Merlin Gerin, Mitsubishi, Reyrolle, Siemens and Toshiba. The nozzles on these circuit breakers can be classified into two categories — long and short. There is no evidence available to show that at 525 kV the dielectric performance of the long-nozzle is superior to that of short-nozzle, since most of the designs have achieved 50/63 kA ratings at 420/525 kV. It is, however, very clear that the rate of nozzle ablation very much depends upon the choice of material, which could be either pure (virgin) PTFE or filled PTFE.

Pure PTFE is white in colour and is most commonly used because of its reasonable price. The rate of ablation of pure PTFE is relatively high and very much depends upon grain size, moulding or compacting pressure, cintering procedure and quality of machining and surface finish. It has also been observed that the radiated arc energy penetrates deep in the body, producing carbon molecules. To overcome this, some manufacturers use coloured PTFE which absorbs the radiated arc energy on the surface and prevents the deep penetration.

To ensure consistent performance with reduced rate of ablation and long life, most manufacturers use filled PTFE for high short-circuit currents interruption (in the region of 63 kA and above).

There are three types of filling: Boron Nitride (cream colour), Molybdenum (blue colour) and Aluminium Oxide (white).

Since the ablation rate of the filled-nozzles is low, the change in nozzle throat diameter after about 20 full short-circuit interruptions is normally very small. The pressure rise characteristic of the interruption hardly changes and therefore the performance of the interrupter remains consistent, giving a long, satisfactory service life.

Filled PTFE material is slightly more expensive than pure PTFE, but its consistent performance and extra-long life justifies its use on high current interrupters. Some examples of PTFE nozzles are shown in Fig. 7.4.

7.3 Arc interruption

7.3.1 Fault current

Present day transmission circuit breaker designs are based upon single pressure puffer type SF_6 interrupters. The gas flow is produced by the self-generated pressure difference across the nozzle, either by the movement of the cylinder against the stationary piston, or by the simultaneous movement of both the piston and cylinder. The magnitude and the rate of rise of the no-load pressure rise depends upon the interrupter design parameters. These are: diameter of the cylinder, nozzle geometry, throat area, swept volume and opening speed.

During fault current interruption, an arc is drawn between the moving and fixed arcing contacts or between the two fixed arcing contacts. The throat area of the nozzle is partially or completely choked by the diameter of the arc, while the arc energy heats up the SF_6 gas in the interrupting chamber, thus causing a substantial pressure rise (Fig. 7.5).

The total pressure rise inside the interrupting chamber consists of the no-load pressure-rise and the arcing pressure rises which produce a considerable pressure difference between the upstream and downstream regions of the nozzle. This pressure difference causes a sonic flow of relatively cold SF_6 gas across the arc. This fast movement of SF_6 gas makes the arc unstable and removes heat energy and in the process cools the arc.

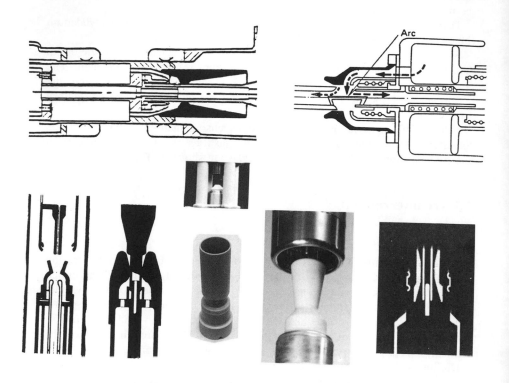

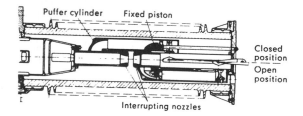

Fig. 7.4   Examples of PTFE nozzles

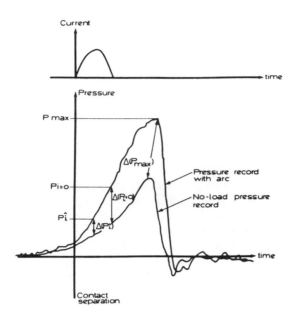

Fig. 7.5   *Pressure-rise characteristic in a puffer type interrupter*

If the rate of recovery of the contact gap at the instant of current zero is faster than the rate of rise of the recovery voltage (RRRV), the interruption is successful in the thermal region (i.e. first 4 to 8 μs of the recovery phase), followed by successful recovery voltage withstand in the dielectric region (above 50 μs) and then full dielectric withstand of the a.c. recovery voltage. This whole process is known as successful fault current interruption.

If, however, the rate of rise of the recovery voltage (RRRV) is faster than the recovery of the gap, then failure occurs either in the thermal region, or in the dielectric region after clearing the thermal region.

Over the past 30 years, researchers worldwide have carried out very useful work which has brought better understanding of the physical processes associated with the arc interruption at and near current zero. Computer models and programmes have been developed which can accurately predict the performance of an interrupter at and near current zero. However, in a real interrupter the recovery process in the dielectric region (i.e. above 50 μs) is very complex, since it is influenced by many factors. These include contact and nozzle shapes, arc energy, gas flow, rate and nature of ablation of nozzle material, rate and amount of metal vapour present in the contact gap, dielectric stress of the gap, dielectric stress on the contact tips and the proximity effect of the interrupter housing.

The dielectric aspect of interruption is not yet fully resolved and to the author's knowledge accurate prediction is not possible at present.

7.3.2 Capacitive and inductive current switching

When a puffer type interrupter switches small inductive (transformer and reactor) or capacitive (line, cable, capacitor bank) currents, it relies entirely upon its no-load pressure rise characteristic, since there is practically no contribution of pressure from these small load current arcs.

The magnitude of the no-load pressure rise and the gas flow across the nozzles determines whether the current is interrupted at, before (on falling-current) or after (on rising-current) current zero.

If the current is interrupted at current zero, the interruption is normal and the transient recovery voltages are within the specified values. However, when premature interruption occurs due to current chopping the interruption is abnormal — it causes high frequency reignitions and overvoltages. If the interrupter chops the peak current, the voltage doubles instantaneously. If this process is repeated several times due to high frequency reignitions, the voltage doubling continues with rapid escalation of voltages (Fig. 7.6). When these overvoltages exceed the specified dielectric strength for the switchgear, the interrupter and/or other parts of the switchgear may be damaged.

The phenomena of chopping and reignition is attributed to the design of an interrupter. Most of the EHV interrupters are designed to cope with high fault currents, up to 63 kA in the UK and up to 80 kA in some other countries. If a design is concentrated only on performance of high currents with high, no-load pressure rise in the interrupting chamber, it will be too efficient for small current and will try to interrupt before its natural current zero. This efficiency sometimes works against it and produces the phenomenon of current chopping and reignitions with adverse consequences.

The interrupter design should therefore incorporate features which cope equally well with small as well as high currents (i.e. softer interruption). It is sometimes desirable to have a softer interrupter to give satisfactory performance for all conditions.

7.3.3 Reactor switching

In a high voltage system, reactors are used for system VAr compensation. These are connected either directly onto the high voltage system with EHV circuit breakers, or to the low voltage delta, tertiary windings of the auto-transformer (at 12 or 33 kV) by MV circuit breakers. The shunt reactors are frequently switched at least two to three times per day. The SF_6 circuit breakers for this duty are

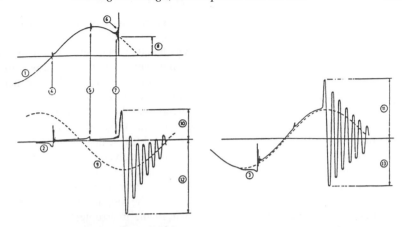

Chopping phenomena in a single phase circuit

1 current to interrupt
2 voltage across circuit-breaker
3 voltage across inductive load
4 failed interruption due to reignition (short contact distance)
5 influence of ard voltage
6 current instability oscillation leading to current chopping
7 arc voltage oscillation
8 effective chopping level
9 main power frequency voltage
10 suppression peak, first voltage maximum across circuit-breaker
11 first voltage maximum across inductive load
12 recovery voltage peak, second voltage maximum across circuit-breaker
13 second voltage maximum across inductive load

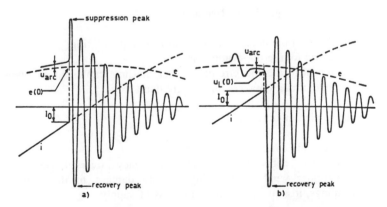

Fig. 7.6   *Overvoltages from inductive current chopping and reignitions.*
Lower diagrams show current chopping (a) before and (b) after
natural current zero

expected to carry out about 5000 satisfactory switching operations on one set of contacts and nozzles.

Service experience worldwide has shown that this switching duty can cause difficulty for some circuit breaker designs which were earlier tested satisfactorily, on available standard circuits at the major testing stations. It is generally acknowledged that those circuits did not truly represent the site conditions.

The mechanism of reactor current interruption, the phenomena of current chopping and multiple reignitions and the generation of high frequency oscillatory overvoltages are well understood. It has been established that the high frequency oscillations are governed by the electrical circuit in a given system, configuration and the interrupter design (i.e. load side capacitance, load side inductance, inductance of the busbars, value of parallel capacitance across the interrupter (including the grading capacitance), inductance of the loop formed by the grading capacitors, no-load pressure rise characteristic of the interrupter).

Since one or all the above parameters could be variable, the issue becomes extremely complex. IEC Subcommittee, CIGRE Working Groups and ANSI have examined various aspects to produce universally acceptable switching test guidelines and a test circuit. The most recent IEC Document 17A(Sect.)358 — May 1992 — gives a circuit for testing and application guide for assessing the general performance of a circuit breaker.

Until IEC provides a final solution, it is recommended that site measurements and system studies should be carried out to ensure that as realistic a circuit as possible is used for testing a circuit breaker on that site. In addition, where possible "R" and "C" damping circuits and metal-oxide surge arresters should be used to ensure safe operation. Failing that, there could be very costly and serious consequences for the users. Some examples are described below.

The overvoltages produced by chopping and high frequency reignition can result in the following:

- **For extra high voltage circuit breakers**:

  — Tracking on the nozzle surface and nozzle puncturing;
  — Dielectric failure on switchgear;
  — Flashover inside the circuit breaker causing explosion in open terminal circuit breakers and severe damage.

- **High voltage circuit breakers**:

  — Catastrophic damage to circuit breakers;
  — Resonance between the high frequency overvoltages and transformer windings possibly causing damage to the tertiary windings, loss of transformer and loss of supply.

7.3.4 Arc interruption: gas mixtures

Pure SF_6 gas is very efficiently used in the present day commercial circuit breakers to interrupt high currents up to 63 and 80 kA. It is, however, an expensive gas and at common operating pressures of 6 bar(g), it starts to condense at -20 °C and in its pure form it is not suitable for operation at lower ambient temperatures. When the ambient temperature becomes too low (i.e. -50 to -60 °C) the gas starts to liquefy and the interrupter contains a mixture of SF_6 gas, liquid droplets and fine mist. This mixture affects the gas flow conditions in the nozzle and transfer of thermal energy from the arc, thus impeding the performance of the breaker.

The need for more efficient and economical transmission switchgear which can operate satisfactorily at temperatures down to -50 °C in Canada, Scandanavian countries and Russia has stimulated the search for new gases and gas-mixtures, as a potential replacement for pure SF_6 gas in puffer type circuit breakers. Extensive research in USA, Japan and Europe has confirmed that at present no single gas tested is found to be superior to pure SF_6 gas in all aspects of dielectric withstand capability and arc interruption.

Several gas-mixtures have been studied with the objective of exploiting the properties of the component gases so that they could be used effectively in switchgear. A survey of some of the published work on various gas mixtures has shown that the performance of several gas mixtures appears very promising.

One such mixture is SF_6/Nitrogen (N_2). Its dielectric strength at about 15 % increased pressure equals that of pure SF_6 gas at normal pressure. The mixture is less sensitive to strong localised electric fields and can be operated at higher pressures up to 6 bar or at lower temperatures down to -50 °C, with potential cost savings up to 35 %. While the dielectric strength of the clean SF_6/N_2 gas mixture is extremely encouraging, its interrupting performance does not compare favourably with pure SF_6 gas. The short circuit rating of a SF_6/N_2 circuit breaker at -50 °C is normally degraded by one level of the IEC standard rating (i.e. a 50 kA circuit breaker is generally used for 40 kA rating).

Work is in progress worldwide to find a suitable mixture which can be used efficiently down to -50 °C ambient temperature without penalising its short circuit or dielectric performance. I understand that some manufacturers have achieved this and hopefully soon the breakers with such gas mixtures will be commercially available.

7.4 Third generation interrupters

Puffer-type interrupters require drive mechanisms to provide energy for moving the cylinder of the interrupter at relatively high speeds of 6 to 9 metres per second. The fast movement of the cylinder compresses the SF_6 gas. The high pressure rise upstream of the nozzle due to compression and arc heating of the gas is required for quenching the longest possible arcs associated with both the single-phase to earth fault and the last phase to clear condition of the three-phase fault. This results in very complex and powerful drives which exert high reaction forces on dashpots, seals, joints, structures and foundation, and affects the reliability and cost of a circuit breaker. The experience over the last fifteen years has shown that the majority of failures on site are mechanical. Therefore switchgear manufacturers have concentrated their efforts on producing simple interrupting devices and reliable and economical mechanisms. To achieve this, they have addressed the fundamental issue of reducing the retarding forces on the drive mechanisms during an opening stroke. This work has led to the development of third generation interrupters which have the following improved design features and economies compared with first and second generation interrupters:

- 10 to 20% reduction in energy has been achieved by optimising the puffer type interrupter designs, which ensures that the maximum arc duration for the highest current does not exceed 21 milliseconds;
- 50 to 60% reduction in drive energy has been achieved by skillfully utilising the arc energy to heat the SF_6 gas, thus generating sufficient high pressure to quench the arc and assist the mechanism during the opening stroke.

At least two switchgear manufacturers have successfully used this principle to produce low-energy circuit breakers (Fig. 7.7). The design criteria depend upon optimising the volumes of the two chambers of the interrupter:

- The expansion chamber, to provide the necessary quenching pressure by heating the gas with arc energy;
- The puffer chamber, to provide sufficient gas pressure for clearing the small inductive, capacitive and normal load currents.

The optimum sizes of these chambers are determined by detailed computer studies of the arc input and output energies, temperature profiles, gas flow, the quenching and total pressures.

The main advantages of this design concept are as follows:

- Softer interruption producing low overvoltages for switching small inductive and capacitive currents;
- Low energy mechanisms, lighter moving parts, simpler damping devices and reduced loads on foundation and other switchgear components;
- Long service life, at least 10,000 trouble free operations;
- Increased reliability and low cost circuit breakers.

7.5 Dielectric design and insulators

On EHV circuit breakers for voltages up to 550 kV, the number of interrupters per phase has reduced from six to one. Therefore the dielectric performance has become extremely important. To optimise the designs, it is essential to carry out

Arcing chamber with optimised quenching principle

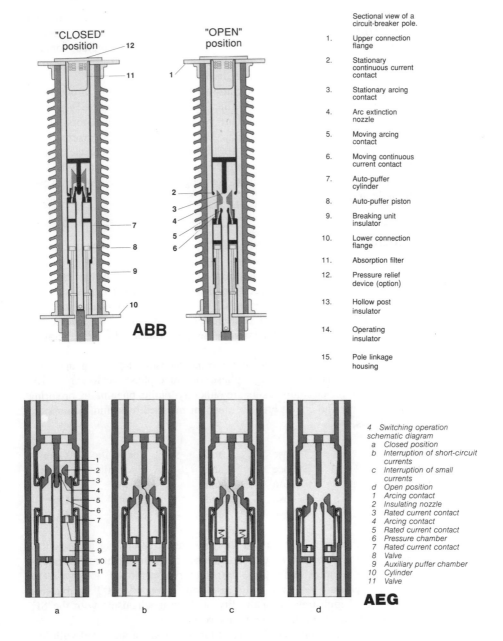

"CLOSED" position — 12

"OPEN" position

ABB

Sectional view of a circuit-breaker pole.

1. Upper connection flange
2. Stationary continuous current contact
3. Stationary arcing contact
4. Arc extinction nozzle
5. Moving arcing contact
6. Moving continuous current contact
7. Auto-puffer cylinder
8. Auto-puffer piston
9. Breaking unit insulator
10. Lower connection flange
11. Absorption filter
12. Pressure relief device (option)
13. Hollow post insulator
14. Operating insulator
15. Pole linkage housing

4 Switching operation schematic diagram
  a Closed position
  b Interruption of short-circuit currents
  c Interruption of small currents
  d Open position
  1 Arcing contact
  2 Insulating nozzle
  3 Rated current contact
  4 Arcing contact
  5 Rated current contact
  6 Pressure chamber
  7 Rated current contact
  8 Valve
  9 Auxiliary puffer chamber
  10 Cylinder
  11 Valve

AEG

a   b   c   d

Fig. 7.7   *Third generation interrupter circuit breakers*

detailed stress analysis studies for all critical components, using sophisticated computer programmes and CAD systems.

Most of the major manufacturers have developed their own computer programmes and their designs are based on stress levels from past experience. These CAD techniques are used to optimise the shapes of stress shields, contacts and insulators. They also optimise stress levels on stress shields, gaps, grading capacitors (if any), support insulators and drive rods.

In addition to the proper stressing of components, it is extremely important to choose the correct insulating material, since the presence of the degradation products inside the circuit breaker can damage the silica base insulators. Long term resistance of insulating material to SF_6 degradation products is essential and most manufacturers have chosen alumina-filled cast resin insulators or the designs which prevent direct contact of degradation products with the surface of silica base insulators. Field computation relating to switchgear design has been covered elsewhere in detail [3].

7.6 Mechanism

The operating mechanism is a very important component of a circuit breaker. When it operates, it changes the circuit breaker from a perfect conductor to a perfect insulator within a few milliseconds. A failure of the mechanism could have very serious consequences. It is therefore essential that the mechanism should be extremely reliable and consistent in performance for all operating conditions.

The circuit breaker may have one or three mechanisms depending upon the operational requirements either single-phase or three-phase reclosing. The mechanisms fitted to the circuit breakers are either hydraulic, pneumatic, or spring or their combination. The circuit breaker mechanisms used by manufacturers are grouped as follows:-

Pneumatic	Close	and	Pneumatic	Open
Hydraulic	Close	and	Hydraulic	Open
Spring (motor charged)	Close	and	Spring	Open
Hydraulic	Close	and	Spring	Open
Pneumatic	Close	and	Spring	Open

The number of operating sequences and the consistency of closing and opening characteristics generally determines the performance of the mechanism. Although IEC-56 type tests require only 2000 satisfactory operations to prove its performance, the present tendency is to carry out extended 5000 trouble-free operations tests to demonstrate compatibility of these mechanisms with the SF_6 circuit-breakers which are virtually maintenance-free.

The task of the third generation SF_6 circuit breakers, which are fitted with low energy mechanism and light weight moving parts, becomes much easier. They satisfactorily perform 10,000 trouble-free operations without any stresses and excessive wear and tear on the moving and fixed parts of the breaker.

7.7 SF$_6$ live- and dead-tank circuit breakers

In live-tank circuit breakers the interrupters are housed in porcelain insulators. The interrupter heads are live and mounted on support insulators on top of a steel structure to conform with the safety clearances (Fig. 7.8a). In dead-tank circuit breakers the interrupters are housed in earthed metal tank, usually of aluminium, mild or stainless steel, depending upon the current rating. The GIS circuit breakers could be of either vertical or horizontal configuration. Four switchgear manufacturers in the world produce horizontal EHV circuit breakers while the rest have vertical designs (Fig. 7.8b).

The live-tank circuit breakers are generally used in open-terminal outdoor substations (Fig. 7.9a) while the dead-tank circuit breakers are used in GIS indoor and outdoor substations (Fig. 7.9b). Both types of circuit breakers have been developed for ratings up to 63 kA at 525 kV and have given satisfactory service all over the world during the past two decades. The choice of the type of circuit breaker, however, depends upon many factors. Some of these are:

—   Cost of the switchgear;
—   Atmospheric pollution;
—   Potential environmental restrictions;
—   Price and availability of the land;
—   Individual preference;
—   Security against third party damage.

In locations where the price of land is high (i.e. in the centre of cities) the GIS option becomes very attractive because it drastically reduces the overall dimensions of a substation. For example at 400 kV, the ratio of land required for a GIS installation to that necessary for an open-terminal, substation is about 1:8.

7.7.1 Basic GIS substation design

GIS substations are of two types; outdoor and indoor. The basic design of both indoor and outdoor switchgear is the same but the switchgear for outdoor substations requires additional weather-proofing to suit climatic conditions. Both substation types have been in service throughout the world over the last two decades and have given satisfactory performance.

In Japan, most SF$_6$ GIS installations are located outdoors without a protective building and have been in service since 1969. In the UK, GEC and Reyrolle SF$_6$ GIS installations have been in use outdoors since 1976, notably at Neepsend (1976) and Littlebrook (1979). To the author's knowledge there have been no major failures in either country.

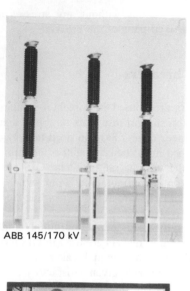

ABB 145/170 kV

ABB 525/550 kV

145 kV

GEC-ALSTHOM

420 kV

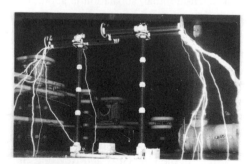

SIEMENS 800 kV circuit-breaker

REYROLLE 420/525 kV

Fig. 7.8(a)   Examples of SF$_6$ live-tank (GIS) circuit breakers

145/170 kV

145/170 kV

ABB **SF₆ GIS, Type ELK 3**

420/525 kV

Line feeder with double busbars

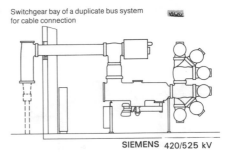

Switchgear bay of a duplicate bus system
for cable connection

SIEMENS 420/525 kV

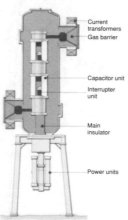

Current transformers
Gas barrier

Capacitor unit
Interrupter unit

Main insulator

Power units

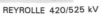

REYROLLE 420/525 kV

GEC-ALSTHOM 420/525 kV

Fig. 7.8(b) *Examples of SF₆ dead-tank (GIS) circuit breakers*

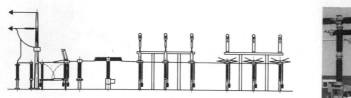

REYROLLE   420 kV

SIEMENS   525/550 kV

GEC-ALSTHOM

420 kV

SIEMENS
800 kV

⬛Ⓖ MERLIN GERIN   765 kV

Fig. 7.9 (a)   Open terminal substations

550 kV Switchgear

MERLIN GERIN

'SPD' 420 kV Gas Insulated Switchgear

420 kV FLUOBLOC

The ALPHA GIS substation 800 kV GIS

420 kV outdoor substation

Fig. 7.9 (b)   GIS substations

Recently in the UK, there has been a trend in favour of indoor GIS substations for technical and environmental reasons. SF_6 GIS equipment requires clean, dry and particle free assembly for safe operation. The assembly and dismantling of outdoor GIS equipment in the UK has been carried out in the past under a portable tent and internal drying of the GIS chambers has been achieved by circulating dry nitrogen. This process can take some days. Therefore it has become usual practice in the UK to install GIS substations indoors in purpose built buildings so that assembly, installation and maintenance can be undertaken in controlled conditions.

Since a GIS installation consists of an assembly of pipework and steel structures, it is usually regarded as unattractive and unsuitable for environmentally sensitive locations. The visual impact is reduced by housing the GIS in a building.

The switchgear equipment used at open terminal and GIS substations includes:

— Circuit breaker (already discussed in detail);
— Disconnector*;
— Switch disconnector*;
— Earth switch*;
— Current transformer*;
— Voltage transformer*;
— Closing resistors*;
— Surge arresters*;
— Busbars*.

The basic designs and operations of these devices* have been discussed elsewhere in detail [5, 6] but the following important issues will be discussed here with case examples:

— Closing resistors/metal-oxide surge arresters — to control overvoltages on long line switching;
— Disconnector switching;
— Ferroresonance;
— Monitoring on site.

7.8 Closing resistors/metal-oxide surge arresters

The switching and re-energising of long EHV transmission lines can generate very high overvoltages which can overstress equipment insulation and sometimes cause dielectric breakdown. The magnitude of the overvoltages depends upon circuit breaker characteristics, circuit breaker grading capacitors, circuit parameters and line lengths. The system is normally protected against excessive overvoltages by damping the transient recovery voltage. This is achieved by inserting a specified value of resistance in the line circuit just before the electrical contact is made. For safety, the resistor contacts are then fully open before the travel of the circuit breaker main contacts is complete. These precise movements are achieved by mounting the closing resistors on the interrupter assembly in parallel with the main contacts. The moving contacts of the closing resistors are directly connected to the main drive of the interrupters, so that the relative movements can be accurately set. Since the pre-insertion time of the resistor before the main contact touch is very critical, the mechanical drive has to be very precise and positive. The closing resistor drive thus becomes complex, requiring high energy operating mechanisms for the circuit breaker.

The pre-insertion time for the closing resistors is determined by detailed network analysis, which takes into consideration all circuit parameters and point-on-wave switching techniques. In most cases the ideal pre-insertion time for optimum overvoltage control is in the region of 10 to 12 milliseconds.

If we examine a circuit breaker with closing resistors and compare it with the performance of a standard SF_6 circuit breaker without closing resistors, we can see that resistors add complexity to the drive mechanism, increase the drive energy requirement of the circuit breaker mechanism, reduce the reliability of the SF_6 circuit breaker and increase the cost by 30 to 40%.

All the above complexities are introduced just for the duration of 12 milliseconds while the circuit breaker is closing, otherwise the closing resistors remain in the open position.

The author understands that several major utilities have experienced difficulties with failures of closing resistors on air-blast circuit breakers and have had problems with the long term reliability of closing resistor drives. Some have already implemented alternative solutions for controlling the switching overvoltages on long lines, by applying metal-oxide surge arresters (at both line ends and in the middle of the line) [8] and by point-on-wave switching.

Surge arrester technology has undergone a radical change in the last ten years. The undoubted simplicity of the metal-oxide arrester, in which the overvoltage is controlled basically by the arrester's internal non-linear resistance, was initially attractive but there were reservations about the ageing of the resistor "blocks". The technology has now advanced to such a stage and sufficient experience has been gained that the metal-oxide arrester is now fully accepted in the industry and is now applied in cases where overvoltage protection is required. This is evidenced by the fact that all manufacturers of surge arresters in the world have changed over completely to the metal-oxide arresters.

7.8.1 Main features of metal oxide surge arresters (MOA)

MOAs are continuous acting, their response time is short, they reduce switching overvoltages as voltages start to build up and, because there are no gaps in the assembly and no arc products, they have prolonged life.

Because of increased reliability and trouble-free service experience of metal-oxide surge arresters (MOA) over the last ten years, the availability of low protective levels and high discharge energy capabilities of the resistor blocks, several utilities have started to replace the closing resistors with the simple and more economical MOA devices for controlling overvoltages during energising and de-energising of long lines. ABB have installed the first 500 kV GIS circuit breakers without closing resistors in China. They have been in service for about two years without any trouble. The switching overvoltages at this installation are controlled by MOAs [8, 9].

7.9 Disconnector switching

Disconnectors in GIS installations are used mainly to isolate different sections of busbars either for operational reasons or for safety, during maintenance and refurbishment. They are also used for certain duties, such as load transfer from one busbar to another, off-load connection and disconnection of busbars and circuit breakers. The switching duties imposed on disconnectors have caused difficulties on some GIS designs in service. These difficulties have resulted in dielectric failure to earth on GIS equipment or dielectric failures on power transformer windings.

At the 1982 Discharge Conference (GD82), Yanabu *et al.* highlighted the switching problems associated with slow moving contacts of a disconnector. Some utilities have observed bright glow on GIS flanges during the disconnector closing and opening operations and reported a few dielectric failures on GIS equipment. Since then, switchgear manufacturers and utilities have continued their investigations. IEC and CIGRE have also taken a very active interest to achieve a better understanding of the disconnector switching phenomena.

During the past six years, the techniques for high frequency measurements have improved considerably. With the present sophisticated measuring and recording techniques, the very high frequency (VHF) transient voltages produced during the disconnector switching operations can now be very accurately recorded and analysed. They also help to explain the switching phenomena.

When the slow moving contacts of a disconnector close or open, hundreds of pre-strikes or re-strikes occur between the contacts. These restrikes generate steep-fronted (4 - 15 ns) voltage transients which last for several hundred milliseconds. The magnitude and frequency of the VHF transient voltages depend upon:

—   Contact speed of the disconnector;
— SF_6 gas pressure in the disconnector;
—   Dielectric stresses on contact tips, stress shields and contact gap;
—   Circuit parameters, voltage offset, polarity and trapped charge.

This subject has now been extensively discussed worldwide during the past six years and the results of investigations have been reported in numerous publications (e.g. from IEE, IEEE, ISH, CIGRE).

The pre-strikes and re-strikes during the closing and opening operations of the disconnector generate very fast, increased voltage transients locally across the contact gap and to the earth giving rise to transient ground potential rise (TGPR). The VHF transient voltages propagate on both sides of the disconnector as very fast travelling waves into the GIS installations, sometimes causing failure of GIS switchgear and the transformer.

Most switchgear manufacturers now have sufficient experience to incorporate design features which avoid failures on GIS equipment. The latest IEC 17A(Sect.)338 and 17A(Sect.)101 — July 1992 — document provides further guidelines for testing and evaluating the switching performance of disconnectors.

In an installation where the high voltage side of the transformer is directly connected to the SF_6 metal clad circuit breaker by the GIS busbars and disconnectors, a surge arrester is normally connected near the transformer for protection against most of these overvoltages (Fig. 7.10). However, even the present-day fast acting metal-oxide surge arresters cannot cope with the VHF transient voltages generated by disconnector switching. They let these fast transients through to the high voltage windings of the transformer. The continuous overstressing of the transformer windings causes deterioration of the winding insulation. This ultimately can cause dielectric failure to earth inside the transformer tank with very severe consequences such as loss of the transformer, loss of supply and expensive repair.

Discussions with the manufacturers of surge arresters have confirmed that the present day metal-oxide surge arrester technology may not be able to provide adequate protection against these VHF transient voltages in the foreseeable future.

The switchgear manufacturers on the other hand have had sufficient experience with these switching processes to accurately measure the amplitude and the rate of rise and the durations of these VHF transient voltages. The author firmly believes that by careful design, they will be able to dampen or eliminate altogether these overvoltages so that other switchgear equipment such as transformer on the substation will not be damaged.

7.10 Ferroresonance

Ferroresonance is a well-known phenomenon and it is defined in Chambers' dictionary as follows:

> "A special case of paramagnetic resonance, exhibited by ferromagnetic materials. It is explained by simultaneous existence of two different pseudo-stable states for the magnetic material B-H curve each associated with a different magnetisation current for the material. Oscillation between these two states leads to large currents in associated circuitry."

In a switchgear installation where electromagnetic voltage transformers are fitted, the ferroresonance can occur if the conditions are conducive. Ferroresonance depends upon the capacitive coupling between the unearthed, disconnected parts of the switchgear and the remainder of the plant. This happens when the sections of busbar fitted with voltage transformers are left energised through the grading capacitors of the open circuit breaker, the value of the resultant voltages depends on the values of coupling capacitance and voltage transformer characteristics. The

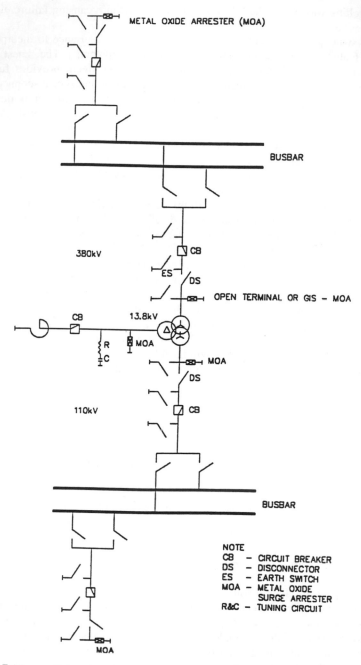

Fig. 7.10   Reactor switching by EHV and MV circuit breakers

occurrence of ferroresonance is a statistical phenomenon which depends upon the switching instants and the remanence effects of the voltage transformer.

The duration for which a voltage transformer may be left energised through the grading capacitors is critical. It is determined by the duty on the voltage transformer and the thermal capacity of its primary windings. In an ideal situation when a bus zone is de-energised, it should be immediately isolated and earthed. This may not always be possible in practice. Therefore, as soon as the switching sequence permits, the disconnectors nearest to the circuit breaker should be opened and the earthing switches closed.

It is essential to analyse the network switching sequence to determine the optimised switching sequences which minimise the coupling effects and the maximum energising time for the VTs. It is recommended that ferroresonance damping devices must be fitted to the secondary windings of the VTs on a "fit and forget basis" for safe operating of the system. The consequences of not carrying out the above recommendations could be quite serious. Several installations have experienced burnt out voltage transformers and flashovers on busbar insulating barriers.

7.11 System monitoring

7.11.1 Monitoring during installation and in service

The designs of EHV switchgear are becoming simpler and the number of interrupters per phase for the highest system voltages and fault currents (550 kV up to 63 kA) are getting fewer and fewer. There were six interrupters per phase for 420 kV in 1976 and only one in 1985/92. Consequently the reliability of present day SF_6 circuit breakers has improved and they are now virtually maintenance-free.

Most SF_6 circuit breakers are capable of interrupting 20 to 25 full short circuit currents and of performing over 10,000 trouble free mechanical operations. This is the result of:

— Improved arc interruption techniques employed in SF_6 gas and SF_6 gas-mixtures;
— Availability of low erosion rate nozzle materials;
— Reduced operating energies with low mechanical stresses on switchgear components;
— Computer-aided dielectric stress analysis techniques to optimise the shapes of the critical components and to obtain low dielectric stresses on contact tips, SF_6 gaps (across the contacts and to earth), stress shields, insulators and drive rods;
— Reduced number of moving parts and dynamic and static seals.

Therefore the practice of conventional regular maintenance will have to be re-examined. Because the present day SF_6 circuit breakers can perform a large number of mechanical operations and have longer contact and nozzle service life, it is not necessary to open a GIS circuit-breaker every six months. Instead of regular maintenance, it is recommended that essential parameters listed below should be monitored, some continuously and others periodically, so that the assessment of switchgear condition can be made.

7.11.2 Continuous monitoring

The necessary parameters to be monitored by fibre-optic diagnostic techniques are:

—   Current;
—   Voltage;
—   Arcing time;
— SF_6 gas pressure;
—   Circuit breaker contact travel characteristics.

Fibre-optic monitoring equipment has now been fully proven in service. It is stable over a long service period, robust, maintenance-free, easily installed and replaced and easily accessible. These measuring devices are now commercially available. They are accurate, reliable and reasonably priced. Fibre optic instruments which can be used to see inside the circuit breaker tank when the switchgear is live are also available.

7.11.3 Periodic monitoring

During the assembly and installation of GIS switchgear on site, care must be taken to ensure that all joints are correctly tightened, all loose particles are removed and all gas chambers are properly cleaned. After assembly, a UHF partial discharge technique may be used to ensure that the whole GIS installation is free from loose particles. This technique has been very successfully used to locate loose particles within a few hundred millimetres (Reyrolle and Strathclyde University who developed the technique have published several papers on this subject). The other techniques of partial discharge measurement have been evaluated by CIGRE WG 15-03 (CIGRE paper 15/23-01, 1992 Paris). These are not discussed here.

The advantage of this technique is that measurements can be made at a relatively low voltage level without overstressing the switchgear insulation. Once the installation is found to be free of any loose particles, it can be safely energised. After the energisation of GIS substation, there is, in the author's view, no need for continuous partial discharge monitoring. Several GIS installations in this country and abroad have been continuously energised for 10 to 15 years without Ultra High Frequency Partial Discharge (UHF-PD) monitoring devices and have given trouble-free service. The long term reliability of these sophisticated monitoring devices are still to be proven. However, a periodic check, say every two years, can be made so that the signature prints of the spectrum of discharges can be compared with those obtained just before energisation and any deterioration in the dielectric integrity of the GIS installation can be detected.

7.12 Insulation coordination

To ensure the safety and reliability of a GIS open-terminal or a hybrid switchgear installation, it is necessary to carry out proper insulation coordination of switchgear equipment and complete installation, so that the switchgear assembly shall be able to withstand all overvoltages imposed on the system during its service. The possible sources of overvoltages are:

- Atmospheric overvoltages — caused by direct lightning strikes and back-flash;
- Transient overvoltages — caused by inductive (reactor), capacitive (line, cable) loads and out-of-phase switching;
- Temporary overvoltages — caused by the resonance of the network and power transformer windings and by ferroresonance in electromagnetic VTs.

The choice of insulation level of equipment is very critical. The design of switchgear should ensure that flashover cannot occur across the open contact gaps with impulse or other overvoltages on one terminal and the out-of-phase a.c. peak voltage on the other terminal (i.e. the gap sees the sum of the two over-voltage peaks). The design should be complemented by detailed system studies of the switchgear installation to optimise the switching sequence and the number and locations of suitable metal-oxide surge arresters.

7.13 Conclusion

Modern SF_6 circuit breakers are simple, reliable and virtually maintenance-free and some designs have now achieved ratings up to 63 kA at 522 kV with one break per phase. Circuit breaker designs with third generation SF_6 interrupters have reduced the driving energy by 50 to 60 % and have brought increased reliability and further reduction in costs.

Reactor switching causes difficulty for some circuit breaker designs. Until a realistic test circuit is available to verify the performance of the circuit breaker, for reactor switching duties it is recommended that metal-oxide surge arresters and R-C tuning circuits (where possible) should be used for added safety.

Owing to increased reliability, high energy discharge capability and trouble-free service over a decade, metal-oxide surge arresters are extensively used for insulation coordination on the substation and gradually replacing the closing resistors for switching EHV transmission lines [8, 9]. On GIS installations, where electromagnetic voltage transformers are used, ferroresonance can occur and some protection is afforded by fitting ferroresonance damping devices to the secondary windings of the voltage transformers.

7.14 Acknowledgements

The author wishes to thank the assistance of colleagues at Merz and McLellan, in particular Mrs E. Adamson, in preparing this chapter and the directors of Merz and McLellan for permission to publish it. The views expressed do not necessarily represent the views of Merz and McLellan Limited.

Acknowledgement is also made to the following organisations for providing technical information and illustrations: ABB, AEG, GEC-Alsthom, MG, NGC, Reyrolle, Siemens and Scottish Power.

7.15 References

1   BARNEVIK, P.: "Electrifying experience: ASEA Group of Sweden" 1983-93
         CLOTHIER, H.W.: "Switchgear stages"
         KAHNT, R.: "The development of high-voltage engineering — 100 years of AC power transmission". Siemens
         ROWLAN, J.: "Progress in power". The contribution of Charles Merz and his associates to sixty years of electrical development, 1899-1959
         RYAN, H.M. and JONES, G.R.: "SF_6 switchgear" (Peter Peregrinus Ltd, 1989)

         Technical information on switchgear, from ABB, AEG, GEC-Alsthom, MG, NGC, Reyrolle, Siemens and Scottish Power

2   ALI, S.M.G., RYAN, H.M., LIGHTLE, D., SHIMMIN, D.W., TAYLOR, S. and JONES, G.R.: "High power short circuit studies on a commercial 420kV-60kA puffer type circuit breaker". IEEE - PES 1984 Summer meeting, Seattle (USA) Paper 84, SM 643-3

3   ALI, S.M.G. and RYAN, H.M.: "Further application of field computation strategies to switchgear design". ISH-89, Sixth International Symposium on High Voltage Engineering, New Orleans (USA), Paper 27.37

4   SUZUKI, K., TODA, H., AOYAGI, A., IKEDA, H., KOBAYASHI, A., OHSHIMA, I. and YANABU, S.: "Development of 550kV 1-break GCB: Part I — investigation of interrupting chambers performance". IEEE-PES 1992 Summer meeting, Seattle (USA), Paper 92 SM 577-7 PWRD

5   ALI, S.M.G. and GOODWIN, W.D.: "The design and testing of gas insulated metal-clad switchgear and its application to EHV substation". *IEE Power Engineering Journal*, January 1988

6   GOODWIN, W.D. and WILLS, A.S.: "The design of outdoor open-type EHV substation". *IEE Power Engineering Journal*, March 1987

7   ALI, S.M.G.: "Field computation in switchgear development". 8th CEPSI Conference 1990, Singapore

8   ERIKSSON, A., GRANDL, J. and KNUDSEN.: "Optimised line switching surge control using circuit-breakers without closing resistors". CIGRE 1990, Paris

9   SCHMIDT, W., RICHTER , B. and SCHETT, G.: "Metal oxide surge arresters for GIS-insulated iubstations". CIGRE 1992, Paris

Appendix 7.1

Since the 1830s, electricity has been used commercially in the telegraph industry, and many firms in the UK, Europe and America supplied various types of low voltage equipment to suit.

In Germany, Siemens & Halke was the largest company which became interested in higher voltage technology. The others which followed were Allgemeine Elektrische Gesellschaft (AEG), established in 1883 as Deutsche Edison Gesellschaft, Schuckert & Co. and the Union Company. In Switzerland, the leading companies were Oerlikon (1882) and Brown Boveri (1891), and in Sweden, ASEA (1883).

The use of alternating current started in lighting plant at the end of 1880s. Edison, among others, was opposed to the use of alternating current, which at the time was considered hazardous. Crompton was one of the leading advocates of direct current while Ferranti supported the cause of alternating current. However, the research continued and engineers in different countries had a breakthrough almost at the same time in the use of polyphase alternating current. Later, great progress was made in this technology, in particular with the invention of the transformer and alternating current (a.c.) motor.

This subsequently led to the concept of the present day transmission system, in which electric power at high voltage and low current was transmitted to another place at a distance and transformed back to a reasonable voltage and distributed for local consumption (i.e. generation at one point and consumption some distance away).

The first successful long distance transmission of electrical power employing three-phase alternating current was from Lauffen hydroelectric station. The Lauffen Frankfurt transmission line in Germany, 175 km long, transmitting at 15 kV, 40 Hz with an overall efficiency of 75%, was inaugurated on 24 August 1891.

Other important dates on the transmission calendar worldwide [1] are:

1911 : 110kV Transmission line — Lauchhammer, Riesa (Germany)

1929 : 220kV Transmission line — Brauweiler, Hoheneck (Germany)
1932 : 287kV Transmission line — Boulder Dam, Los Angeles (USA)
1952 : 380kV Transmission line — Harspranget, Halsberg (Sweden)
1965 : 735kV Transmission line — Manicouagan, Montreal (Canada)
1985 : 1200kV Transmission line — Ekibastuz, Kokchetau (USSR)

The main reason for using ever higher voltages was economy of transmission.

In the UK electrical industry, Merz and McLellan's contribution stretches back to 1889 when they were the consulting engineers to the North Eastern Electric Company, the pioneer in its field of generation and distribution. This company was regarded as a model of an efficient private utility not only in the UK but throughout the world.

The foresight of Charles Merz brought about fundamental developments in electrical power supply in the North East, elsewhere in Britain, its overseas dominions and in the USA. He was instrumental in establishing the first 20kV integrated transmission system of the North East in 1907 and the standardising of the British Supply frequency at 50Hz instead of 24 and 40Hz used in different parts of the country.

In 1907 Merz predicted a saving of 55 million tons of coal each year if UK power supply was operated and managed as an integrated whole. In 1924, he proposed the establishment of a super-tension transmission network for linking up the existing supply areas and developing new ones and to allow interchange of power at one frequency, similar to those already existing at that time in USA, Canada, Sweden, Japan, France, Germany and the North East coast of England.

Merz's suggested form of super-tension network was endorsed by the similar IEE-proposal twenty eight years later for the British 275kV supergrid. His dream became a reality when on April 1 1948 the British Electricity Authority (BEA), the largest utility in the western world, was created. In 1954 it became the Central Electricity Authority (CEA) in Britain and at the same time the formation of South of Scotland Electricity Board (SSEB). In 1957 Merz's vision was completed by the formation of the central electricity generating board (CEGB) with the 400kV interconnected supergrid in the UK.

From 1890 to 1960, the transmission voltages increased from 2kV to 400kV. The important dates in the UK transmission calendar [1] are:

1890 - 2.0 kV
1905 - 5.5 kV
1907 - 20 kV
1924 - 60 kV
1926 - 132 kV
1953 - 275 kV
1963 - 400 kV

The switchgear industry worldwide has kept pace with the increasing demands of both currents and voltages during this period, by developing reliable switchgear for controlling and protecting the electrical networks.

Fig. 7.11   *One-break 420/550kV SF$_6$ circuit breakers (1985-1992)*

Merz once again played an important role in bringing Parson, Clothier and Reyrolle together on Tyneside who jointly brought about the electrical revolutionin alternating current (a.c.) generation, transmission and control. Merz, Clothier and Reyrolle pioneered the concept of bulk oil, compound filled switchgear in UK and jointly developed the first iron-clad switchgear. This was a bulk-oil double plain-break, metal clad, 5.5 kV circuit breaker with compound filled busbars in 1905. This brought new standards of safety to the high voltage distribution system. With continuous improvement in switchgear technology and the efficient use of SF_6 gas, the ultimate goal in circuit breaker design has now been achieved in the development of one-break 420 kV (Ali *et al.*, 1984, [2, 3, 8] Fig. 7.11) and 550 kV (Suzuki *et al.*, 1992, [4]) circuit breakers.

SF_6 circuit breakers in the UK

Early transmission in the UK at 132 kV employed bulk-oil circuit breakers in open-terminal substations but from the mid-1940s air-blast breakers were made in increasing numbers, particularly driven by the establishment of the 275 kV super-grid system. The earliest 420 kV open-terminal circuit breakers in the UK were first commissioned in the early 1960s and had twelve series air-blast interrupters per phase. SF_6 insulated current transformers were produced over the range 145 kV to 420 kV in 1950s.

Following the development and validation of synthetic testing techniques, the recovery voltages available for tests were no longer governed by the maximum direct output of the short-circuit testing stations. This allowed the development of interrupter units with higher breaking capacity. In 1971 two-cycle 420 kV 35 GVA air-blast breakers with six series breaks per phase were installed by GEC and Reyrolle.

SF_6 open-terminal circuit-breakers appeared in Europe from the mid-1960s onwards and the first SF_6 GIS installations were commissioned in Europe in the late 1960s. In the early to mid-1970s in the UK, 300 kV GIS installations were supplied to CEGB by both GEC and Reyrolle followed by 420 kV GIS substations towards the end of that decade.

The first high power SF_6 interrupters utilised air-blast technology, modified to give a closed two-pressure system. The relatively high cost and complex mechanisms of the system led to the development of single-pressure puffer type interrupters which were first applied in EHV circuit breakers in the early 1970s for both open terminal and GIS installations.

Chapter 8

High voltage bushings

J. Graham

8.1 Introduction

A bushing is a device for carrying one or more high voltage conductors through an earthed barrier such as a wall or a metal tank. It must provide electrical insulation for the rated voltage and for service overvoltages and also serve as mechanical support for the conductor and external connections. The requirements for bushings are specified in IEC137 1984 [1].

8.2 Types of bushings

Bushings are used to carry conductors into all types of electrical apparatus e.g. transformers, switchgear and through building walls. Their form depends on the rated voltage, insulating materials and surrounding medium. Bushings can be broadly grouped into two types, non-condenser and condenser graded bushings.

8.2.1 Non-condenser bushings

In its simplest form a bushing would consist of a conductor surrounded by a cylinder of insulating material, porcelain, glass, cast resin, paper etc. as shown in Fig. 8.1. The radial thickness a is governed by the electric strength of the insulation and the axial clearance b, by that of the surrounding medium.

As shown in Fig. 8.2, the electric stress distribution in such a bushing is not linear through the insulation or along its surface. Concentration of stress in the insulation may give rise to partial discharge and a reduction in service life. High axial stress may result in tracking and surface flashover. As the rated voltage increases, the dimensions required become so large that this form of bushing is not a practical proposition.

Partial discharge can be reduced by including in the bushing design a stress control mechanism. With cast resin insulation a control electrode, electrically

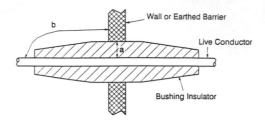

Fig. 8.1   *Non-condenser bushing*

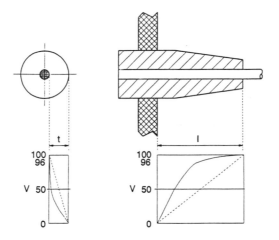

Fig. 8.2   *Stress distribution in non-condenser bushing*

connected to the mounting flange, can be embedded in the insulation reducing the stress at the flange/insulation interface.

Stress control methods have been developed for power cable terminations using heat-shrinkable stress control tubing which can also be applied to bushings. The heat-shrinkable stress grading tube is installed over the exposed solid insulation and overlapping the flange. The tube reduces the voltage gradient at the flange and along the surface of the bushing (Fig. 8.3). It is important that air is eliminated from the interface using a void filling mastic to prevent partial discharge.

8.2.2 Condenser bushings

At rated voltages over 52 kV, the condenser or capacitance graded bushing principle is generally used, as shown in Fig. 8.4. The insulation material of such a bushing is usually treated paper with the following the most common:

—   Resin bonded paper (RBP);

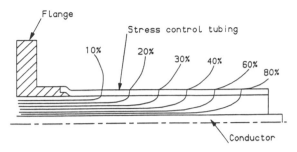

Fig. 8.3   Stress control using heat-shrinkable stress control layer

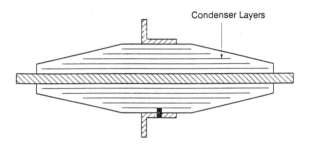

Fig. 8.4   Condenser bushing

— Oil impregnated paper (OIP);
— Resin impregnated paper (RIP).

As the paper is wound onto the central tube, conducting layers are inserted to form a series of concentric capacitors between the tube and the mounting flange. The diameter and length of each layer is designed so that the partial capacitances give a uniform axial stress distribution and control radial stress, within the limits of the insulation material (Fig. 8.5).

8.2.2.1 Resin bonded paper bushings

RBP bushings were previously used extensively, up to 420 kV, for transformer applications but are now limited to low voltage use, particularly in switchgear, due to technical limitations. In RBP bushings, the paper is first coated with a phenolic or epoxy resin then wound into a cylindrical form under heat and pressure, inserting conducting layers at appropriate intervals. The use of RBP bushings is limited by the width of paper available and by the danger of thermal instability of the insulation due to the dielectric losses of the material. RBP bushings are designed to operate in service at a maximum radial stress of approximately

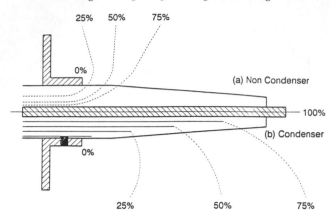

Fig. 8.5   Field distribution in non-condenser amd condenser bushings

20 kV/cm.

The RBP insulation is essentially a laminate of resin and paper. The bushing therefore contains a considerable amount of air distributed between the fibres of the paper and at the edges of the grading layers.

Where internal faults exist, as the voltage applied to the bushing is raised, partial discharge inception can occur at the layer ends where the stress is greater than the radial stress between the layers. During manufacture, incorrect winding conditions may result in circumferential cracking produced by shrinkage or weak resin bonding. Here, in the voids produced, the electric stress is enhanced and partial discharge may occur at low levels of stress.

In service, ingress of moisture into RBP bushings can cause delamination and increased and unstable dissipation factor or tangent delta. Discharges at the layer end produce carbon treeing, extending axially, while discharges in voids may produce breakdown between layers. Both forms of discharge are progressive and ultimately lead to failure over a long period by overstress or thermal instability in the residual material [2, 3].

Overvoltages which occur in service are usually surges produced by switching or lightning. Breakdown of a bushing under this type of stress would normally be initiated axially from the ends of the layers due to breakdown of air in the winding. Complete failure may be a combination of axial and radial breakdown.

8.2.2.2 Oil impregnated paper bushings

OIP insulation is widely used in bushings and instrument transformers up to the highest service voltages. OIP bushings are made by winding untreated paper inserting conducting layers at the appropriate positions and impregnating with oil after vacuum drying.

The paper used is generally an unbleached kraft which is available in widths of up to 5 m. This width is adequate for most applications but, for ultra high voltage bushings, various methods of extending the condenser length by multi-piece construction or paper tape winding have been used. It is important that the paper be sufficiently porous to allow efficient drying and impregnation while maintaining adequate electric breakdown strength.

The oil used is a mineral oil as used in power transformers and switchgear [4]. Prior to impregnation, processing is carried out to ensure low moisture and gas content and high breakdown strength [6]. In certain applications, other properties may be important; for example low pour point for low temperature installations and resistivity and fibre content for d.c. bushings.

Processing of the bushing may be carried out by placing the whole assembly in an autoclave or by applying vacuum directly to the bushing assembly before impregnation. Manufacturing defects are generally detected in routine tests. In the case of properly processed OIP bushings, there are no gaseous inclusions in the material. Internal discharge inception therefore occurs at much higher stress levels than with the RBP bushings. IOP bushings are therefore being designed to operate at radial stresses of typically 40 kV/cm. Discharges can occur at the layer ends (due to misalignment) of the layers at the high stress levels associated with lightning impulse and power frequency tests. If this stress is maintained, gassing of the oil and dryness in the paper can be produced, and eventually carbonisation at the layer ends may occur which, due to the high radial component of the stress, tends to propagate radially leading to breakdown.

8.2.2.3 Resin impregnated paper

RIP insulation was developed in the 1960s for use in distribution switchgear and insulated busbar systems. In recent years, development has increased its utilisation to 525 kV.

In the manufacturing process, creped paper tape or sheet is wound onto a conductor. Conducting layers are inserted at predetermined positions to build up a stress controlling condenser insulator. The raw paper insulator is dried in an autoclave under a strictly controlled heat and vacuum process. Epoxy resin is then admitted to fill the winding. As a 525 kV bushing may have a core greater than 6 m in length, it is important that the resin has low viscosity and long pot life to ensure total impregnation. During the curing cycle of the resin, shrinkages must also be controlled to avoid the production of cracks due to internal stresses. The resulting insulation is dry, gas tight and void free giving a bushing with low dielectric losses and good partial discharge performance.

During manufacture, the conducting layer follows the shape of the creped paper. The spacing between individual layers varies between the peaks and troughs of the creping. The layer spacing with RIP is therefore coarser than with RBP and OIP bushings and full advantage of the high intrinsic strength of the resin cannot be taken. RIP bushings are designed to operate with a radial stress of about 30 kV/cm.

8.3 Bushing design

It is essential that a bushing be designed to withstand the stresses imposed in test and service. These are summarised in Table 8.1.

Table 8.1 Likely stresses for bushings

Electrical	Lightning Impulse Voltage (BIL)
	Overvoltages caused by switching operation (SIL etc.)
	Power frequency voltage withstand
Thermal	Conductor losses
	Dielectric losses
	Solar radiation
Mechanical	Loads due to external connections
	Self loads due to angle of mounting
	Earthquake forces
	Short circuit forces
Environment	Wind loads
	Nature of surrounding medium (air, oil, gas)
	Pollution

Condenser type bushings are used predominantly at high voltages and their design and application will now be considered.

Electrical stresses act both radially through the insulation and axially along its surface. The maximum allowable stresses for each material have been determined by experience and test to give a minimum service life of 30 years.

The condenser design controls these stresses to a safe level. The stress distribution is dependent on five principle factors (Fig. 8.6):

r_o radius of conductor
r_n radius of outer layer
l_1 length of first condenser
l_n length of last condenser
U_r rated voltage

A typical OIP bushing, one end operating in air and the other in oil, as shown in Fig. 8.7, will be considered to describe the procedures of bushing design [6]. Fig. 8.8 shows 420 kV OIP bushings installed on a transformer supplied by NEI Peebles, Edinburgh, UK to NTPC Rihand, India.

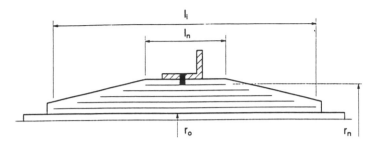

Fig. 8.6   Major dimensions of condenser bushings

The condenser winding (1) is enclosed in air side (2) and oil side (3) insulators which are pressed against the mounting flange (4), the underside of which may be extended to provide current transformer accommodation, by springs contained in the head of the bushing (5) acting to tension the central conductor or tube (6). The assembly is sealed by gasket to prevent oil leakage.

8.3.1 Air end clearance

For indoor use with moderate pollution and humidity, resin based insulating materials need no further protection from the environment. Oil or gas filled or impregnated bushings always require an insulating enclosure. This is commonly porcelain but modern glass reinforced plastic with rubber coatings (composite insulators) are also used.

The length of the insulator is governed by the lightning impulse and switching impulse requirements. The design of the bushings produces uniform axial stress along the surface of the insulator and the length can therefore be less than for a simple air gap. The length of the insulator is also affected by the service environment. In polluted atmospheres, resistance to flashover under wet conditions even at working voltage is dependent on the surface creepage distance, i.e. the length of the insulating surface between high voltage and earth, and the proportion of it protected from rain. IEC 815 [7] gives guidelines on the design of insulator profiles for use in polluted atmospheres.

From information on the site pollution severity, a minimum nominal specific creepage distance l is specified for each of the pollution levels indicated in Table 8.2. The minimum total creepage distance L is given by:

$$L = K_D \, U_r \, l$$

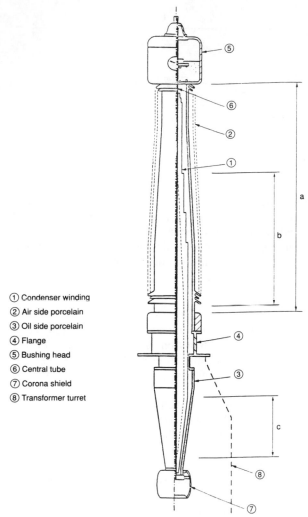

① Condenser winding
② Air side porcelain
③ Oil side porcelain
④ Flange
⑤ Bushing head
⑥ Central tube
⑦ Corona shield
⑧ Transformer turret

Fig. 8.7   *Section of typical transformer/air bushing*

where

K_D = diameter correction factor to increase the creepage distance with average diameter of insulator D_m

D_m < 300 mm $K_D = 1$
300 < D_m < 500 mm $K_D = 1.1$
D_m > 500 mm $K_D = 1.2$
U_r = rated voltage of bushing (kV)

Fig. 8.8   420 kV oil impregnated paper transformer/air bushing

In certain desert areas, a combination of adverse climatic conditions, long periods without rain, frequent fog, sand storms etc. lead to the accummulation of conductive pollutants and insulator flashover. To combat these conditions, enhanced creepage distances of 40 mm/kV or more have been specified.

Modern porcelain insulator designs generally use an alternate long/short (ALS) shed profile which gives superior performance to the previous antifog type. ALS profiles allow easier cleaning under natural rain and wind conditions and can be produced economically by modern turning techniques. Typical profiles are shown in Fig. 8.9. To avoid bridging of the sheds by rainfall, it is recommended that dimension C be equal to or greater than 30 mm, S/P_1 equal to or greater than 0.8 and $P_1 - P_2$ equal to or greater than 15 mm.

The manufacture of porcelain insulators is limited by the tendency to bend during firing if the piece has a high ratio (typically > 6) of height to bore. This is overcome by bushing manufacturers using epoxy adhesive to bond together several sections. The outdoor porcelain of a 420 kV bushing may typically be assembled from 3 sections to give an overall height of 3.5 m with a height to bore ratio of approximately 10. The adhesive produces a high strength, oil-tight joint and the assembly is considered to be a single piece.

Table 8.2 Pollution levels, equivalent ambient severities and minimum specific creepage distance

Pollution levels and typical environments	Equivalent ambient severities (reference values)			Minimum specific creepage distance, *l* (mm/kV)
	Salt-fog method	Solid-layer methods		
		Steam-fog	Kieselguhr	
	Salinity (kg/m³)	Salt deposit density (mg/cm²)	Layer conductivity (μS)	
I Light - frequent winds and/or rain falls - agricultural - mountainous (> 10 km from sea, no sea-winds)	5 to 14	0.03 to 0.06	15 to 20	16
II Average - industries without pollution smoke - dense housing with wind and/or rain, exposed to wind from sea (but not too close)	14 to 40	0.1 to 0.2	24 to 35	20
III Heavy - high density of industries - suburbs of large cities (close to sea)	40 to 112	0.3 to 0.6	36	25
IV Very heavy - conductive dust, smoke - sea-spray, strong sea winds - desert	> 160			31

The application of composite insulators is limited at present due to concerns over durability. They have advantages over porcelain being lighter, explosion proof and having a greater hydrophobicity due to the nature of the silicone rubber moulded shed profile.

Having determined the height of the air side insulator (dimension *a*), the length of air side grading (dimension *b*) can be determined. It is not necessary to grade 100 % of the air side insulator length; in practice 60 % internal grading or less gives adequate surface grading for large bushings.

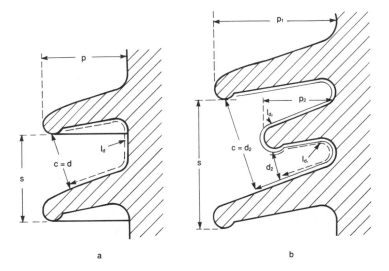

Fig. 8.9   *Typical porcelain insulator shed profiles*
(a) normal sheds
(b) alternate long/short (ALS) sheds

8.3.2 Oil end clearance

The oil side insulator (3) is usually a conical porcelain or cast resin shell. The internal axial grading over the condenser is dependent on the power frequency test voltage at a stress of approximately 10 kV/cm, this determines dimension c. It is normal that the oil side of a transformer bushing be conservatively stressed due to the consequences of a flashover within the transformer. Dimensions b and c together with the physical requirements of the mounting flange and current transformer determine the lengths l_1 and l_n of the condenser.

8.3.3 Radial gradients

Whilst it is possible to design a bushing with a constant radial gradient, this can only be achieved at the expense of a variable axial gradient. In most cases a constant axial gradient is desirable and the radial gradient may be allowed to vary and will be a maximum at either the conductor or the earth layer.

The values are given as follows:

$$E_o = \frac{V\,(a + 1)}{2\,a\,r_o\,\log b}$$

Earth layer stress

$$E_n = \frac{V(a + 1)}{2\,a\,r_{n-1}\,\log b}$$

where $a = l_l/l_n$ and $b = r_n/r_o$.

E_o is maximum when $a < b$ and E_n is maximum when $a > b$. A minimum insulation thickness is achieved when radial gradients at the HT layer and the earth layer are approximately equal, i.e. when $a = b$; however, this cannot always be achieved.

The radius r_o is dependent on the current rating, the method of connection between the bushing and the transformer winding and on the bushing construction. An optimum value for r_n can then be calculated.

Having determined the limiting dimensions, the positions of the intermediate layers can be calculated. The detailed calculation method may vary but the object is to achieve acceptable radial stress on each partial capacitor and uniform axial stress, with the minimum number of layers.

Since, as stated, the radial gradient varies throughout the insulation thickness, the layer spacing for constant voltage per partial capacitor will also vary. In this way, a 420 kV bushing may contain about 70 layers.

8.4 Bushing applications

8.4.1 Transformer bushings

Transformers require terminal bushings for both primary and secondary windings. Depending on the system configuration the outer part may operate in air, oil or gas.

At distribution voltages up to 52 kV, non-condenser type bushings are generally used. In the case of dry-type transformers, the bushings form an integral part of the cast resin winding. With liquid insulated transformers, porcelain insulated bushings are commonly used for outdoor applications and cast resin for connections inside cable boxes or with separable connectors. These types of bushing are covered by the European standard HD506 [8].

Condenser-type bushings have been developed for rated voltages up to 1600 kV [9]. Transformer bushings are not exclusively of the OIP type. RIP and some RBP are used, particularly at ratings up to 245 kV. RIP has some advantages over OIP for certain applications.

In many cases, the flexible cables from the transformer winding are drawn through the bushing and terminated at the head of the bushing within the bushing tube. This "draw lead" type of connection is limited to approximately 1250 A rating due to the dimensions of the flexible cable required. In cases of higher current ratings, connections may be made at the lower end of the bushing and the bushing tube itself used as the conductor as shown in Fig. 8.10.

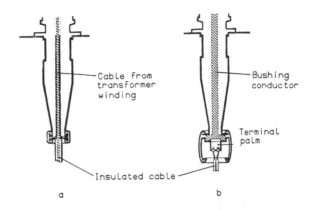

Fig. 8.10   Transformer bushing connections
(a) draw lead type
(b) bottom connection type

With RIP insulation the layers are embedded in a solid material of sufficient strength that there is no need for a supporting tube. In the case of draw lead connection using paper insulated cables, a transformer end stress shield is unnecessary which allows a reduction in the transformer turret diameter.

The oil end of the bushing may take two forms: conventional or re-entrant type. The conventional type is considered above. From a comparison of the two forms shown in Fig. 8.11, it can be seen that the re-entrant type is shorter and, as no stress shield is required, the transformer turret diameter can be reduced. The re-entrant form has been used extensively in the UK on power transformers but has been replaced by the conventional type. Re-entrant bushings cause difficulties with their installation. The transformer lead must be insulated with paper to approximately 30 % of the service voltage and it is possible for gases to become trapped on the inner surface.

At the mounting flange of the bushing a connection to the last layer of the condenser is brought through a test tapping. This tapping is used during partial discharge and capacitance measurement of the bushing and the transformer. As the capacitance of the tapping is low it is essential that it is connected directly to earth when in service to prevent generation of high voltage and sparking at the tapping terminal. In certain cases, particularly in North America, a potential tapping may be required. In this case, extra layers are included in the winding to provide a capacitance voltage divider. This type of tapping has high capacitance compared to the main bushing and can be used in service, connected to a bushing potential device, to provide a voltage source of up to 5 kV and an output power, typically 100 VA. This power may be used to supply relays and measuring equipment.

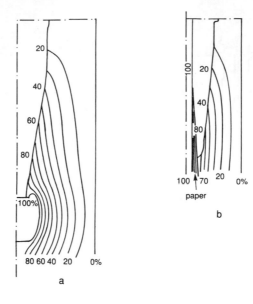

Fig. 8.11   *Field plots of the transformer side of HV bushing connections*
(a) conventional type
(b) re-entrant type

8.4.2 High current bushings

Bushings used on the low voltage side of generator transformers require special consideration due to their operating condition. Bushings of this type are often required to operate with their outdoor side enclosed in a phase isolated bus duct. This arrangement can produce ambient air temperatures as high as 90 °C around the bushing, differing greatly from the standard conditions. It is essential that the bushing and the connections are designed to reduce conductor losses and dissipate heat efficiently. At service currents of up to 40 kA, local heating due to poor connections can cause serious damage. To facilitate cooling, a multi-palm configuration is often used at the end terminals. Fig. 8.12 shows an RIP condenser type bushing having an aluminium conductor. However, copper conductors and non-condenser types are also widely used.

Where low voltage, high current bushings are mounted in close proximity, consideration should be made of distortion of the current path in the bushing due to magnetic effects.

Fig. 8.12   36 kV 31500 A resin impregnated paper transformer/air bushing

8.4.3 Direct connection to switchgear

Due to advantages of space saving given by gas insulated switchgear (GIS) operating with gas (usually sulphur hexafluoride SF_6) at a pressure of about 4 bar(g), it is increasingly common for transformers and switchgear to be directly connected. Direct connection also reduces pollution problems in coastal and industrial areas giving increased system reliability. Oil to gas bushings, used to provide the interconnection, generally have a double flange arrangement for connection to the transformer turret and the GIS duct. The bushing design therefore needs to be flexible to cater for the requirements of different equipment manufacturers. Work is underway within IEC to standardise dimensions on the gas side of the interconnection.

It is important that escape of gas from the GIS is minimised. Precautions must be taken with the bushing design to effectively seal the conductor and flange interfaces to prevent leakage of gas into the transformer. Fig. 8.13 shows typical arrangements where double seals are provided at each position, the effectiveness of which can be tested by applying high pressure between the seals [10].

RIP bushings provide an ideal solution (Fig. 8.14), and are available up to 525 kV. Due to the gas-tight nature of the insulation an additional porcelain shell is unnecessary. The dry insulation can be mounted at any angle without any need for oil expansion devices as would be required with OIP. The high electric strength of the resin also allows reduced axial dimensions, particularly of the gas part.

Electrical tests for oil to gas bushings require special arrangements. The gas side is tested in gas instead of oil to prevent contamination of the bushing seals and the gas duct in service. Draw lead type bushings are generally not used as the

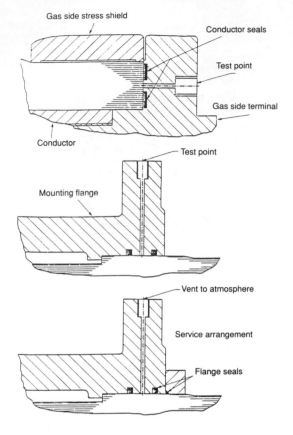

Gas side stress shield

Conductor seals

Test point

Gas side terminal

Conductor

Test point

Mounting flange

Vent to atmosphere

Service arrangement

Flange seals

Fig. 8.13 Typical sealing arrangements for transformer/gas bushings

risk of leakage of gas through this site-made joint would be undesirable.

8.4.4 Switchgear bushings

Entrance bushings for high voltage gas insulated switchgear often utilise pressurised porcelain. The gas within the bushing is common with the duct. Stress control is achieved by profiled electrode screens between the flange and the conductor. The porcelain must be dimensioned to withstand the full pressure of the system and presents an obvious danger if damaged in service. An improvement of this technique is the so called "double pressure" bushing where a glass reinforced plastic tube is used as a liner and the gap between the tube and porcelain is at reduced gas pressure.

An alternative solution is to use condenser bushings as shown in the installation in Fig. 8.15. RIP gas to air bushings have been manufactured up to 525 kV. The

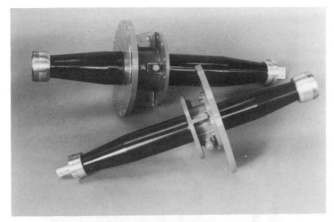

Fig. 8.14 Resin impregnated paper transformer/gas bushings

RIP condenser seals the GIS and the porcelain may be filled with a compound material or low pressure gas. This enables a lightweight porcelain to be used and operation at any angle without modification. Developments of this type of bushing are being made to replace the porcelain by a composite insulator or to mould silicone rubber sheds directly onto the RIP surface. As the gas side may be used directly into, or close to, a circuit breaker, the components of the bushing must exhibit resistance to the decomposition products of SF_6, particularly hydrogen fluoride (HF). This is achieved by coating the RIP with a special alumina-rich varnish.

In GIS, very fast transients (VFTs) generated by disconnector switching are recognised to present a problem to the internal connection bushings. Due to the speed of propagation of the VFT, it is possible to develop a high voltage between the conductor and the first layer of the condenser. At present, no test exists within IEC137 to demonstrate acceptability of the bushing design. However, tests have been proposed which apply lightning impulse chopped within the gas duct at approximately 70% of the rated BIL.

8.4.5 Direct current bushings

Direct current bushings require special consideration. HVDC schemes are becoming increasingly popular for the transmission of power over long distances and also for the connection of separate a.c. networks. These so called back-to-back schemes may be used on systems of different frequency and a sychronous operation or to increase operational stability and operate at typically ±80 kV d.c. while long distance transmission occurs at up to ±600 kV d.c.

The design of a d.c. bushing is influenced by the resistivities of the various materials used as opposed to their permittivity in the a.c. case. While permittivities of paper, oil, porcelain etc. are of a similar order, their resistivities

Fig. 8.15   420 kV 2000 A resin impregnated paper gas/air bushings installed in Saudi Arabia

vary by up to 10000:1. It is therefore important to study the voltage distribution in the core and the surrounding area [11]. The effect on field distribution is shown in Fig. 8.16. The upper plot shows the a.c. field in a typical oil-impregnated paper transformer bushing. A concentric field is produced in the oil gap between the paper insulated stress shield and the transformer turret wall. In the centre plot, the d.c. field of the same arrangement is illustrated, where the high resistivity of the paper, compared to that of oil, concentrates the stress in the shield insulation. To reduce this stress concentration and the stress of the surface of the porcelain, concentric cylinders of pressboard are placed around the bushing. This technique is illustrated in the lower plot. In the practical case a greater number of cylinders and conical barriers may be required to achieve suitable stress control. As the resistivity ratios vary with temperature, studies of the field are made across the operating range of the transformer. It is important therefore that the transformer and bushing manufacturers co-operate in this detail of the design.

In a d.c. scheme, pollution and fire risks are a major concern. To reduce both, bushings have been developed to operate horizontally to project directly into the d.c. converter building. Alternative solutions are contrasted in Fig. 8.17, which are comparable to the situations at EdF Les Mandarin and at NGC Sellindge,

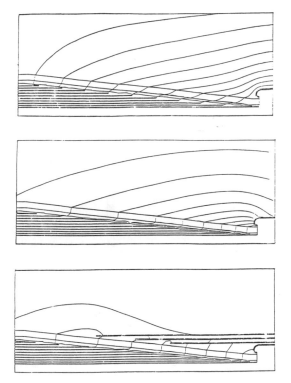

Fig. 8.16   Field plots of HVDC bushings

respectively, at either end of the 280 kV d.c. cross-Channel Link.

In France, more conventional OIP bushings are used in the transformer connected to the converter equipment by dry type RIP wall bushings. Fig. 8.18 shows a 500 kV reactor manufactured by ABB Transformers, Mannheim, Germany for the Pacific Intertie Extension in the USA. The bushings shown are OIP type, mounted in service as Fig. 8.17(b). The angled through-wall bushing is designed to limit the loss of oil from the bushing and reactor in the event of damage to the outboard porcelain insulator.   At Sellindge, due mainly to space restrictions, horizontally mounted through-wall bushings were used on the convertor transformer.   These are dry type RIP bushings and, being indoors, require no porcelain weather shell and admit no oil to the building.

In d.c. schemes pollution induced flashovers [12] have occurred and, where mounted close to a building, non-uniform wetting of bushing insulator has caused problems. Where a polluted insulator is partially protected from rainfall by the building, the difference in surface resistivity in the dry and wet parts reduces the flashover voltage. Various methods of improvement using booster sheds [13] and improving hydrophobicity of the insulator [14] surface have been examined.

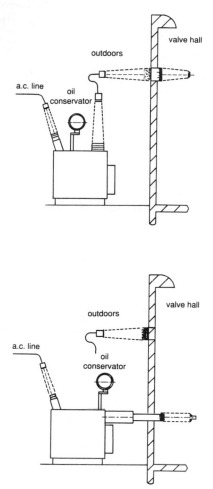

Fig. 8.17   Alternative arrangements of converter transformers
  (a) Outdoor bushings
  (b) Through wall transformer bushing

8.5 Testing

Adequate testing is essential to ensure reliable operation over the required service life. Routine and type tests are performed in accordance with IEC137 and IEC60 [15]. Fig. 8.19 shows a transformer bushing installed in an oil-filled tank being

Fig. 8.18   500 kV d.c. 1250 A oil impregnated paper bushings

prepared for test.

8.5.1 Capacitance and dielectric dissipation factor measurement

This test is probably the most universally applied of all tests on high voltage bushings and insulation systems. Measurements are made by Schering Bridge, or similar equipment, and give an indication of the quality of the brushing processing. Dissipation factor, or tangent delta, is identical to power factor in the range of values obtained. Tangent delta is a measure of the losses in the insulation and indicates the degree of cure of resinous materials or the moisture content of RBP and OIP. The typical tangent delta/voltage curve for a correctly processed bushing is flat up to at least rated voltage. An increase in tangent delta, particularly below working voltage, would almost certainly cause deterioration in service, due to increased dielectric losses or internal partial discharge.

Fig. 8.19 Bushing tests at NEI Reyrolle Ltd Clothier Laboratory

8.5.2 *Power frequency withstand and partial discharge measurement*

Although classed as separate tests, power frequency withstand and partial discharge are often combined. Partial discharge is a major cause of failure in bushings and, as discussed earlier, may occur in voids, cavities or inclusions in solid or liquid impregnated insulation and surface discharges at material boundaries.

In earlier times, an audible "hissing" test was used to assess the quality of RBP insulation. A trained ear could detect discharge of about 100 pC. RBP bushings with this limit of discharge at $1.05 U_r/\sqrt{3}$ have given years of satisfactory service and this limit has been retained in recent specifications.

Discharges have a more damaging effect on OIP and a limit of 10 pC at $1.5 U_r/\sqrt{3}$ is agreed. In general, well-processed OIP and RIP insulation are free from detectable discharge at this level.

Modern discharge detection equipment has been developed to improve the sensitivity of the measurement. Most systems monitor current from the bushing test tapping and display on an oscilloscope in the form of an ellipse. Partial discharges appear as pulses, the magnitude of which can be compared to a calibrated pulse. Much work has been published on partial discharge interpretation

[16, 17]. From the position of the discharge pulse on the ellipse, it is possible to recognise certain types of fault. When combined with the routine power frequency withstand test, normally applied for 1 min, the stress dependence of any discharge can provide further information on the discharge site.

For transformer bushings, proposals have been made that bushing withstand tests be carried out at 10 % above that of the transformer for which they are destined. In some cases, transformer manufacturers specify bushings coordinated one level above the transformer to avoid the possibility of internal fault during transformer test and the ensuing cost of rebuild.

8.5.3 Impulse voltage tests

Lightning and switching impulses represent transients occurring naturally in a high voltage system under operation. Tests with impulse voltage are designed to demonstrate the response of equipment to transients over a wide frequency range. Dry lightning impulses are applied to all types of bushing as a type test and wet switching impulse tests on bushings above 300 kV rating. Fifteen impulses of positive polarity and 15 impulses of negative polarity are applied to the bushing. As the internal insulation is considered non-self restoring, a maximum of two flashovers are allowed in air external to the bushing insulator with no internal fault permitted. To ensure that transformers, complete with bushings, can be safely subjected to impulse test, it is increasingly common for dry switching and chopped lightning impulses to be applied to bushings as a routine test. Negative lightning impulse tests have been carried out routinely by some bushing manufacturers for a number of years and this will be incorporated in the next issue of IEC137.

8.5.4 Thermal stability test

This test is particularly applicable to bushings for transformers of rated voltage above 300 kV and is intended to demonstrate that the dielectric losses do not become unstable at the operating temperature. The test is carried out with the bushing immersed in oil heated to 90 °C. A voltage is applied equal to the maximum temporary over voltage (usually $0.7 U_r$) seen by the bushing in service. By continuously measuring the capacitance and tangent delta of the bushing, the dielectric losses are calculated. Should the bushing be incapable of dissipating these losses, the tangent delta would increase and thermal runaway would occur, resulting in breakdown of the insulation.

Due to the inherently low value of tangent delta of OIP and RIP bushings, thermal stability is not normally a problem. In certain applications such as oil to gas bushings where the cooling is restricted, special attention should be paid to thermal stability.

It is the intention of the specification IEC137 that dielectric and conductor losses be applied to the bushing simultaneously. This is not always possible due to design restrictions of the bushing, and conductor losses are considered separately during the temperature rise test.

8.5.5 Temperature rise test

This test is intended to demonstrate the ability of the bushing to carry rated current without exceeding the thermal limitations of the insulation. OIP and RIP are restricted to a maximum temperature of 105 °C and 120 °C, respectively. The higher thermal rating of the RIP material does not necessarily mean that smaller conductors can be used. RIP is a good thermal insulant and the design of OIP bushings more readily allows cooling of the conductor by convection within the oil of the bushing. The service condition of different types of bushing, particularly high current bushings used in phase isolated bus ducts, must be carefully considered. In a typical test, a bushing achieved a rating of 10 kA under the standard test conditions laid out by the specification, while with an increased ambient air temperature, equivalent to the duct, the maximum current was reduced to 7 kA. This causes obvious difficulty in the specification and use of this type of bushing.

8.5.6 Other tests

In addition to the major electrical tests discussed, tests or calculations are usually required to demonstrate the suitability of the bushing. These include:

(1)   Leakage tests, resistance to leakage by internal or external pressure of oil or gas;
(2)   Cantilever test, demonstration of the ability of the bushing to withstand forces imposed by connections, short circuit, self loads etc.;
(3)   Seismic withstand, usually demonstrated by static calculations, the effect of the stiffness of the equipment to which the bushing is mounted is also important;
(4)   Short circuit, again usually demonstrated by calculations, given in IEC137 1984, to prove adequate thermal capacity to prevent overheating and insulation damage during short circuit events.

8.6 Maintenance and diagnosis

Bushings are hermetically sealed devices generally operating in service under low electrical and mechanical stress. Ingress of moisture, however, due to gasket defects is a major cause of insulation deterioration. Internal partial discharge can result from moisture ingress, system overvoltages or inadequate stress control. External contamination build-up and the risk of pollution flashover can be reduced by periodic washing or the use of silicone rubber or grease coatings.

Dielectric diagnosis techniques can be applied to installed bushings [18]. On-line infrared scanning and radio influence voltage (RIV) measurement can detect thermal problems and corona. Off-line measurement of capacitance and tangent delta can be made on bushings and compared with factory results and other similar equipment. Information on bushing insulation should include ageing, moisture content and condenser breakdown.

A continuous tangent delta monitor for online transformer bushings has been developed [19]. Signals derived from test tapings on all bushings within a substation are compared and abnormal changes activate an alarm. The system is claimed to be cost-effective in high-risk situations.

With OIP insulation, dissolved gas analysis (DGA) offers an established technique for the assessment of insulation condition. Most published work refers to transformer insulation [20, 21] comparing the relative concentrations of fault gases. The Rogers ratios give an indication of the type of fault as shown in Table 8.3. These ratios are particular to transformers and their direct application to bushings has not yet been established. The different proportions of oil, paper and copper in transformers and bushings may alter the ratios significantly. By experience, admissible levels of gases have been proposed [23], see Table 8.4.

Today the trend is for the development of online diagnosis techniques to minimise the need for periodic line diagnosis in assessing insulation condition and thereby predicting fault development and extending equipment life.

8.7 References

1   IEC137 1984. "Bushings for alternating voltages above 1000V", (equivalent BS223: 1985)

2   DOUGLAS, J. L. and STANNETT, A. W.: "Laboratory and field tests on 132kV synthetic resin bonded paper condenser bushings". *IEE Proc. A*, 1985, **105**

3   BRADWELL, A. and BATES, G. A.: "Analysis of dielectric measurements on switchgear bushings in British Rail 25kV electrification switching stations". *IEE Proc. B*, 1985, **132**

4   IEC296 1982. "Claused mineral oils for transformers and switchgear", (equivalent BS148: 1984)

5   KALLINIKOS, A.: "Electrical insulation" (Peter Peregrinus, London, 1983, Chapter 11)

6   BARKER, H.: "High voltage bushings". Harwell high voltage technology course, 1968

Table 8.3 Interpretation of DGA

Gas ratio	Range	Code
	(<0.1)	5
$\dfrac{CH_4}{H_2}$	(>0.1,<1)	0
	(>1,<3)	1
	(>3)	2
$\dfrac{C_2H_6}{CH_4}$	(<1)	0
	(>1)	1
$\dfrac{C_2H_4}{C_2H_6}$	(<1)	0
	(>1,<3)	1
	(>3)	2
$\dfrac{C_2H_2}{C_2H_4}$	(<0.5)	0
	(>0.5,<3)	1
	(>3)	2

$\dfrac{CH_4}{H_2}$	$\dfrac{C_2H_6}{CH_4}$	$\dfrac{C_2H_4}{C_2H_6}$	$\dfrac{C_2H_2}{C_2H_4}$	Diagnosis
0	0	0	0	normal deterioration
5	0	0	0	partial discharges
½	0	0	0	slight overheating - < 150°C(?)
½	1	0	0	overheating - 150-200°C(?)
0	1	0	0	overheating - 200-300°C(?)
0	0	1	0	general conductor overheating
1	0	1	0	winding circulating currents
1	0	2	0	core and tank circulating currents, overhead joints
0	0	0	1	flashover without power follow through
0	0	½	½	arc with power follow through
0	0	2	2	continuous sparking to floating potential
5	0	0	½	partial discharges with tracking (note CO)

Table 8.4 DGA for bushings

	Gas	H_2	CO	CO_2	CH_4	C_2H_6	C_2H_4	C_2H_2
Max. admissable values in service (from - to)	at delivery	13 18	90 130	180 260	4 7	4 7	1.8 2.6	0.9 1.3
	after 5 years	200 300	315 435	720 1040	20 30	20 30	9 13	2 4
	after 10 years	100 500	500 650	1300 2200	30 80	28 30	12 15	5 15
repeat measurement	within 1 year	100 500	270 390	900 1300	30 500	30 100	10 20	4 8
(from - to)	Take out of service	300 1000	900 1300	1800 2600	70 200	70 200	25 40	10 15

7   IEC185 1986. "Guide for the selection of insulators in respect of polluted conditions"

8   HD506. "Bushings for liquid filled transformers above 1kV up to 36kV"

9   BOSSI, A. and YAKOV, S.: "Bushings and connections for large power transformers". Report No. 12-15, CIGRE, 1984

10   GALLAY, M., FOURNIER, J. and SENES, J.: "Associated problems, inspection and maintenance of two types of interface between EHV transformer and metalclad electric link insulated with SF_6" Report No. 12-01, CIGRE, 1984

11   MOSSER, H. P. and DAHINDEN, V.: "Transformer board II" (H. Weidmann AG, Rapperswil, Switzerland)

12   LAMPE, W., ERIKSSON, K. A. and PEIXOTO, C. A. O.: "Operating experience of HVDC stations with regard to natural pollution". Report No. 33-01, CIGRE, 1984

13   LAMBETH, P. J.: "Laboratory tests to evaluate HVDC wall bushing performance in wet weather". IEEE 90 WM 167-7 PWRD

14   LAMPE, W., WIKSTROM, D. and JACOBSON, B.: "Field distribution on an HVDC wall bushing during laboratory rain tests". IEEE 91 WM 125-5 PWRD

15   IEC60-1 1989. "High voltage testing techniques"

16   NATTRASS, D. A.: "Partial discharge detection in high voltage equipment" (Butterworth & Co., 1989)

17   KREUGER, F.: "Partial discharge detection in high voltage equipment" (Butterworth & Co., 1989)

18   "Dielectric diagnosis of electrical equipment for AC applications and its effect on insulation co-ordination" State of Art Report SC15/33, CIGRE, 1990

19   ALLAN, D. *et al.*: "New techniques for monitoring the insulation quality of in-service HV apparatus". *IEEE Trans. on Elect. Insulation* (1992), **27**, (3)

20   ROGERS, R. R. *et al.*: "CEGB experience of the analysis of dissolved gas in transformer oil for the detection of incipient faults". IEE Conference Pub. No. 94, 1973

21   IEC599 1978. "Interpretation of the analysis of gases in transformers and other oil filled electrical equipment in service", (equivalent BS5800: 1979)

22   "Reliability, surveillance and maintenance of high voltage insulation in power systems", State of Technique Report SC15, CIGRE Symposium, Montreal 1991

Chapter 9

Design of high voltage power transformers

A. White

9.1 Introduction

The tremendous development and progress in the transmission and distribution of electrical energy during the past 100 years may not have been possible but for the capability of linking the generator, the transmission line, the secondary distribution system and a great variety of loads, each part operating at its most suitable voltage [1]. This linking of systems at different voltages has relied upon a simple, convenient and reliable device — the power transformer. This unique ability of the transformer to adapt the voltage to the individual requirements of the different parts of the system is derived from the simple fact that it is possible to couple the primary and secondary windings of the transformer in such a way that their turns ratio will determine very closely their voltage ratio as well as the inverse of their current ratio, resulting in the output and input volt-amperes and the output and input energies being approximately equal.

9.2   Transformer action

A transformer essentially comprises at least two conducting coils having mutual inductance. The primary is the winding which receives electric power and the secondary is the one which may deliver that power as shown in Fig. 9.1. The coils are usually wound upon a core of laminated magnetic material and the transformer is then known as an iron-cored transformer. The modern iron-cored transformer has so nearly approached perfection that for many calculations it may be considered a perfect transforming device. In the simplest form of the theory of the transformer it is assumed that the resistances of the windings are negligible, the core loss is negligible, the entire magnetic flux links all the turns of the windings, that the permeability of the core is so high that a negligible magnetomotive force produces the required flux, and that the capacitances of the windings are negligible.   That is, the transformer is assumed to have characteristics approximating to those of an ideal transformer with no losses, no magnetic leakage and no exciting current. Thus, the instantaneous terminal voltage, v_1, is numerically identical to the instantaneous voltage, e_1, induced by the time-varying flux linkages which in turn is equal to the number of turns in the coil N_1 multiplied by the rate of change of flux linkages. Thus for the primary:

$$v_1 = e_1 = N_1 \frac{d\phi}{dt} \qquad (9.1)$$

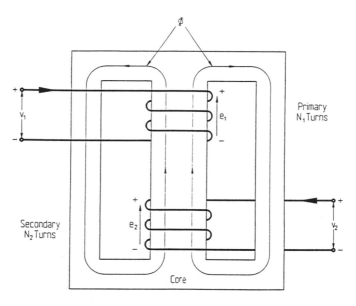

Fig. 9.1   *Schematic diagram of a transformer showing positive directions of voltages and currents*

For the secondary circuit similar criteria apply:

$$v_2 = e_2 = N_2 \frac{d\phi}{dt} \qquad (9.2)$$

The parameter ϕ is the resultant flux produced by the simultaneous actions of the primary and secondary currents and therefore:

$$\frac{v_1}{v_2} = \frac{N_1}{N_2} \tag{9.3}$$

Thus for an ideal transformer the instantaneous terminal voltages are proportional to the number of turns in the winding and their waveforms are identical.

The net magnetomotive force required to produce the resultant flux is zero and the net magnetomotive force is the resultant of the primary and secondary ampere turns; hence, if the positive direction of both primary and secondary currents are taken in the same direction about the core, then:

$$N_1 I_1 + N_2 I_2 = 0 \tag{9.4}$$

that is for an ideal transformer

$$\frac{I_1}{I_2} = -\frac{N_2}{N_1} \tag{9.5}$$

The minus sign indicates that the currents produce opposing magnetomotive forces.

The performance of a transformer depends upon a time-varying flux and therefore in the steady state a transformer operates on alternating voltage only. The transformer therefore is a device which transforms alternating voltage or alternating current or even impedance. It may also serve to insulate one circuit from another or to isolate direct current whilst at the same time maintaining alternating current continuity between circuits.

9.3 The transformer as a circuit parameter

In the simple theory it may be assumed that the transformer is electrically perfect. In a more comprehensive theory of its electrical characteristics, however, account must be taken of some "aspects which are not ideal" which occur in iron-cored transformers. First of all the windings have resistance; secondly the windings cannot physically occupy the same space and there is magnetic flux leakage between the coils of the transformer; thirdly an exciting current, albeit small, is required to produce the flux and, fourthly, there are hysteresis and eddy current losses in the core. Furthermore, where rapid rates of voltage change occur, then the capacitances relating to the windings can no longer be neglected.

When a transformer is supplied with power through a transmission circuit whose impedance is relatively high, the primary terminal voltage may vary with changes in load over an undesirably large range, because of changes in the impedance drop in the transmission circuit. Therefore in order to maintain the secondary voltage at its desired value under varying conditions of load and power factor, it is often necessary to provide tappings to vary the turns in one of the windings (Fig. 9.2). Large tapchanging power transformers are also frequently used to control the flow of reactive power between two interconnected power systems or between component parts of the same system, whilst at the same time permitting the voltages at specified points to be maintained at desired levels. Often the tapchanging apparatus is designed so that the ratio of transformation can be adjusted whilst the unit is on-load without interruption of the load current. Various types of on-load tapchanger are available, the most commonly used type being high speed resistor equipment. More will be related on this topic later in this chapter.

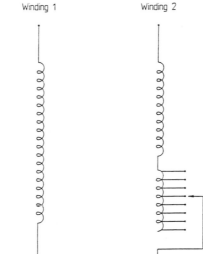

Fig. 9.2   Voltages regulation diagram

The voltage transformation effected by a transformer is accomplished at the expense of an exciting current. This has a power component which corresponds to the no-load loss which is practically all confined within the core, and there is a reactive component which corresponds to the reversible magnetisation of the core and is wattless. At no-load also there is a loss due to current travelling through the copper conductors. This is usually small. There is also a loss in the insulation which is usually minute.

The in-phase and reactive components of the excitation current depend upon the quality and thickness of the core steel used, upon the flux density, upon the frequency, upon the quality of the joints and upon the overall length of the magnetic circuit. It is totally independent of coil design. The desirable quality in a core is minimum values of both the power and reactive components. Of these two, the former is by far the more important consideration.

9.4 The core

One of the most significant improvements in transformer manufacture has been the ongoing development and introduction of higher grade core steels [2]. Prior to the turn of the century, soft magnetic iron materials were used for transformers (Table 9.1). Early in the century however, silicon steels were produced which gave very much reduced losses. Prior to the Second World War, metallurgists were able to produce steels in which the grains were oriented in the direction of rolling making them much easier to magnetise and thus reduce losses. By the early 1970s, Japan was beginning to lead the world in steel production and this included core steels. By clever metallurgical methods they were able to introduce a variety of low-loss steels which are known universally as "Hi-B". The technology was licensed to other countries including the UK. "Hi-B" steel is now in common usage throughout the world. In 1980 various steel companies, led principally by the Japanese, were producing ever thinner core steels. Whereas in 1960 most core steel used in the UK was 0.35 mm thickness, the Japanese have perfected 0.18 mm thick materials and even thinner materials are under development. However, the cost of the very thin materials, together with cutting and handling, becomes almost prohibitive and alternative methods of loss reduction have proved to be more economic. These methods include the use of laser or other scribing methods which break up the long grains of the "Hi-B" steels, thus allowing easier rotation of those grains, and make the steel much more easy to magnetise.

Table 9.1   Core material development

Date	Material
1885	Soft magnetic iron
1900	Non-oriented silicon steel
1935	Grain-oriented silicon steel
1970	Hi-B
1980	Thin Hi-B
1983	Laser scribed Hi-B

There was a further benefit noted in respect of the noise level of the transformers which indicated a 2 – 3 dBA reduction. Such a reduction is extremely important in Japan and is becoming increasingly important in Europe especially in densely populated areas.

The introduction of the grain-oriented materials has certainly reduced both the magnetising current and the loss components of current day cores. However, by making it easier for core materials to be magnetised along the grain, it usually means that it is more difficult to magnetise the materials across the grain. Inevitably, in any transformer design the flux must be persuaded to turn through angles, usually of 90°. Adjacent to joints between plates, the flux must change direction. This introduces localised power and quadrature components which can be substantially higher than those components which appear away from the joint. The method used to reduce the effects of change of flux direction is to mitre the joint at an angle. The optimum angle for conventional grain-oriented materials is normally recognised as being of the order of 55°. However, cutting at this angle increases the amount of waste material as the angle on the corresponding plate must be cut at 35°. The economic optimum angle of mitre is 45°, as the loss difference between 55° and 45° is small, and the waste is small.

As has already been stated, change of direction of flux introduces additional losses. Thus if cores are clamped by bolts passing through holes in the core plates, the flux must change direction to circumvent the holes and this introduces higher local losses in the bolt regions (Fig. 9.3). A large proportion of the transformer industry has therefore developed from bolted cores to banded cores. In general, this move has tended to reduce noise levels of transformers in addition to about a 3% improvement in the no-load loss.

Thus the transformer designer is faced with a large number of core steel grades (Table 9.2) but the better the grade the higher the price; the choice depends upon economics (Table 9.3). When the high grade, thin laser-etched "Hi-B" material first came onto the market, the author's company manufactured 4 transformers which were identical in design with one exception. That exception was that 2 transformers were produced with high 0.30 mm thick "Hi-B" steels which up to that time had been widely used, and the other two cores were built from 0.23 mm thick laser-scribed "Hi-B" material. The test results (Table 9.4) demonstrated reduction of 27% or thereabouts in the no-load loss.

Transformers are frequently of 3-phase construction although trackside transformers and those transformers for which weight or dimensions are critical features of the specification may require single phase designs. For 3-phase cores, a 3-limb design with each limb carrying windings may be used (Fig. 9.4). In this case the leg section and the yoke section area may be identical. Where transport height becomes critical or where the zero phase sequence impedance of the transformer is required to be equal to the positive phase sequence impedance, a 5-limb construction is often used. In this case the 3 inner legs of full section area are equipped with sets of windings and outer legs are provided to recirculate some of the flux. This reduces the magnitude of flux which must be transferred directly between phases and permits reduction in the yoke section areas. Economics provide the control thereafter. Some core designs will have the inner yokes of area 60% of that of the main wound limbs with the outer yokes and outer legs of 40% area. Other variations of 58/44%, 50/50%, 60/60% main to outer yoke section areas are commonly found amongst various manufacturers. For single phase designs, there are a large number of possible variations. These include a

Flux paths - general

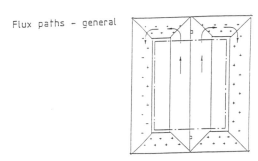

Flux paths around
bolt holes

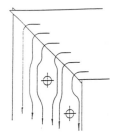

Fig. 9.3   *Core construction — mitres and bolt holes*

3-limb construction with only the centre limb wound, a 2-leg construction with both legs wound, a 4-limb construction with 2 limbs wound, and a complex cruciform core construction with the windings linking one leg on each of the cores forming the cruciform.

The losses in the core materials can be subdivided into two principal groups. Firstly there is the hysteresis loss which is inherent in the material and is a function of the metallurgy and of the flux density. The second component is the eddy loss which results from currents circulating within the laminations due to the flux which is passing through them. This is a function of the resistivity of the material, the dimensions, and of course, the flux density. The core steel manufacturers have worked upon the metallurgy to reduce the hysteresis components and on the thickness of the material to reduce the eddy components. The ultimate aim is to achieve high permeabilities and low losses at high flux densities.

The coreplates are usually cut on high speed, high accuracy automatic machines using sharp tools in order to ensure that burrs and slivers are eliminated so far as is possible. The plates must be built layer by layer, usually no more than two plates at a time, with alternate layers displaced in order to give a rigid mechanical system when the whole core is fully clamped. The core constitutes by far the heaviest component of the transformer.

Table 9.2   *Grades and losses of electrical steels*

Type	Grade	Loss, W/kg 1.7 T, 50 Hz	Relative loss, %
Conventional	35M6	1.44	127
	28M4	1.26	112
	27M3	1.18	104
	23M3	1.08	96
Hi-B	30M2H	1.13	100
	30MOH	1.04	92
	27MOH	1.02	90
	23MOH	0.94	83
Etched Hi-B	30ZDKH	0.97	86
	27ZDKH	0.92	81
	23ZDKH	0.85	75

Table 9.3   *Typical net worth of core steels for a single phase, 50 Hz generator transformer operating at a flux density of 1.7 Tesla. Base material 30M2H*

Type	Grade	Capitalisation at £3000/kW
Conventional	35M6	-900
	28M4	-390
	27M3	-130
	23M3	150
Hi-B	30M2H	0
	30MOH	265
	27MOH	330
	23MOH	370
Scribed Hi-B	30ZDKH	240
	27ZDKH	400
	23ZDKH	510

Table 9.4   *Comparison of tested parameters of 45 MVA transformers manufactured from two grades of core steel (Courtesy GEC Alsthom Transformers Ltd.)*

	Original transformer (30M2H)	Duplicate transformer (Z3ZDKH)
Load loss		
Unit 1	280 kW	280 kW
Unit 2	281 kW	281 kW
No-load loss		
Unit 1	20.1 kW	14.3 kW
Unit 2	19.7 kW	14.2 kW
Noise		
Unit 1	66 dB	64 dB
Unit 2	68 dB	65 dB

9.5 The windings

Windings also form an essential feature of any transformer. Various winding constructions may be used and in the UK, and indeed in most of Europe, the arrangement used is the core-type transformer with concentric windings. Essentially a core-type transformer is one in which the windings are assembled over a core leg. The second type of transformer is known as the shell-type in which the windings are normally of rectangular pancake construction which are sandwiched and encapsulated by the core. This type of transformer finds its greatest support in the USA although a number of other manufacturers throughout the world tend to construct to this principle. This chapter will concentrate on the core-type transformer which has virtually exclusive utilisation in the UK.

The type of winding used depends upon the current and the voltage. A very high current winding will normally be associated with a low voltage and hence few turns. The principal design intent is to accommodate as much of the conducting medium in the smallest physical space possible. In order to achieve this, the most straightforward and simple winding is the helical type in which each turn is wound successively against its neighbour (Fig. 9.5). As the windings become larger it becomes more difficult to remove the heat generated due to the losses within the winding and cooling ducts inevitably are required. The so-called spaced spiral winding therefore becomes necessary. This winding type is

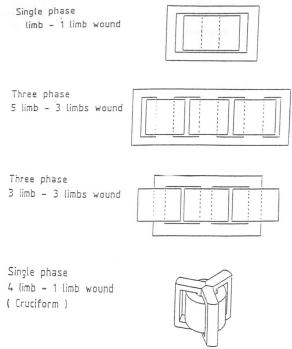

Single phase
limb – 1 limb wound

Three phase
5 limb – 3 limbs wound

Three phase
3 limb – 3 limbs wound

Single phase
4 limb – 1 limb wound
(Cruciform)

Fig. 9.4   Core types

essentially the same as the helical winding except that cooling ducts are introduced between the turns. As the voltage increases, the current tends to decrease and therefore the windings tend to be of smaller section conductors but of many more turns. Eventually, the economics of space utilisation demand a change from the simple helical or spiral construction to a disc arrangement. As the voltage increases further, the insulation between the conductors, especially the discs adjacent to the line end of the winding, reach the point where additional insulation becomes uneconomic to apply. The dielectric control within the winding may then be achieved by introduction of electrodes of controlled potential inserted within the winding. This latter type is known as the intershielded disc type winding. Ultimately further complications have to be introduced in the design and manufacture of the disc winding, and a feature known as the interleaved disc winding, invented by George Stern of English Electric in the early 1950s, now has worldwide appeal (Fig. 9.6). It must be noted that not all of the core-type transformer manufacturers utilise disc windings. A number of companies have continued the principle, once used by some UK manufacturers, of multiple layer construction.

Considering next the conductor material, it is recognised that the best electrical conductor is silver. However, the use of such material in transformers is totally

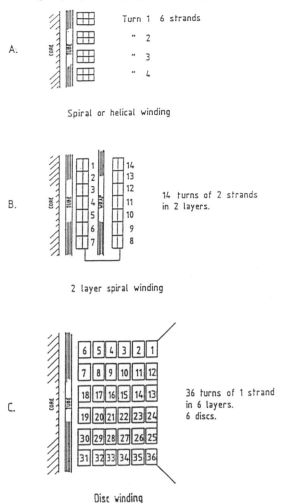

Fig. 9.5   *Types of transformer windings*

uneconomic and copper is the next best alternative.   There are also certain financial and sometimes technical reasons for the use of aluminium instead. The conductors themselves may be round, particularly where a very large number of turns with low voltages between them are involved. Power transformers, however, tend to use rectangular conductors. These conductors may come in a variety of forms, either individually as strip conductors, or in paired or triple format with reduced insulation between the individual conductors of the bundle whilst maintaining a high value of insulation between adjacent turns of the winding. The conductors may be laid side by side or end to end.   A further alternative is continuously transposed cable which comprises a number of rectangular copper

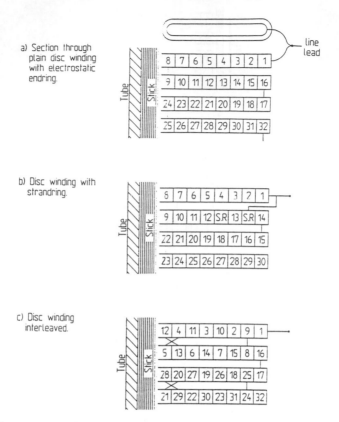

Fig. 9.6 *Methods of increasing series capacitance in disc windings*

conductors each insulated with an enamel, transposed with one another at regular intervals along the cable length, and the whole of the bundle wrapped with an insulating material to provide the interturn insulation (Fig. 9.7). Where very few turns are involved, the use of foils, either of copper or aluminium, may find preference.

Essentially a transformer requires two windings. However, there are frequently regulating windings which are usually associated with the high voltage winding. These regulating coils may on occasions simply be tappings brought out from the main high voltage winding or alternatively wound as a completely separate winding for reconnection with the high voltage winding at some point outside the winding assembly. Such a separate tapping winding may be wound concentrically outside the HV winding or concentrically inside the·LV winding (Fig. 9.8). In addition, the high voltage winding may be split into two parallel components with the tapping winding also split to permit the high voltage line lead to pass between the two halves. There are sometimes requirements to split the HV into series components in order to achieve a specific impedance or to reduce some of the

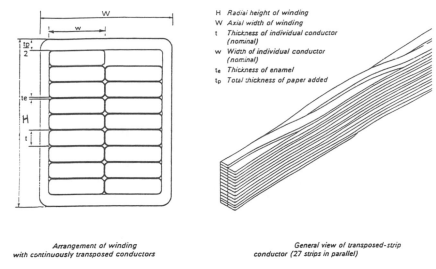

H *Radial height of winding*
W *Axial width of winding*
t *Thickness of individual conductor (nominal)*
w *Width of individual conductor (nominal)*
t_e *Thickness of enamel*
t_p *Total thickness of paper added*

Arrangement of winding
with continuously transposed conductors

General view of transposed-strip
conductor (27 strips in parallel)

Fig. 9.7   *Continuously transposed conductors*

force components under short circuit current flow conditions. This arrangement is known commonly as "double concentric". The arrangements may be further complicated by the addition of other windings and a locomotive transformer may typically have four primary high voltage windings, four secondary low voltage windings and five tertiary windings, all having specified impedances between each pair of windings thus controlling the physical location of each of the many windings with respect to any other [3].

9.6 Cooling systems

A winding carrying current will be subject to a loss in the conductor simply due to the resistance of that conductor. This component of loss is known as the I^2R loss. The current passing through the coil creates a magnetic field, and eddy currents are created in the conductors within that field, giving increased loss. This eddy loss is not uniformly distributed because the field is not uniform. Furthermore, if each conductor does not physically occupy a space in an identical field to that of any other conductor to which it is ultimately connected, the voltage induced in each interconnected conductor will be different from its neighbour and circulating currents will result. All of these load and parasitic currents create loss and this loss must be extracted and carried away from the windings to a position at which it can be dissipated. The commonly used cooling medium is transformer oil, although other synthetic fluids, gases, or even air, are found in a range of power and distribution transformers.

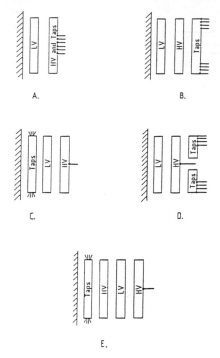

Fig. 9.8   Winding arrangements

In fluid systems there is a choice of flow conditions. Flow may be natural, i.e. the oil flow results from convection. Alternatively, a pump may be used to induce a higher rate of flow than simple convection will permit. This forced system will provide greater efficiency of heat transference. Within the winding there is the possibility of having nondirected flow or directed flow.

In the former case the flow is again determined solely by convection and in the latter case, the flow paths are predetermined by the designer to further improve the efficiency of heat transfer from the winding conductors to the fluid. The choice of natural or forced flow lies with the purchaser who selects the type of flow according to the local conditions. The choice of directed or non-directed flow within the windings of the transformer is usually the choice of the designer.

The oil flow velocities are also usually the choice of the transformer design engineer. There are no mandatory limits on oil velocity although each manufacturer will impose his own velocity limitations. Very high oil velocities especially with very dry oil conditions induce electrostatic charge which may be concentrated in local areas of insulation causing partial discharge and tracking and may ultimately lead to dielectric failure.

9.7 The insulation

The life of transformer insulation is generally determined by the highest temperature which appears within the winding — this highest temperature being known as the hotspot temperature. This usually appears near the top of a winding and probably in the first or second disc from the top. Of course, the actual hotspot position is totally a function of the design and manufacture. Only the transformer manufacturer has access to all the information to determine the position of the hottest spot, at the present time, although direct measurement methods are under consideration.

Just as it has been identified that transformer oil is the most common cooling fluid used in transformer manufacture, it is also an excellent insulant although dielectrically inefficient in large volumes. Conductor insulation is usually paper. Paper itself is not mechanically strong and when dealing with the heavy weights of copper involved in a winding or in the short circuit forces, a much more substantial insulation in the form of board is required. Major inroads have been made over the years in relation to the quality of insulation materials. The most commonly used insulating board material is made completely from wood. A number of companies, however, mix wood with cotton. By using the relatively long cotton fibres the oil impregnation of the pressboard material is usually easier to accomplish. The conservationists will also argue that cotton is an annual product and therefore its use is environmentally friendly and reduces the rate of depletion of the forests. Unfortunately, however, harvesting the cotton is a very labour-intensive operation and costs are rising. Other systems at lower ratings may use cast resin as the insulant and air as the coolant. Higher temperature systems are available using synthetic fluids which include silicone and are usually associated with higher temperature grade solid insulations such as Nomex and resin or polyester glasses [4, 5]. These materials are very much more expensive than paper and pressboard. In general however, they are mechanically stable in moist air, whereas paper and pressboard systems are subject to moisture absorption and dimensional changes.

9.8 The tank

Having considered the core, the windings, the insulation and the cooling fluid, it is obvious that another principal feature of the transformer must be the container used to encapsulate the fluid. As well as being called upon to meet the rigours of nature in its allocated environment for periods for 30, 40 or perhaps even 50 years, this tank is required to sustain the vacuum necessary for oil filling, to withstand the pressure exerted by the oil in normal operation, and to withstand the forces imposed during transportation at least twice in its lifetime. In addition, many of the forces which appear within the transformer under short circuit current conditions are effectively restrained to some degree by the tank. Tanks are therefore usually of mild steel but on occasions it may be necessary to introduce non-magnetic steels or other materials in order to reduce stray losses or the temperatures resulting therefrom.

9.9 The bushings

Connections are required between the system on the primary side and the primary winding of the transformer and between the secondary winding of the transformer and the secondary system. This interconnection requires a device which will permit the high voltage leads to pass through the tank which is normally at earth potential for safety reasons. This interconnection is achieved by the provision of bushings. Lower voltage bushings comprise a through stem which is centrally located in conjunction with an insulating system which has an external surface and which is mechanically coupled to a mounting flange which is then affixed to the tank opening. Such a bushing may be of the non-condenser type and for in-air use, may have epoxy resin major insulation or alternatively have oil contained within a shedded porcelain.

For higher voltages, a higher level of control of the voltage distribution is often required and condenser bushings are used. The major insulation is either oil impregnated paper or epoxy resin bonded paper, with the oil or the resin being introduced under vacuum. Very thin metallic foils are inserted at regular intervals within the layers of paper making up the major insulation. The inner foil is often the longest and is electrically bonded to the current-carrying conductor. The outer foil is usually the shortest and is electrically bonded to the fixing flange of the bushing. One or two of the intermediate foils may be electrically connected to bushing tappings which may be used either for check power factor measurements or to drive low power, voltage-measuring circuitry. In normal operation the power factor tapping is connected to earth.

A number of termination possibilities are available — oil/air bushings, oil/SF_6 bushings, oil/oil terminations for ongoing connection to a cable, oil/semi-fluid compound-type terminations or a moulded-type bushing which makes protected but direct connection to a cable.

9.10 On-load tapchangers

On-load voltage regulation has been referred to above. The device which performs this on a transformer is known as an on-load tapchanger. In the USA, reactor-type tapchanging is still available and is frequently used, but the rest of the world uses high speed resistor-type tapchanging. The on-load tapchanger is a device which includes a series of switches which operate in a pre-set sequence to permit the load current to flow throughout the operation [6]. The tapping which is adjacent to the one which is in service is preselected by the tap selector. The service tapping selector switch carries the load current. A switching device transfers the current flow from a direct connection to one which permits the current to be bypassed through a resistor. The switching movement then continues such that the selected and the preselected tapping are both permitted to share the load current and an additional circulating current created by the voltage difference between the selected and preselected taps is restricted in magnitude by two resistors in the circuit. The switch then continues to move to break contact with that resistor which is connected to the previously selected tapping. Finally, by further switch action, the second resistor is taken out of the circuit. The switch is designed to operate very quickly and the resistors are therefore only required to have short time ratings. It is therefore essential that the operation is not interrupted in order that the tapchanging equipment is fully protected. The complete sequence is therefore actuated by a motor drive mechanism which operates through geneva gears to operate the tap selector switches and simultaneously to wind up an energy spring mechanism which actuates the load current breaking or diverter switch. The diverter switch operating cycle is carried out only when the spring is released, independently of the motor drive mechanism. The actual current transfer time is of the order of 40 to 60 milliseconds, but the total time for one tapchange between initiation of the motor drive mechanism and the completion of the tapchange operation amounts to between 3 and 10 s depending upon the type and manufacture of the on-load tapchanging equipment.

High speed resistor tapchangers are classified into two groups, the first being the separate tank tapchanger and the second the in-tank tapchanger. As the name implies, the separate tank tapchanger divorces oil of both the selector and diverter tank from the oil in the main tank of the transformer. For the in-tank type tapchanger the selector switch operates within the main tank oil but the diverter oil is maintained in a separate system. Whilst the UK has had a general preference for the separate tank tapchanger, it must be recognised that the in-tank type tapchanger has a very much greater world population. The principle of in-tank tapchanger design offers greater flexibility in matching the tapchanging equipment to the transformer than does the separate tank tapchanger. Furthermore, the number of solid insulation surfaces which are under high dielectric stress in the in-tank design are limited with the critical surface being associated with the insulating drive tube. The major insulation is thus oil whose quality is easier to maintain and check. With the in-tank tapchanger, however, a certain amount of gas is produced by capacitance current sparking and many believe that this will disguise any gases which may emerge from the transformer core and windings. Analysis of hydrocarbon gases dissolved in the transformer oil is an important part of condition monitoring.

9.11 Design features

Transformer design is a complex process taking into account dielectric, electromagnetic, thermal and mechanical aspects, many of which are interrelated. The design engineer is, therefore, equipped with a number of calculation tools to confirm the capability of the transformer which s/he has designed.

9.12 Dielectric design

Dielectrically, the winding is designed to have certain duties. It must be satisfactory for the voltage conditions which appear in normal operation, and must also be capable of meeting any overvoltage conditions which are impressed upon it. This capability is demonstrated in the final test of the transformer as an induced overvoltage test at power frequency. Normally, the voltage which is induced in this test is at least twice the normal rated voltage between any parts. The transformer in service also has imposed upon it transient voltages in the form of lightning and switching impulses. The lightning impulse, in particular, is of very short duration (Fig. 9.9). Because of this, the capacitances of the windings have a much more predominant effect than at power frequencies. The voltage distributions within the winding are therefore determined by capacitance networks, at least for the first two or three microseconds of the impulse wave arriving at the terminal. The capacitance network of the transformer winding assembly gives a non-uniform voltage distribution throughout the coils (Fig. 9.10). The highest voltage drops appear at the turns of the winding which are closest to the impinging lightning impulse wave. If the voltage wave was a step function then ultimately a linear voltage distribution would occur. Between the initial distribution and the final distribution, therefore, considerable variations can appear at various parts of the winding. The effect is similar to that which would be achieved by holding a piece of elastic between the live and neutral terminals on a graph and then stretching the elastic to follow the curve of the initial distribution. On letting go of the elastic, oscillations are produced which eventually die to give the linear stretch between the two fixed points. The voltage distributions within the winding under stepped wave application give similar effects.

Usually, control of the amount of insulation on the conductors is determined by the capacitive distribution. Thus, over-insulation of the line end turns may be a cheap and viable solution. However, the feasibility of this method decreases with increasing voltage level. Other solutions then become necessary. The first is the utilisation of shields within the winding. These may either be of the form of covered rectangular conductors inserted between turns of the winding or alternatively as toroids having a metallic coating which are inserted between the discs. By interconnection of the shields with a part of the coil closer to the line end, the capacitance network may be modified and a more linear initial distribution obtained. The distribution is characterised by the value of the square root of the ratio of the shunt to series capacitances in the network. The closer this value is to zero, the more linear is the voltage distribution. In the winding design, variation of the series capacitance is often easier to accomplish than variation of the shunt capacitance. In the intershielded construction, the shields effectively increase the series capacitance especially at the line end. There are other methods of increasing the series capacitance, the most notable of which is the interleaved disc construction to which reference has already been made. The interleaving is accomplished by reconnecting the conductors so that the charge circulates through

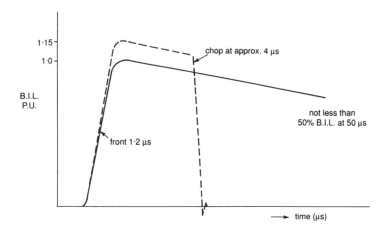

Fig. 9.9   *Standard impulse waveshape*

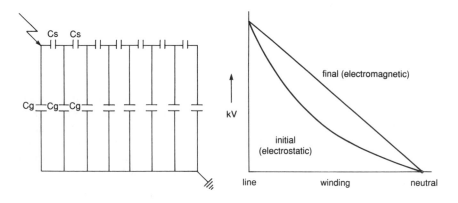

Fig. 9.10   (a) *Equivalent capacity network*
(b) *Impulse volts distribution*

each disc twice. The interconductor capacitances which act in parallel give a very high series capacitance and improve the degree of non-uniformity of the initial voltage distribution.

The enhancement may be taken further if there are a number of conductors electrically in parallel. It is then possible to achieve various combinations of capacitance throughout the coil. This is known as grading and by careful grading from fully interleaved through partial interleaving to no interleaving, i.e. continuous disc, the initial distribution may be even further improved.

The windings are required to be separated electrically and often mechanically. This separation is essentially by solid insulation materials but in order to introduce cooling oil to the windings, it is common for the insulation to be provided as a combination of solid and oil. In this respect, advantage is taken of the phenomenon whereby a small oil volume is capable of withstanding a greater stress than a large volume. In most major insulation, therefore, the solid insulations serve as barriers to break up the oil spaces into smaller volumes. Nevertheless, it is normally the oil which provides the weakest link. Under power frequency and also lightning impulse conditions, the voltage distribution within the major insulations is controlled by capacitive effects and hence by the relative permittivities of the insulating materials. For transformers which are connected to d.c. equipment and are therefore subject to d.c. voltages, it is the relative resistivities of the materials which provide the voltage control. Whereas the permittivity ratio of paper to oil is of the order of 2 to 1, the ratio of resistivities can achieve 100 to 1. The effect of this is that the stress is concentrated within the solid insulation with generally very low stresses appearing in the oil. However, where the solid insulation terminates abruptly, very high stresses may appear at the boundary between the solid and the oil. This could lead to breakdown commencing in the area. The d.c. field is thus very different from the a.c. field within a transformer. D.C. testing in the field on an a.c. transformer is therefore not a viable alternative to repeating the factory test conditions.

9.13 Electromagnetic design

A transformer is, of course, an electromagnetic device and in operation electromagnetic fields are induced. The effects of these fields are various. Eddy losses are induced in the conductors of the windings when carrying current. Electromagnetic fields exist outside the windings as well as within them and flux is attracted to, and perhaps concentrated within magnetic materials, whether these be the core structural steelwork, the tank or devices which are deliberately installed to affect the field.

The leakage flux produced by the equal and opposite primary and secondary ampere turns spreads out at the ends of the windings and circulates, some through the core and some between the outer winding and the tank and some through the tank and other metalwork (Fig. 9.11). The flux which impinges upon or travels within metallic components induces eddy currents and gives rise to loss and temperature increase. In order to control either temperature rise or the level of loss, or both, knowledge is required of the electromagnetic field. Analogue methods of field solution are possible but are extremely limited in application and inaccurate in operation. Rapid developments in computation have resulted in considerable flexibility, accuracy and improvements in response time. Mathematical solutions may be derived either by finite difference or finite element methods [7]. A considerable number of finite element analysis software packages are currently available. GEC Alsthom Transformers Limited at Stafford is a licensee of that which it considers the best available, and which can be used for very complex three-dimensional problems (Fig. 9.12).

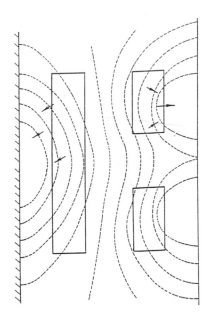

Fig. 9.11   Electromagnetic flux and force distribution

Knowledge of the electromagnetic field forms only part of the requirement; the capability of controlling such a field is also extremely important. Control of electromagnetic fields takes two forms. The first is flux attraction and the second is rejection. Each plays its part in the manufacture of large power transformers.

It has already been mentioned that the leakage field associated with load currents spreads out at the ends of the winding. The introduction of magnetic shields above and below the windings will effectively straighten the field at the ends (Fig. 9.13) and usually give some reduction in the eddy losses within the windings themselves. In addition, as these magnetic flux shunts are located between the windings and the metallic structural framework, they effectively shield the framework from impinging flux and reduce stray loss in the structural materials (Fig. 9.14). As the shunts provide an attractive sink for the flux, less flux impinges upon the tank, with resulting reduction in tank wall loss and tank heating. It is also possible to provide magnetic shunts on the tank wall. This has the purpose of directing the flux through laminated material giving lower losses than would result from direct impingement upon the tank wall.

The second method of electromagnetic flux control is by rejection. The provision of conducting shields, usually of aluminium or perhaps copper, has the

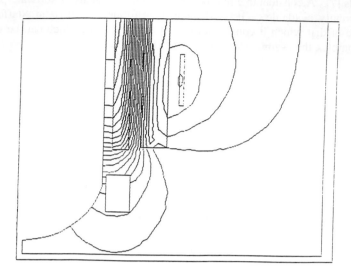

Fig. 9.12   Field plot of a typical transformer without flux shunts or rejectors

effect of preventing flux penetration, provided a sufficient thickness of material is available. Before the advent of sophisticated finite element modelling software, the use of rejectors presented the difficulty of determining where the rejected flux was actually transferred and consequential effects.

9.14 Short circuit forces

As every electrical engineer knows, when a conductor which is carrying current lies within a changing magnetic field, it experiences a force. This force is proportional to the square of the current. The transformer is required to be capable of withstanding through fault currents and hence the forces resulting from the peak value of those currents must be resisted and the design prepared accordingly. The action of the forces is complex but, for simplicity, they may be resolved into axial and radial components. The axial component of the force results from the radial component of the flux and vice versa. It may be shown that for a simple two winding transformer the axial force on a conductor is greatest at the ends of the coil and acts towards the centre of that coil. The summation of the forces gives rise to a compressive stress which is maximised at the coil centre. The radial component of the force acts on the inner winding radially inwards and is greatest at the centre of the coil. On the outer winding, it acts outwards thereby imposing a hoop stress on the conductors. Furthermore,

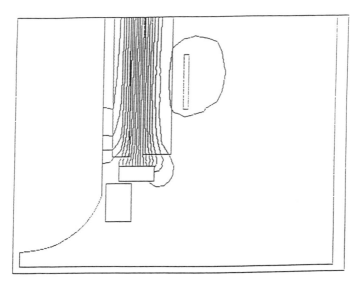

Fig. 9.13   Field plot of a typical transformer with winding end flux collectors

any unbalance of the ampere turns of the two windings, either by design intent or by manufacturing tolerances, will give rise to forces on the ends of the windings. Mechanical engineering takes over from electrical engineering in the design of the transformer in these respects.

9.15 Winding thermal design

The heat generated by the currents passing through the winding conductors must be extracted and transferred to some other part of the cooling system for dissipation. In order to achieve this, the winding is equipped with a labyrinth of channels which permit oil to come in contact with the covered conductors, extract the heat and, by induced convection or forced flow, carry it from the winding assembly (Fig. 9.15). The heat flow is affected by the proximity or otherwise of conductors to a cooling oil duct, the thermal conductivity through the paper, and the oil velocity over the exposed surfaces of the winding. For disc-type winding constructions, the bulk of the exposed surface of a coil usually lies in the horizontal plane. In order to achieve greater efficiency of heat removal, it is necessary, especially under forced oil conditions, to persuade the oil to flow past these surfaces. This is achieved by inserting oil restriction washers alternately at the inside and outside diameters of the coil. The number of horizontal ducts which are located between these restriction washers has an effect on the winding

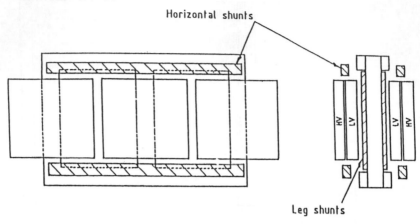

Fig. 9.14   Leakage flux shunts

to local oil gradient (Table 9.5). For forced oil flow conditions, the rate of flow also has an effect on the gradient. Increasing the flow beyond a certain level (which depends upon the particular design) has little effect on the gradient (Table 9.6).

High oil flow rates, as stated previously, may give rise to charge build-up with a phenomenon known as streaming electrification [8].

Table 9.5   Temperature effects of varying the number of disc coils per pass

Coils per pass	9	18
Winding hot spot rise above ambient air, °C	55.4	59.6
Winding average rise above ambient, °C	45.7	47.0
Hot spot to average difference	9.7	12.6
Winding oil rise difference	4.4	4.4
Winding over local oil	5.3	8.2

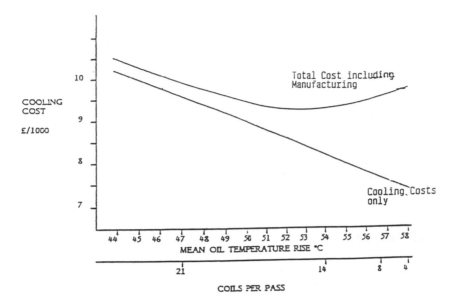

Fig. 9.15   Relationship between cost of cooling a typical transformer and the number of disc coils between oil restriction washers

Table 9.6   Temperature effects of varying oil flow rates in a typical transformer

Oil flow	+ 20%	Nominal	- 20%	- 50%
Average winding rise above ambient, °C	45.4	45.7	46.1	47.1
Hot spot winding rise above ambient, °C	54.1	55.4	56.9	61.5

9.16 Conclusion

The power transformer is a device which has been in use for more than 100 years. Its principles have not changed in that time, but the efficiency of material utilisation and the quality of those materials most certainly have. Core steels, solid insulation materials and insulating fluids have shown considerable strides in their development to fulfil the demands of the market which requires a very high level of reliability, more power per unit volume and weight and economy in operation. Transformer design has kept in step with these developments, increasing the efficiency of material utilisation by a more accurate and more effective calculation for every aspect of transformer manufacture, test and service. Continued investigation into transformer features and the wider application of sophisticated calculation methods confirm that transformer design is not a mystic art but a science. Experience is vital in leading the on-going development, but the benefits of new ideas emanating from unopinionated minds is essential in stepping up the gearing to produce ever more reliable units in shorter times and at reduced prices. Transformers may be an old idea but new brains will always be needed and welcomed in the industry.

9.17 References

1   ALLAN, D J.: "Power transformers — the second century". *IEE Power Engineering J.*, 1991, **5** (1)

2   DANIELS, M R.: "Modern transformer core materials". *GEC Review*, 1990, **5**

3   BENNETT, P C.: "Aspects of the design, construction and testing of BRB's class 90 and 91 locomotive transformers". International Conference on Main Line Railway Electrification, IEE Conference Publication No. 312, September 1989

4   SADULLAH, S. and WILLRICH, M.: "Performance characteristics and advantages of insulation systems in GEC cast resin transformers". Conference Publication of 6th BEAMA International Electrical Insulation Conference, May 1990

5   WHITE, A. and HOWITT E. L.: "Low flammability insulation systems for fluid filled transformers". Conference Publication of 6th BEAMA International Electrical Insulation Conference, May 1990

6   BREUER, W. and STENZEL, K.: "On-load tapchangers for power and industrial transformers — a vital apparatus on today's transformers". TRAFOTECH 82, International Conference on Transformers, New Delhi, India, February 1982

7   DARLEY, V.: "The practical application of FEM techniques in transformer design and development". ISEF 1991 Proceedings, International Symposium on Electromagnetic Fields, Southampton, September 1991

8   DAVIES, K.: "Static charging phenomena in transformers". Conference Publication of 6th BEAMA International Electrical Insulation Conference, May 1990

Chapter 10

Transformer user requirements, specifications and testing

J.A. Lapworth

10.1 Introduction

Transformers are required to transform the power output from electrical generators up to the voltages used in the transmission system (400 and 275 kV in the UK), to interconnect parts of the transmission system and to step down the voltage at bulk supply points and at various points in the distribution network before reaching the consumer. In addition, special transformers are required for a.c./d.c. converter stations and for transmission control devices such as quadrature boosters and static VAr compensators.

For a utility, large power transformers are major capital items, costing up to £2 M, with construction lead times up to 18 months. Although generally very reliable, when problems occur they are often difficult to diagnose and expensive to correct. For example, just to handle the oil from a transformer to allow an internal inspection, which very often can be inconclusive, can cost over £100,000 for a large transformer.

A utility is therefore concerned, as a customer, to ensure that as far as possible every new transformer purchased is capable of performing to requirements and will continue to do so for a service life of at least forty years, over the specified operating conditions and without being damaged by the inevitable occasional system abnormalities. The user requires a transformer which will be effective, efficient, reliable and also economical. Key activities in ensuring this are the specification for the transformer, quality assurance during manufacture, effective testing of the transformer before it leaves the manufacturer's works, and appropriate maintenance and diagnostic testing in service.

Several commercial and technical factors combine to colour the transformer procurement scene. First, transformers are generally not ordered in large batches, and each customer has different requirements, so there is seldom an opportunity to benefit from economies of scale. Secondly, it is usual for each new transformer order to go out to competitive tender, unless there are over-riding reasons why a repeat order is necessary, e.g. to meet a very short delivery time. This further reduces the possibilities of extended production runs. Thirdly, even though the basic technology of building transformers has changed little over the last 50 years, transformer manufacturers continue to refine their designs to take into account changing requirements, material advances or technical design improvements, with the result that even though a sizeable number of transformers may have been purchased to the same basic specification over a decade, there will be several different designs involved.

The overall outcome of all these factors is that transformers are effectively custom made rather than purchased off the shelf. In such situations the onus is even more than usual on the customer to ensure that his particular requirements are identified and met.

10.2 User requirements

10.2.1 Specific requirements

The fundamental specific performance parameters for transformers are:

Primary and secondary voltages and ratios	e.g. 400/132 kV
Tapping range	e.g. +15 % to -5 % in 14 steps
Rated power	e.g. 240 MVA
Impedance	e.g. 20 %
Connection	e.g. Auto (YNa0)

The principal parameters of the main types of power transformers connected to the UK Grid system are given in Table 10.1.

10.2.1.1 Voltage ratios
The voltages specified are for the no-load condition. Voltage ratios must be met to within ±0.5 % at every tap position. This is particularly important where transformers have to operate in parallel, otherwise large circulating currents can arise.

10.2.1.2 Impedances
The magnetising flux set up in the core of a transformer (Fig. 10.1) generates the voltage and current transformations. Winding currents also generate leakage fluxes which do not link both windings, and these are responsible for winding impedances.

The impedance of a transformer is largely reactive, the resistive components being very small. Impedance is useful in that it limits currents that will flow under short circuit currents, thereby providing protection from short circuit damage and enabling switchgear to operate within rating. The disadvantage of impedance is that it results in a reduction of secondary voltage with load (Fig. 10.2), an effect referred to as regulation. Following standard electrical engineering practice, the

Table 10.1   Classification of UK power transformers

Voltage ratio kV	Rating MVA	Type	Impedance % on rating	Tap range
Generator transformers				
18, 22 or 23.5/300 or 432	Up to 600 or 800	Delta/star	20	+5 to -15% of HV
Transmission transformers				
400/275	1000	Auto	16 or 20	-
"	750	"	12 or 20	-
"	500	"	12	-
400/132	240	"	20	+15 to - 5% of 132 kV
400/66	180	Star/delta	22.5	+5 to -15% of 400 kV
275/132	240	Auto	20	±15% of 132 kV
"	180	"	15	"
"	120	"	15	"
275/66	120	Star/star	20	±15% of 275 kV
275/33	120	Star/delta	20	±20% of 275 kV
Distribution transformers				
132/66	90	Star/star	17	+10 to -20% of 132 kV
"	60	"	12.5	"
"	45	"	"	"
132/33	120	Star/delta	22.5	"
"	90	"	"	"
"	60	"	15	"
132/11/11	60	Star/star	15	"
132/11	30	"	22.5	"

impedance is expressed on a per unit basis, in terms of the percentage volt drop caused.

During operation, depending on the loading and impedances involved, it may be necessary to change tap position to maintain the desired secondary voltage. Since impedance can vary with tap position, it is important that a change to a tap position with a higher secondary voltage does not introduce too much additional impedance, otherwise the effectiveness of the tap change can be severely reduced.

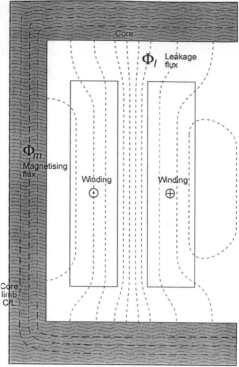

Fig. 10.1   Transformer magnetic fluxes

Also, and perhaps of greater importance in practice, it is necessary to ensure that where transformers operate in parallel, their impedances have similar characteristics, to avoid unequal power sharing between them. For these reasons it is now common practice to specify impedance envelopes, i.e. minimum and maximum impedances at the extremes of the tap range. Fig 10.3 illustrates the impedance envelope currently specified for NGC 275/66kV transformers, together with the characteristics of some new and old transformer designs.

For three phase systems, the zero phase sequence impedance of the transformer is also of importance, since this determines the magnitude of fault currents flowing between the neutral of a star-connected winding and earth if a single phase to earth fault occurs. Zero phase sequence impedance depends on whether the transformer has a three- or five-limb core, and whether a delta-connected tertiary winding is fitted. The user requires to know what the zero phase sequence of a transformer will be in order to select appropriate circuit breakers.

10.2.1.3 Tertiary windings
It has become common practice to specify that star-auto and star-star connected transmission transformers are fitted with delta-connected tertiary windings to provide one or more of the following:

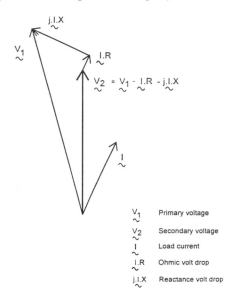

Fig. 10.2   *Transformer regulation phasor diagram*

- Stabilisation of the neutral point of unearthed transformers, or line voltages for earthed neutrals;
- Reduction of triple harmonic voltages in the lines and currents in the neutral/earth circuits;
- Reduction of zero phase sequence impedance to a known level, to have some control of earth fault currents;
- Reduction of current imbalance in primary phases resulting from unbalanced secondary loading;
- An auxiliary supply for loading or for reactive compensation.

Whatever the primary purpose, it has become practice to specify that all tertiary windings be capable of supplying an external load of 60 MVA at 13 kV, without detracting from the rating of the secondary winding, whatever the cooling state in operation, and of limiting the fault level at the tertiary terminals to 1250 MVA, assuming fault infeeds from the other terminals at their designed system maximum levels.

If the tertiary winding is not to be used for external loading, then only two tertiary terminals are brought out so that the delta can be closed and earthed at one corner externally to the tank.

10.2.1.4 Tap windings and tap-changers
When voltage tappings are required, there are often several possible arrangements of tap winding (e.g. line end, neutral end or linear, and reversing or coarse/fine). In the main there are no specific requirements limiting this choice. If neutral end

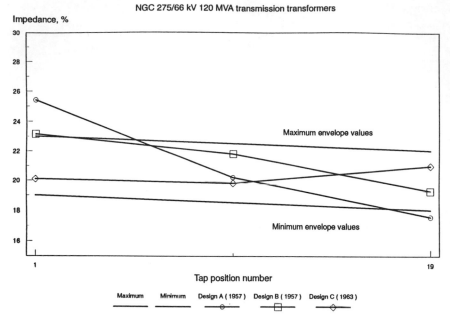

Fig. 10.3   Transformer impedance characteristics

taps are used in the design, then there may be a requirement to provide some boosting arrangement to ensure that tertiary winding voltages do not vary with tap position because of varying volts per turn on the core.

Where taps are required, a tapchanger must also be provided, and this is usually an on-load high speed device. Several designs are available to the transformer manufacturer. The utility may have a preference, based on previous experience or a need to restrict the number of different sets of spares held. One specific requirement that is common in the UK is for the tapchanger to be housed in a separate tank from the windings, or at least separated from them by an oil-tight compartment. The reason for this is that UK utilities rely heavily on dissolved gas analysis of main tank and tap selector oil for condition monitoring.

10.2.1.5 Rated power, temperatures and cooling

Load currents flowing in windings cause ohmic and stray loss heating. Maximum MVA is often determined by the necessity to limit winding temperatures to levels which will result in an acceptable insulation life. Since maximum MVA is temperature dependent, the rated power required by the user is defined for particular ambient conditions, typically 10 °C.

It is usual for the rated power to refer to steady state loading, i.e. continuous maximum rating, CMR. However, particularly for transmission and distribution transformers, the user is more concerned with variable loads. For this reason, in addition to the rated power required, a user usually specifies that the transformer also be capable of operation for specific load cycles and emergency overloads with limited periods above nameplate rating.

When the transformer is intended for operation at constant load, e.g. generator transformers, then only one cooling regime is normally required. For transmission and large distribution transformers the loading is often well below nameplate rating because of daily and seasonal variations and because more than one transformer normally operates in parallel for security of supply. In such cases, two or more cooler stages are required. It has become practice in the UK for transmission transformers to be required to have an "un-cooled" ONAN (oil natural circulation, air natural circulation) rating of 50% of their main OFAF (oil forced circulation, air forced circulation) rating. Switching between the two cooler stages is automatic.

10.2.1.6 Capitalisation of losses

Transformers are highly efficient devices, but nevertheless over the life of the unit the cost of the losses (up to 1000 kW for a 500 MVA unit) is significant. Losses can be reduced by increasing the amount of copper and iron, but this increases the capital cost of the transformer.

In order to arrive at the most cost-effective design, a utility calculates the sum required to pay for the expected losses over the life of the transformer, taking into account transformer loading, electricity costs, interest rates and projected movements in these quantities and derives £/kW figures for capitalising load and no-load losses which are specified in tender enquiries. Alternative designs offered by manufacturers are evaluated on the basis of a total capitalised cost.

The cost of losses can reach 30% of the total and is often a decisive factor in determining the most cost effective design.

10.2.2 General requirements

In addition to the specific requirements of a particular installation, there will be requirements common to all installations covering the following aspects:

— Transport limitations on weight and dimensions of the largest indivisible load arranged for transport;
— System fault levels;
— Electrical and safety clearances;
— Maximum core flux density;
— Maximum sound pressure levels;
— Short circuit performance;
— Overload performance;
— Protection practice and CT provision;
— Terminations and bushings;
— Cooler arrangements, oil pipework and valves;
— Ancillary components and fittings such as: winding temperature indicators, dehydrating breathers to limit the moisture uptake of the cooling oil,

pressure relief valves, gas in oil monitors, and marshalling kiosks with control, alarm and trip facilities;
—   Documentation.

10.3 Specifications and Standards

In the procurement process, the user's technical requirements are detailed in a technical specification, which with a contractual specification together form the tender and subsequent contract documentation on customer requirements. Technical specifications make reference to National, European or International Standards.

10.3.1 Standards

Two common dictionary definitions of the word *standard* are:

—   "a measure to which others conform or by which others are judged";
—   "a degree of excellence or minimum performance required for a particular purpose".

In this context neither definition is wholly complete, since the main purpose of transformer standards is to provide a common definition of technical terms and processes, without attempting to set acceptable or target figures for performance.
The main standards relevant to transformers and reactors are:

—   IEC 76 (BS 171) Power Transformers;
—   IEC 214 (BS 4571) On-Load Tapchangers;
—   IEC 289 (BS 4944) Reactors;
—   IEC 354 (CP 1010) Loading guide for oil-immersed transformers;
—   IEC 551 (BS 6056) Determination of transformer and reactor sound levels.

10.3.2 Specifications

Most large power transformers and reactors purchased in the UK in the last 30 years have been ordered to the *British Electricity Boards Specification for Transformers and Reactors (BEBS T2): 1966*, an electrical industry specification based on the practices of the UK electricity supply and transformer industries.
As is the case for any general technical specification, the aims were:

—   To ensure system needs were met;
—   To obtain technical uniformity;
—   To eliminate unsuccessful practices;
—   To satisfy maintenance, safety and environmental requirements;
—   To achieve economies.

In over 30 sections, BEBS T2, within the framework of BS 171 and other relevant standards, details general requirements, both for performance, ancillary components and facilities, provides a standardised format for the user to describe his specific requirements, and specifies test requirements and acceptance criteria.

In addition to providing detailed functional requirements, BEBS T2 also specifies in many key aspects how a transformer is to be constructed. For example, Section IX *Windings, Connections and Terminal Markings* specifies in clause 3.4 that:

> "No radial cooling duct shall be less than 3 mm thick and no axial duct containing strips supporting radial spacers less than 6 mm thick."

This necessity to control manufacturing practices has now largely been obviated by improved understanding of technical issues and the introduction of quality systems such as ISO 9000/BS 5750.

In recent years BEBS T2 has not always been used, either because of greater pressure to achieve a lower initial capital cost since the privatisation of the UK Electricity Supply Industry, or because it is not entirely compatible with EC Procurement Directives.

Because of this, users have now introduced new "functional" or "technical" specifications which cover the requirements of their own particular company. Most companies have reinforced this approach by retaining an approval procedure which includes a design review to ensure product compatibility with the user's needs, ensures that the equipment has been fully tested and that the manufacturer can consistently meet the necessary quality assurance standards.

10.4 Testing

When all manufacturing processes have been completed, tests are performed on transformers at the manufacturer's works as part of the procurement process for one of two purposes:

—   To prove the design meets requirements and to obtain transformer characteristics;
—   To check quality requirements have been met and that performance is within the tolerance guaranteed.

Tests performed for the former purpose are referred to as type tests (TT) and are performed on the first unit of a new design, and on selected repeat units where a large number of units of a given type have been ordered. Tests performed for the latter purpose are referred to as routine tests (RT) and are carried out on every unit manufactured. During works testing, special tests (ST) may also be performed to obtain information useful to the user during operation or maintenance of the transformer.

The tests to be performed on a particular transformer and the acceptance criteria are specified in the customer's purchase order and specification, and appropriate standards.

The following tests are performed at the manufacturer's works and form part
of the customer's acceptance tests:

—   Voltage ratio and vector group (RT);
—   No-load loss and magnetising current (RT);
—   Load losses and impedance (RT);
—   Zero phase sequence impedance (TT);
—   Tapchanger operation (RT);
—   Noise levels (RT);
—   Winding resistance (RT);
—   Temperature rise (TT);
—   Applied and induced over-voltage (RT);
—   Lightning (RT) and switching (TT) impulses;
—   Insulation resistance (RT);
—   Hydrostatic oil pressure (RT);
—   Pressure and vacuum (TT);
—   Ancillary components (RT and TT).

In some cases, the tests involve a simple check that the transformer has passed
some criterion. In others, quantities have to be measured for information or to
qualify for performance payments.

10.4.1 No-load loss and magnetisation current

The no-load losses are essentially the iron losses in the core due to hysteresis and
eddy currents set up by the main core magnetising flux, and are independent of
load current and dependent only on the core excitation.

No-load losses are measured by applying rated voltage at rated frequency to one
of the windings, i.e. HV, LV or tertiary, with all other windings open circuit.
Measurements are made at 90, 100 and 110% of rated voltage on normal tap and,
for auto transformers with neutral end taps, on maximum and minimum tap
position.

10.4.2 Noise levels

The basic cause of transformer noise is the magnetostriction of the sheets of
laminations forming the transformer core. The fundamental frequency is double
the system frequency, 100 Hz in the UK, but there are numerous harmonics.

Noise measurements are made as specified in IEC 551 with the transformer on
open-circuit at rated voltage. Sound level measurements are made at a number of
equally spaced points around the transformer, at a distance of 0.3 m away from a
contour formed by a taught string around all projections on the tank at a height
of 1.2 m above the ground.

Cooler noise, including fans and pumps, is a further sound source, and noise
measurements are made at 2 m from the cooler, and usually at two heights.

10.4.3 Load losses and impedance

Load losses include ohmic losses in the windings, together with winding eddy losses and other stray losses (in framework, tank etc.) induced by the transformer leakage fluxes.

Load losses are measured by short-circuiting the terminals of one winding (usually the LV) and applying sufficient voltage at rated frequency to the other winding to circulate rated current in both windings. The voltage required is a measurement of the impedance between that pair of windings. Measurements are made for HV/LV, HV/Tertiary and LV/Tertiary combinations, on maximum, normal, mean and minimum tapping positions, or on all tap positions for a type test. Load losses are usually measured at room temperature, and have to be corrected to the appropriate reference temperature, usually 75 °C.

10.4.4 Tapchanger operation

An on load tapchanger is operated up and down its full range with rated voltage applied on the LV or tertiary terminals.

10.4.5 Temperature rise

To prove that temperature rises comply to limits specified in IEC 76, and to derive thermal characteristics for the transformer, a heat run is carried out supplying full load losses for sufficient time to ensure that the temperature rises of the winding and oil reach steady state values. The "top oil" temperature is measured by a thermometer in a pocket at the top of the transformer tank, and this is used to verify that steady conditions have been reached. Final winding temperatures cannot be measured directly, and have to be determined by measuring winding resistances for about 15 minutes after excitation has been removed, and extrapolating back the cooling curves to derive the mean winding temperatures at shut-down (Fig 10.4). Standards describe how winding hotspot temperatures are to be calculated from the heat run test data.

Since with large transformers it is not practicable to supply both rated current and volts simultaneously, the accepted practice is to use the same test method as for a load loss test, but at an enhanced current, so as to supply full load losses to the windings until oil temperatures have stabilised. Once steady state oil rises have been determined, the excitation current is reduced to rated value and the test continued for a further hour, after which time the winding rises above oil, referred to as "winding gradient", will have reached their steady state.

Oil samples are taken from the main tank before and after heat runs for dissolved gas analysis (DGA) to detect whether any localised overheating has occurred.

Tank temperature rises are also measured during heat run tests, often with the aid of infra-red thermal imagers, to ensure that there are no excessive hotspots.

Temperature rise

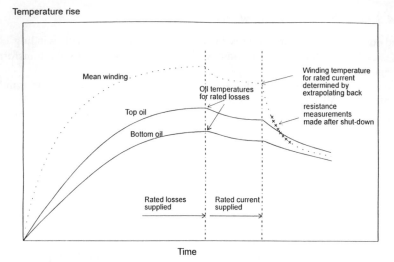

Fig. 10.4   Heat run test temperatures

10.4.6 Induced and applied overvoltage

The induced overvoltage test is intended to verify the power frequency withstand strength between turns and adjacent winding parts, along the winding, between connections, between windings and between phases.

An alternating sinusoidal voltage is applied to the terminals of one phase winding with the other windings open circuit (Fig. 10.5) so as to achieve the test voltages specified. The test is carried out at an increased frequency (up to 400 Hz) to avoid core saturation.

IEC 76 recognises two different methods for performing overvoltage tests. NGC adopts Method 1, in which the duration of the test is between 15 and 60 seconds, depending on the test frequency, with test voltages as given in Table 10.2. During the induced overvoltage test, partial discharge measurements are made at the line terminals of each phase at 1.2 and 1.6 p.u. of the rated phase to earth voltage before reaching the full over-potential voltage, at that level, and again at 1.6 and 1.2 p.u. while reducing the voltage. The intention of the partial discharge measurement is to ensure that the transformer is discharge free at system highest voltage. Any significant discharges detected have to be investigated to the satisfaction of the customer.

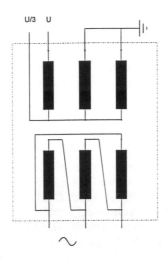

Fig. 10.5   Induced overvoltage test arrangement

Table 10.2   NGC test voltages

Rated voltage between phases (kV)	400	275	132	66	33	13	11
(a) Minimum Lighting Impulse Voltage withstand (Full Wave) kV, peak	1425	1050	550	325	170	150	95
(b) Minimum Induced Overvoltage withstand kV r.m.s.	630	460	230	132	66	26	22
(c) Minimum Applied Voltage withstand kV r.m.s.	38	38	38	140	70	50	38
(d) Minimum Switching Impulse withstand kV, peak	1050	850	-	-	-	-	-

The Method 2 test involves a longer test period of about 35 minutes at somewhat lower test voltages.

Applied over-voltage tests are intended to verify the power frequency withstand strength of a winding under test to earth and to other windings. A sinusoidal single-phase voltage of not less than 80 % rated frequency is applied to the winding for 60 seconds. The test is passed if no collapse of the test voltage occurs.

10.4.7 Lightning and switching impulses

Lightning and switching impulses are intended to simulate the types of high frequency transient overvoltages a transformer is expected to withstand in service. Transients such as these produce high stress concentrations at the winding ends.

Lightning impulse tests are made using a standard 1.2/50 μs waveshape (Fig. 10.6), which may be chopped to simulate the effect of a protective gap flashover. A switching impulse has a 175/2500 μs waveshape. Tests are made by direct application of the impulse to each line terminal in turn, with the transformer set up as in service (but without attached busbars) (Fig. 10.7), on the most onerous tap position, as previously determined by RSO tests.

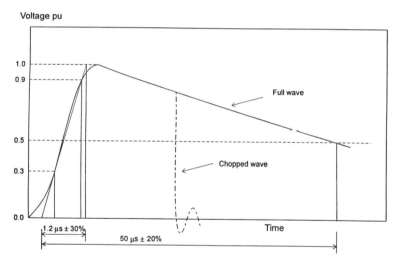

Fig. 10.6   Lightning impulse wave shape

The test sequence required by NGC for routine tests on every unit, at the voltage level specified in Table 10.2, is:

(i)   1 Reduced full wave (between 50 and 75 %);
(ii)   1 Full wave (100 %);
(iii)   2 Chopped waves (115 %);

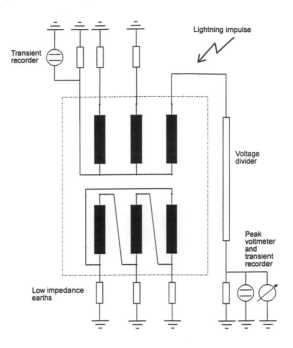

Fig. 10.7   Lightning impulse test arrangement

(iv)   2 Full waves (100%);
(v)   1 Reduced full wave (at same level as first impulse).

Transient recordings of the applied voltage and of the current at the neutral end of the winding under test are taken.

The transformer passes the test if there is no evidence of complete or incipient failure from audible indications or changes in the voltage or current records.

10.5 Concluding remarks

Current UK practice concerning specifications and testing is based on experience gained over almost 50 years on high voltage transformers of 275 kV and over. During this time there have been few major changes in the transformer procurement scene. In the main such practice has served very well in that transformers purchased have proved to be very reliable over an extended life.

With the recent changes in the structure of the electricity supply industry, and new EC procurement directives there are now new pressures to review current practice to achieve immediate capital savings. The challenge is to achieve this without endangering longer term benefits.

Chapter 11

Basic testing technique

E. Gockenbach

11.1 Introduction

Generally, tests are necessary to show that the equipment under test fulfils certain specified requirements. Depending on different factors the tests can be routine tests or type tests, and the following parameters have to be taken into account:

— Regulations;
— Recommendations;
— Mutual agreements on technical specifications;
— Experiences;
— Economy.

This chapter deals only with high voltage testing technique, requirements and recommendations, without consideration for different types of test, experience or economic factors.

11.2 Insulation coordination

An IEC recommendation, Publication 71: "Insulation Co-ordination", describes the relationship between the different test voltages:

— Alternating, at power frequency;
— Lightning impulse;
— Switching impulse.

The general definitions and requirements for the relevant test voltages are given in "Recommendations for High Voltage Test Technique"; IEC Publication 60 - 1, and, when the general conditions cannot be fulfilled, in relevant Apparatus Recommendations, e.g. for transformers, cables. A transformer may influence the standard switching impulse, due to its saturation effect, and so a suitable switching impulse for transformer testing is recommended. The normally large capacitance of a cable causes a longer front time, and so the tolerances are increased for such a case. It is necessary to define the main parameters regarding the test techniques in order to make the tests clear and to ensure that the test results are comparable and reproducible.

The following definitions have been adopted for the purpose of insulation coordination:

Nominal voltage of a three-phase system
The r.m.s. phase-to-phase voltage by which the system is designated and to which certain operating characteristics of the system are related.

Highest voltage of a three-phase system
The highest r.m.s. phase-to-phase voltage which occurs under normal operating conditions at any time and at any point of the system.

Highest voltage for equipment
The highest r.m.s. phase-to-phase voltage for which the equipment is designed in respect of its insulation as well as other characteristics which relate to this voltage in the relevant equipment standards.

External insulation
The distances in air and the surfaces in contact with open air of solid insulation of the equipment which are subject to dielectric stresses and to the effects of atmospheric and other external conditions such as pollution, humidity, vermin etc.

Internal insulation
The internal solid, liquid or gaseous parts of the insulation of equipment which are protected from the effects of atmospheric and other external conditions such as pollution, humidity, vermin etc.

Indoor external insulation
External insulation which is designed to operate inside buildings and consequently not exposed to the weather.

Outdoor external insulation
External insulation which is designed to operate outside buildings and consequently exposed to the weather.

Self-restoring insulation
Insulation which completely recovers its insulating properties after a disruptive discharge caused by the application of a test voltage; insulation of this kind is generally, but not necessarily, external insulation.

Non-self-restoring insulation
Insulation which loses its insulating properties, or does not recover them completely, after a disruptive discharge caused by the application of a test voltage; insulation of this kind is generally, but not necessarily, internal insulation.

Type test
A test made on one piece of equipment or on several similar pieces intended to show that all pieces of equipment made to the same specification and having the

same essential details would pass an identical test; it is usually not repeated on different deliveries.

Routine test
A test to which each piece of equipment is subjected.

Overvoltage
Any time-dependent voltage between one phase and earth or between phases having a peak value or values exceeding the corresponding peak value ($U_m \sqrt{2}/\sqrt{3}$ or $U_m\sqrt{2}$ respectively) derived from the highest voltage for equipment.

Switching overvoltage
A phase-to-earth or a phase-to-phase overvoltage at a given location on a system, due to one specific switching operation, fault or other cause, the shape of which can be regarded, for insulation coordination purposes, as similar to that of the standard impulse used for switching impulse tests. Such overvoltages are usually highly damped and of short duration.

Lightning overvoltage
A phase-to-earth or a phase-to-phase overvoltage at a given location on a system, due to a lightning discharge or other cause, the shape of which can be regarded, for insulation coordination purposes, as similar to that of the standard impulse used for lightning impulse tests. Such overvoltages are usually unidirectional and of very short duration.

Statistical switching (lightning) overvoltage
Switching (lightning) overvoltage applied to equipment as a result of an event of one specific type on the system (line energisation, reclosing, fault occurrence, lightning discharge etc.), the peak value of which has a probability of being exceeded which is equal to a specified reference probability. This reference probability is chosen as 2% in the standard of insulation coordination.

Temporary overvoltage
An oscillatory phase-to-earth or phase-to-phase overvoltage at a given location of relatively long duration and which is undamped or only weakly damped.

Temporary overvoltages usually originate from switching operations or faults (e.g. load rejection, single-phase faults) and/or from non-linearities (ferro-resonance effects, harmonics). They may be characterised by their amplitude, their oscillation frequencies, and by their total duration or their decrement.

Statistical switching (lightning) impulse withstand voltage
The peak value of a switching (lightning) impulse test voltage at which insulation exhibits under specified conditions a probability of withstand equal to a specified reference probability. This reference probability is chosen as 90% in this standard. The concept of statistical withstand is at present applicable only to self-restoring insulation.

Rated switching (lightning) impulse withstand voltage
The prescribed peak value of the switching (lightning) impulse withstand voltage which characterises the insulation of an equipment as regards the withstand tests.

Rated short duration power-frequency withstand voltage
The prescribed r.m.s. value of sinusoidal power-frequency voltage that the equipment shall withstand during tests made under specified conditions and for a specified time usually not exceeding 1 minute.

Rated insulation level
(a)   For equipment with highest voltage for equipment equal to or greater than 300 kV: the rated switching and lightning impulse withstand voltages.
(b)   For equipment with highest voltage for equipment lower than 300 kV: the rated lightning impulse and short duration power-frequency withstand voltages.

11.3 Test voltages

A test voltage is normally defined by its amplitude and shape within specified tolerances. The characteristics of some test objects sometimes make it necessary to change or modify the shape of the test voltage, in order to make the tests possible.

11.3.1 Direct voltage

A direct voltage is defined as the mean value between the highest and lowest level within a period. The length of the period depends on the generating system. Fig. 11.1 shows a typical example of a direct voltage with voltage drop ΔU and ripple δU.

The variation of the instant amplitude is given by the ripple, which is defined as the half value of the difference between the highest and lowest values. The maximum value of the ripple is $\pm 3\%$ of U_{gl} according to IEC Recommendation 60. The upper line in Fig. 11.1 represents the peak value of the supply voltage U_T and the lower line the mean value of the direct voltage, which is a fraction of the test voltage.

The main parameters which influence the value of the ripple are load current, frequency of the generating system and the capacitance, assuming that the voltage is generated by a high voltage rectifier. Fig. 11.2 shows a simplified equivalent circuit of an N-stage rectifier.

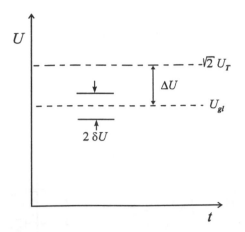

Fig. 11.1   Direct voltage U_{gl} with voltage drop ΔU and ripple δU

The approximate calculation of the expected ripple value can be made using the following equation:

$$\delta U = \frac{I}{(f\,C)\dfrac{N}{4}(N + 1)} \tag{11.1}$$

where

$\quad I \;=\;$ load current
$\quad C \;=\;$ smoothing capacitance per stage
$\quad f \;=\;$ frequency
$\quad N \;=\;$ number of stages.

Fig. 11.3 gives the relationship between the number of stages, the maximum load current and the output voltage. Measures to reduce the ripple can be deduced from equation (11.1) depending on the requirements and conditions of the other parameters.

The voltage drop is another effect of a d.c. generator, which influences the mean value of the voltage. It also depends on the design and parameters of the generator and can be calculated according to the following equation:

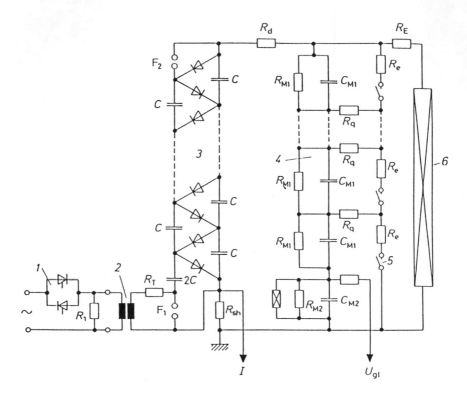

Fig. 11.2   Basic circuit diagram of a simple N-stage testing rectifier
1 - Thyristor controlled voltage regulation
2 - High voltage transformer
3 - N-stage rectifier with diodes, charging and smoothing
  capacitors
4 - Direct voltage divider with grading capacitors
5 - Discharging and grounding device
6 - Test object

$$\Delta U = \frac{I}{(f\,C)\dfrac{N}{3}\left(2N^2 + 1\right)} \tag{11.2}$$

The voltage drop will also be increased by the reduction in the output voltage of the feed transformer due to its short circuit impedance. For test voltage purposes the voltage drop is not a problem, because it can be taken into account by the design of the voltage generator [1].

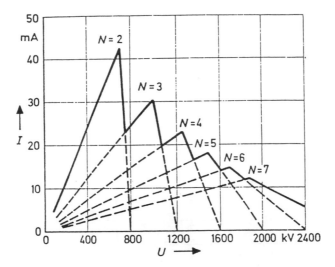

Fig. 11.3 Load diagram of a simple cascade rectifier with 3% ripple, N = number of stages

The d.c. pollution test of an insulator is a very good example to show the practicability of test requirements. For low load currents, the influence of the ripple and the voltage drop is more or less negligible assuming a reasonable number of stages. For high load currents, particularly high current impulses, control of the voltage amplitude is more difficult. The change of current influences the ripple, the voltage drop and the voltage change after breakdown. In order to keep the voltage within a given limit, in this case defined as a mean voltage drop, the control system and the components of the generator must be designed according to the test requirements and the test object performance.

Fig. 11.4 shows the calculated and measured output voltage of a d.c. generator during a pollution test [2].

11.3.2 Alternating voltage

An alternating voltage is defined by its r.m.s. value and/or its peak value. Assuming a pure sinusoidal wave form, the relation between r.m.s. and peak value is given by the square root of 2. This relation is also the base of definition in IEC Recommendation 60, where the deviation of the wave form is given by the equation:

$$\frac{U_{peak}}{U_{eff}} = \sqrt{2} \pm 5 \% \tag{11.3}$$

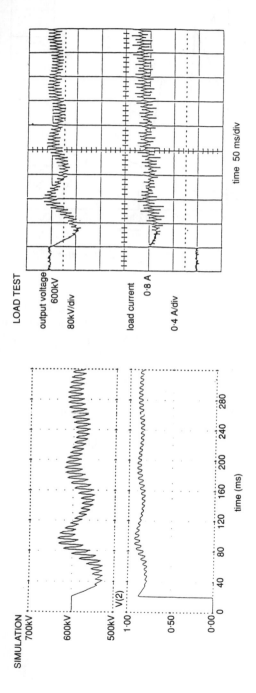

Fig. 11.4 Simulation and measurement of voltage and current during a pollution test

Depending on the test purposes, the r.m.s. or the peak value is more important. For dielectric tests of flashover behaviour the peak value of the alternating voltage is the main parameter, but for load tests, including thermal behaviour, the r.m.s. value is relevant.

The wave form of an alternating voltage can also be defined by the ratio of the basic frequency and harmonics of different order. This may cause some problems for the measuring technique, but with a digital recording system and intelligent software such problems can be solved.

The generation of alternating voltage is usually by a transformer or a cascade transformer. Fig. 11.5 shows a cascade arrangement and the equivalent circuit diagram with the primary windings (1), secondary windings (2) and tertiary windings (3).

The short circuit impedance and the change of harmonics depending on the load are the important parameters for the test transformer. The short circuit impedance, normally given as a percentage, represents the voltage drop under full load conditions, where the resistive part is negligible compared with the inductive part. The short circuit impedance of a cascade transformer cannot be easily calculated, but as a rough approximation it can be mentioned that the impedance of the total system increases more than linearly with the number of stages.

The transformers of each stage are identical and therefore the maximum output power is given by the power of one stage. With a combination of two transformers in parallel at the bottom stage, the output current can be increased at reduced output voltage, assuming three identical transformers. This feature is sometimes helpful for high voltage tests, when different test objects require different voltage and current combinations.

Another important factor is the voltage increase due to a capacitive load. Fig. 11.6 (a) shows the simplified transformer circuit and Fig. 11.6 (b) the vector diagram with capacitive load.

The same physical behaviour of a transformer test circuit can be used to benefit the voltage increase by a so-called resonance test circuit. In this case, the circuit will be changed by a variable inductance or by the feeding frequency at a constant capacitive load in order to get a high output voltage at resonance conditions. This a.c. generating method is very often used for cable testing, where the capacitive load is so high that compensation by an additional inductance is not economical for a normal transformer test set-up.

11.3.3 Impulse voltage

An impulse voltage is defined by its peak value and its time parameters. Fig. 11.7 shows the standardised lightning impulse voltage with the peak value, the front time T_1 and the time to half value T_2.

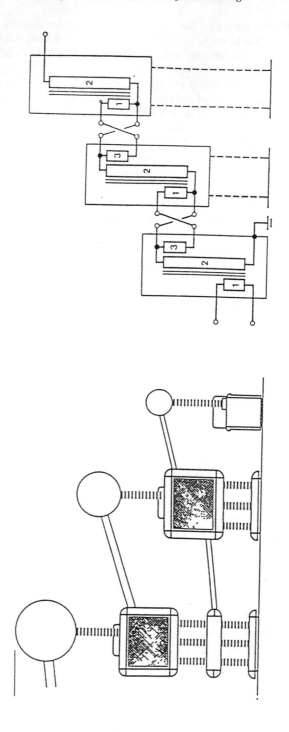

Fig. 11.5 Equivalent circuit diagram for a cascade transformer

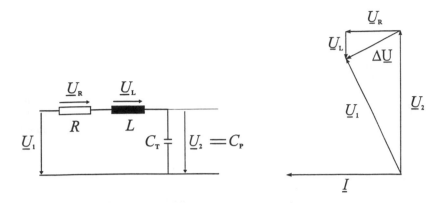

Fig. 11.6 (a) *Simplified transformer circuit*
(b) *Vector diagram of the simple transformer circuit*

U_1 - Input voltage (referred to the secondary side of the transformer)
U_R - Voltage drop across the resistor
U_L - Voltage drop across the inductor
U_2 - Voltage across the capacitive load
I - Transformer current
R - Resistor
L - Inductor
C_T - Capacitor (representing the transformer)
C_P - Capacitor (representing the test object)

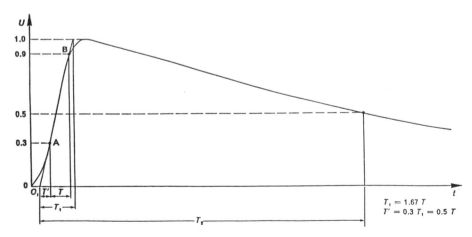

Fig. 11.7 *Standard lightning impulse*

The front time T_1 is 1.67 T, where T is the measured time difference between the 30% and 90% voltage level. The time to half value T_2 is the time difference between the virtual zero point O_1 and the 50% voltage level in the tail of the wave shape.

The tolerances of front time (30%) and time to half value (20%) are quite large, because the test results are not strongly affected by variation of the time parameters. On the other hand, although the adjustment to an impulse generator for different impulse shapes is not difficult it may be very time consuming, if exact time parameters are required. Fig. 11.8 shows the simplified equivalent circuit diagrams without inductance for an impulse test.

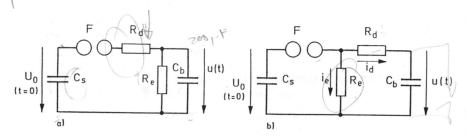

Fig. 11.8   *Simplified equivalent circuit diagrams*
   U_o -   *Charging voltage of impulse capacitor C_s at $t = 0$*
   F -   *Sparking gap*
   R_d -   *Damping resistor*
   R_e -   *Discharging resistor*
   C_b -   *Load capacitor*

The most commonly used configuration is connection (b), because there is no voltage reduction due to connection of damping resistor R_d.

The front time is mainly determined by the resistor R_d and the load capacitor C_b and the time to half value by the resistor R_e and the capacitors C_b and C_s. For a 1.2/50 μs standard lightning impulse calculation of the time constants can be made using the equations:

$$T_f \approx R_e (C_s + C_b) \tag{11.4}$$

$$T_r \approx \frac{R_d \ (C_s C_b)}{(C_s + C_b)} \tag{11.5}$$

The relation between the front time T_1 and the time to half value T_2 and the time constants T_f and T_r depends on the wave shape. For a 1.2/50 μs standard lightning impulse the following relations are valid:

$$T_1 = 2.96 \ T_f \qquad (11.6)$$

$$T_2 = 0.73 \ T_r \qquad (11.7)$$

The generation of an impulse voltage with a high amplitude is normally with a Marx generator. The principle of such a generator is parallel charging and series discharging of the capacitors. Fig. 11.9 shows a schematic diagram of the circuit.

The main problem for reproducing an impulse voltage is the behaviour of the switches, which connect all the capacitors in series. The requirements for the test impulse voltage are to have the same peak value and same wave shape for a given circuit and charging voltage. To reach these requirements two methods of triggering the generator are possible, assuming a simple sphere-gap electrode configuration is used as switch.

With method 1 the distance between the spheres is increased so that no sparkover occurs during the charging period. After reaching the required charging voltage, the spheres are moved and the gap reduced until sparkover occurs at the sphere with the smallest spacing distance. This is normally the sphere gap at the first stage of the generator. Triggered by this flashover, all other sphere gaps will spark over. With method 2 the distance between the spheres is slightly greater than is needed to withstand the charging voltage. When the charging voltage reaches the preselected value and is constant throughout the generator stages, a trigger impulse at the first stage causes a sparkover and the other gaps to breakdown. The second method ensures high reproducibility and reliability, which is really necessary for some tests.

With the same generator a further normalised impulse, the 250/2500 switching impulse can be generated. The time to peak T_p is 250 μs with a tolerance of 20% and the time to half value 2500 μs with a tolerance of 60%. Fig. 11.10 shows the impulse shape with T_p, T_2 and the time above 90% of the peak T_d.

The relationship between the time to peak and time to half value and the time constants is:

$$T_p \approx \frac{(T_r \ T_f)}{(T_r - T_f) \ \ln\left(\dfrac{T_r}{T_f}\right)} \qquad (11.8)$$

$$T_2 \approx T_r \ \ln\left(\frac{2}{\eta}\right) \qquad (11.9)$$

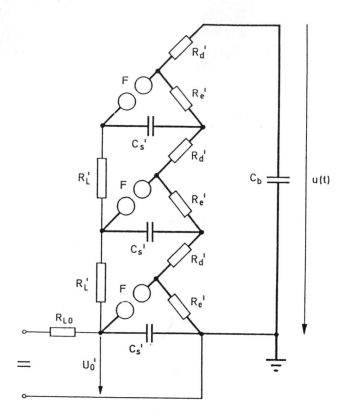

Fig. 11.9   Schematic diagram of an impulse generator

R_{LO} -   *Charging resistor for the first stage*
U_o' -   *Charging voltage per stage*
R_L' -   *Charging resistor per stage except for the first stage*
F -   *Sparking gap*
R_e' -   *Discharging resistor per stage*
R_d' -   *Damping resistor per stage*
C_s' -   *Impulse capacitor per stage*
C_b -   *Load capacitor*

where η = efficiency factor or the ratio between output voltage and charging voltage.

The values of impulse capacitance and load capacitance are important for the efficiency factor of an impulse generator. A good estimation for the output voltage for a lightning impulse is given by the following equation:

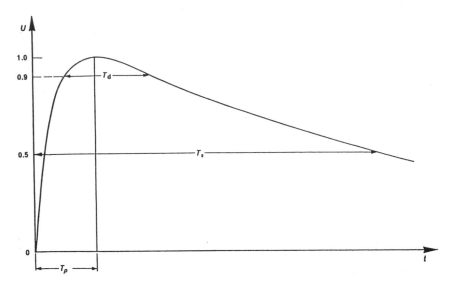

Fig. 11.10   Standard switching impulse

$$U_{out} = U_{charging} \left(\frac{C_s}{C_s + C_b} \right) 0.95 \qquad (11.10)$$

The factor 0.95 takes into consideration the influence of the resistors. For switching impulses, this influence is greater and therefore the factor can be as low as 0.50.

Also of significance are the oscillations at the front or near the peak of an impulse voltage. The oscillations occur only with lightning impulses, when the resistance of the whole test circuit is small. Fig. 11.11 shows some examples of typical lightning impulses, to explain the definition of a mean curve, overshoot and peak value for test purposes according to IEC Publication 60.

The oscillations can be reduced by increasing the resistance, but this leads to an increase in front time. Depending on the inductance of the whole test circuit the minimum value of an aperiodic damped circuit is given by the equation:

$$R \geq 2 \left(\frac{L \left(C_s + C_b \right)}{C_s C_b} \right)^{1/2} \qquad (11.11)$$

For large test circuits or test arrangements with a high inductance the standard lightning impulse cannot be generated. Fig. 11.12 shows the relationship between the resistance and total load capacitance for the 1.2/50 standard lightning impulse

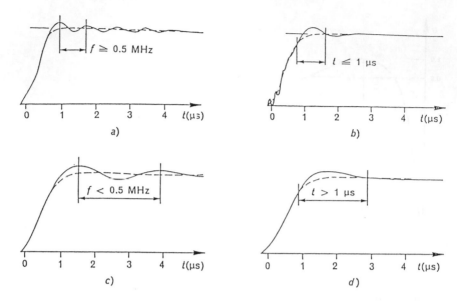

Fig. 11.11   Typical lightning impulses with oscillations and mean curves

with a tolerance of ±30% for front time and 5% for overshoot [3].

A typical example of how the test object can influence the wave shape is the impulse test on large cables. Due to the high capacitance, the standard lightning impulse cannot be generated without very large oscillations. Therefore, the IEC recommendation allows a longer front time, up to 5 μs, in order to get a reasonable and reproducible wave shape. On the other hand, change of the front time has no influence on the dielectric test results on cables tested with lightning impulse voltages.

11.4 Test current

An impulse current is defined by its peak value and time parameters. Fig. 11.13 shows an example of an impulse current; with the definition of the front time T_1, the time to half value T_2 and undershoot i_1 [4].

It should be noted that the definition of front time is not identical with that for impulse voltage. T_1 is 1.25 T, where T is the measured time difference between the 10% and 90% current level. T_2 is the time difference between the virtual zero point O_1 and the 50% current level in the tail of the wave shape. Furthermore, the impulse current typically has an undershoot, which is also defined by its amplitude in order to make impulse and test results reproducible. The generation of an impulse current is similar to that for impulse voltage, but normally the capacitors

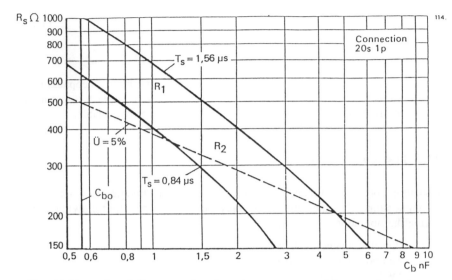

Fig. 11.12   Load diagram of an impulse test system, 20 stages in series

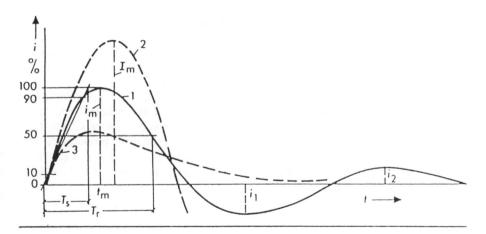

Fig. 11.13   Standard impulse current

are connected in parallel during charging and discharging. More effort is necessary to get the required front time, particularly for short front times, due to the fact that the front time depends on the rate of change of the relevant current.

The steepness at the beginning of the impulse is given by the equation:

$$\frac{di}{dt} = \frac{U}{L}$$ (11.12)

It is clear that a lower inductance of the test circuit gives a higher steepness to the current impulse. A higher voltage is normally related to a larger test circuit and consequently higher inductance, and this is not helpful or reasonable.

11.5 Test conditions

Insulation coordination includes defining the electrical strength of equipment and its application. The dielectric stress may be divided into the following classes during the operation:

— Power frequency voltages;
— Temporary overvoltages;
— Switching overvoltages;
— Lightning overvoltages.

For a given voltage stress the behaviour of internal insulation may be influenced by its degree of ageing, and that of external insulation by the degree of atmospheric conditions. A correction of the test voltage due to ageing is not possible and therefore no correction factor exists. Test under atmospheric conditions should include correction of the test voltage according to the actual atmospheric conditions.

The standard reference atmosphere is a temperature of $20\,°C$, a pressure of $b_0 = 101.3\,kPa$ and an absolute humidity of $h_0 = 11\,g/m^3$. The atmospheric correction factor K_t is divided into two parts, the air density correction factor k_1 and the humidity correction factor k_2. The disruptive discharge voltage is proportional to the atmospheric correction factor K_t that results from the product of k_1 and k_2. The air density correction factor k_1 depends on the relative air density δ and can be expressed by the equation:

$$k_1 = \delta^m$$ (11.13)

where δ is given by the equation:

$$\delta = \left(\frac{b}{b_0}\right)\left(\frac{273 + t_0}{273 + t}\right)$$ (11.14)

and *m* is given in Fig. 11.14 as a function of *g*, which depends on the type of predischarges.

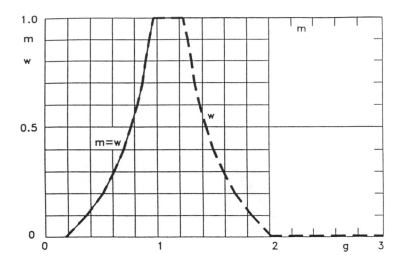

Fig. 11.14   *Values of exponents m for air density correction and w for humidity correction as a function of parameter g*

Equation (11.15) shows the definition of *g* to be:

$$g = \frac{U\,B}{(500\,L\,\delta\,k)} \tag{11.15}$$

where *UB* is the 50 % disruptive discharge voltage at the actual atmospheric conditions (in kilovolt), *L* the minimum discharge path (in metres) with the actual relative air density δ and parameter *k* according to Fig. 11.15.

The humidity correction factor k_2 is expressed as:

$$k_2 = k^{\,w} \tag{11.16}$$

It should be noted that the exponents *m* and *w* are still under consideration (within IEC) and that the values given in Fig. 11.14 are approximations (see also Section 14.6).

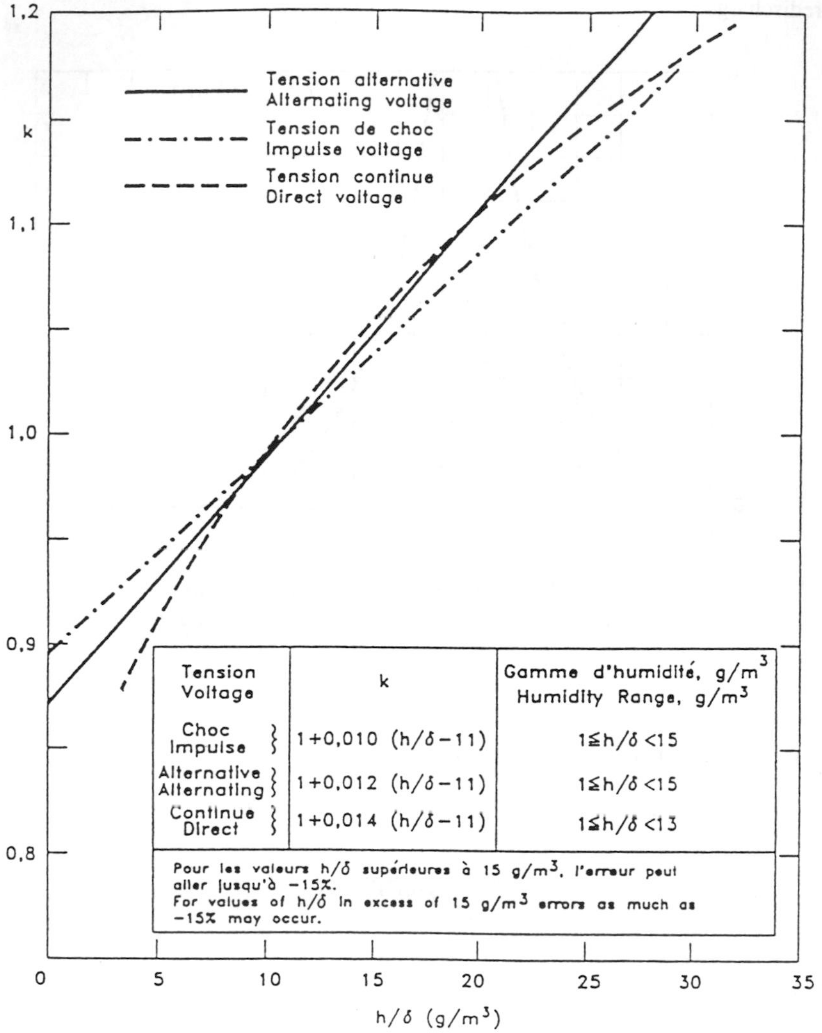
1,2
k
1,1
1,0
0,9
0,8
Tension alternative
Alternating voltage
Tension de choc
Impulse voltage
Tension continue
Direct voltage
Tension
Voltage
k
Gamme d'humidité, g/m3
Humidity Range, g/m3
Choc
Impulse
1+0,010 (h/δ−11)
1≦h/δ<15
Alternative
Alternating
1+0,012 (h/δ−11)
1≦h/δ<15
Continue
Direct
1+0,014 (h/δ−11)
1≦h/δ<13
Pour les valeurs h/δ supérieures à 15 g/m3, l'erreur peut aller jusqu'à −15%.
For values of h/δ in excess of 15 g/m3 errors as much as −15% may occur.
0
5
10
15
20
25
30
35
h/δ (g/m3)

Fig. 11.15 *k as a function of the ratio of absolute humidity h and relative air density δ*

Selection of dielectric tests, according to the recommendations, is different for voltage ranges up to and above 300 kV. Tables 11.1 and 11.2 show the values for a volt[illegible]ange between 52 and 300 kV, where the performance is checked in [illegible] a short duration power-frequency test and a lightning impulse test.

For voltages above 300 kV, performance under power-frequency operating voltage and temporary overvoltages can be checked by a long duration test in order to show that the equipment is suitably designed regarding ageing and pollution. Performance under switching impulse is checked by a switching impulse test, and the performance under lightning impulse by a lightning impulse test. It is important to know that the rated switching impulse withstand voltage, in p.u., goes down with increasing voltage level. The same tendency exists also for the lightning impulse withstand voltage.

Table 11.1   Standard insulation levels for 52 kV $\leq U_m \leq$ 300 kV

1	2	3	4
Highest voltage for equipment Um (r.m.s.)	Base for p.u. values Um $\sqrt{2}/\sqrt{3}$ (peak)	Rated lightning impulse withstand voltage (peak)	Rated power-frequency short duration withstand voltage (r.m.s.)
kV	kV	kV	kV
52	42,5	250	95
72,5	59	325	140
123	100	450	185
145	118	550	230
170	139	650	275
245	200	750	325
		850	360
		950	395
		1050	460

Table 11.2   Standard insulation levels for $U_m \geq 300$ kV

1	2	3	4	5	6
Highest voltage for equipment U_m (r.m.s.)	Base for p.u. values $U_m\sqrt{2}/\sqrt{3}$ (peak)	Rated switching impulse withstand voltage (peak)		Ratio between rated lightning and switching impulse withstand voltage	Rated lightning impulse withstand voltage (peak)
kV	kV	p.u.	kV		kV

1	2	3	4	5	6
				1.13	850
		3.06	750	1.27	
300	245			1.12	950
		3.47	850		
		2.86		1.24	1050
362	296			1.11	
		3.21	950		
		2.76		1.24	1175
420	343			1.12	
		3.06	1050	1.24	1300
		2.45		1.11	
525	429			1.36	
		2.74	1175	1.21	1425
				1.10	
				1.32	
		2.08	1300	1.19	1550
				1.09	
				1.38	
765	625	2.28	1425	1.26	1880
				1.16	
		2.48	1550	1.26	1950
				1.47	2100
				1.55	2400

11.6 References

1   REINHOLD, G.: "Höchstspannungsgleichrichter für große Ströme", *Bulletin SEV*, 1975, **66**, pp. 751 - 756

2   GOCKENBACH, E., *et al.*: "Some aspects of DC pollution tests". 5th International Conference on AC and DC Power Transmission, London 1991 pp. 331 - 336

3   Impulse voltage generator, Bulletin E-114, (E. Haefely & Cie Ltd.), Basel, Switzerland

4   MODRUSAN, M.: "Berechnung von Stroßstromkreisen für vorgegebene Impulsströme". *Bulletin SEV*, 1976, **67**, pp.1237 - 1242

Chapter 12

Basic measuring techniques

E. Gockenbach

12.1 Introduction

The measurement of voltage and current in high voltage tests is difficult because amplitudes are so high that they cannot be measured directly with conventional measuring and recording systems. Furthermore, not only the peak value but also the shape of the measured signal should be measured and evaluated, particularly for impulse voltage and current, and this requires an adequate recording system using either an oscilloscope or a digital recorder.

12.2 Measuring system

A high voltage or high current measuring system generally consists of a converting device, a transmission device and a recording device. An optimised measuring system has components which fulfil the required performance and are similar in their characteristics. The measuring system is as good as the weakest part of the system, and therefore it is not necessary to require exceptional performance from any one of the components.

The converting device should reduce the amplitude of the signal to be measured to a value which is suitable for the transmission and recording devices. The output signal of the converting device should be an exact replica of the input signal in terms of wave shape and time parameters. This requirement is critical in many cases and therefore measurement errors should be estimated and evaluated carefully.

In measuring systems the converting device is usually the key component due to its physical size and transfer behaviour. A simple and typical test to check transfer behaviour is measurement of the unit step response. Assuming the input voltage or current is an ideal unit step, the output signal is deformed by the converting device. Fig. 12.1 shows a simplified output signal of a voltage divider [1].

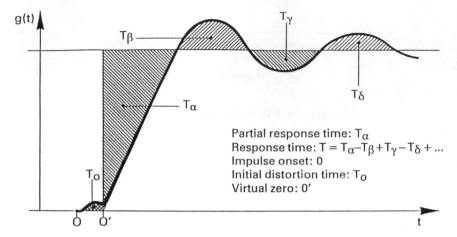

Fig. 12.1 Simplified output signal of a voltage divider for a unit step input

The parameters of the output signal are the partial response time T_α, the response time $|T|$, the settling time t_s and the overshoot β, which are shown in Fig. 12.2 [2]. These parameters describe the transfer behaviour and their values define the performance of the converting device.

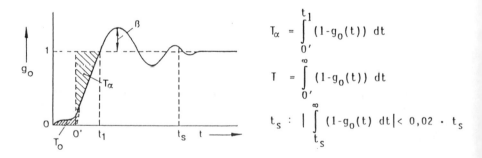

Fig. 12.2 Unit step response parameters

The requirements for the transformation ratio are stability, independence of frequency, linearity and accuracy. Depending on the converting device, not all requirements are fulfilled satisfactorily and therefore an estimation of the measuring uncertainty is necessary. The transmission system is generally a coaxial cable, which does not influence the amplitude but, in particular cases, does influence the transfer behaviour of the system. The recording device very often has an additional converting device on the input side in order to reduce the amplitude further. This converting device is small and normally has sufficiently good transfer behaviour. The transformation characteristic is normally not critical but it should be checked and included in the evaluation of the whole system. Depending on the parameter to be measured the recording device can be a simple peak voltmeter, an oscilloscope or a digital recording system. Because a digital system is characterised by its sampling rate, the bandwidth or sampling frequency can be used to evaluate the performance of the recording device. Fig. 12.3 shows the amplitude content of different components of a high voltage or current measuring system [3].

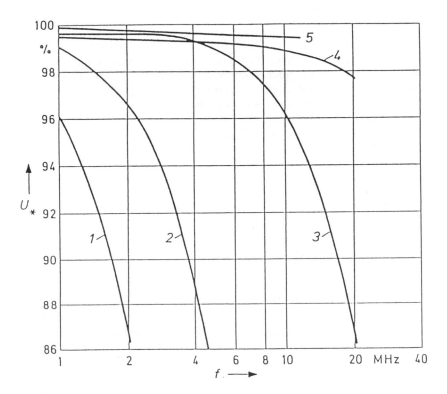

Fig. 12.3   Frequency response of measuring system components
    1 - Shunt
    2 - Divider
    3 - Oscilloscope
    4 - 8-bit recorder
    5 - 10-bit recorder

Because the transfer behaviour of the divider or shunt determines the transfer behaviour of the system, the influence of the recording devices, oscilloscope or digitiser, is negligible, if a reasonable performance is available from the digitiser.

An important problem for all measuring systems, particularly at high voltage or high current, is sensitivity to electromagnetic interference. Some measures to prevent or suppress disturbance are discussed.

The most critical tests are impulse tests, where impulse generation is at the same time a source of radiation of an electromagnetic wave. This wave penetrates the whole system — the converter, the transmission and the recording device — and not all components can be shielded. Therefore, it is necessary to reduce the amplitude of the radiation and to increase the signal level in order to get a very high signal to noise ratio.

The voltages and currents are another reason for electromagnetic interference, because they can be induced by capacitive or inductive coupling or by loops of the measuring cable.

The measures to take against these effects are proper shielding or, in special cases, double shielding of the measuring and control cables and the prevention of loops by star connection of all measuring and control cables. Fig. 12.4(a) shows a simple example of a bad connection and Fig. 12.4(b) a good cable connection.

There are some cases where a cable loop cannot be avoided. A typical example is the residual voltage measurement of a metal oxide arrester. In this case the high magnetic field induces in the loop (consisting of the test object, divider connection, voltage divider and earth potential) a voltage, which is superimposed on the residual voltage. Fig. 12.5 shows the schematic diagram [4].

The earth connection of the divider cannot be changed for certain other reasons and therefore one possibility is to compensate the magnetic field by a suitable connection of the voltage divider. The measuring systems should be designed according to the requirements of the measuring uncertainty concerning the transformation behaviour or amplitude measurement and the transfer behaviour or time parameter measurement.

12.3 Amplitude measurements

The amplitude is the main parameter of a high voltage or high current test, and therefore it should be measured and checked for each test. The time parameters depend mainly on the voltage or current generator configuration and will normally not be changed during one test sequence. Some checks of the time parameter at the beginning and the end of a test sequence may be sufficient under normal conditions.

12.3.1 Direct voltage

The amplitude of a direct voltage can be measured very simply with a high ohmic resistor. The current through the resistor is proportional to the voltage. The measuring instrument is assumed to have a negligible resistance. Fig. 12.6 shows the circuit diagram.

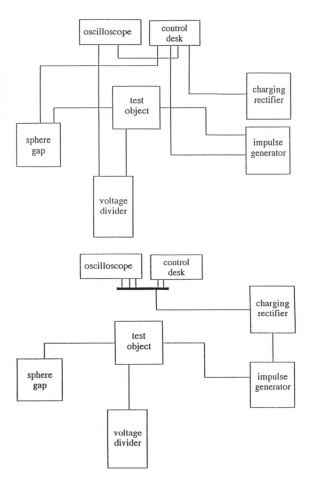

Fig. 12.4 (a) Cable connections with many loops
(b) Cable connections without loops

The surge arrester in parallel with the measuring instrument is very important because, in case of disconnection of the measuring cable or flashover of the high voltage resistor, the full voltage will be applied to the measuring instrument and/or the operator.

The resistor normally consists of many elements connected in series. The whole chain can be fixed on an insulating tube in order to get a reasonable mechanical strength and a certain distance between high and ground potential. To prevent a large measuring error, the minimum required current through the resistor is 0.5 mA according to the IEC 60 Recommendations. A resistance of $10^{12}\,\Omega$ for the insulating material loads the supply at 1000 kV to a current of 1 μA, which is 0.2%

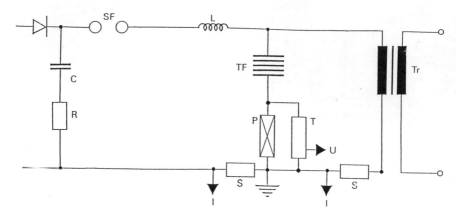

Fig. 12.5   *Schematic diagram for a surge arrester test*

C - *impulse capacitor*
R - *damping resistor*
SF - *sparking gap*
L - *inductor*
TF - *isolating gap*
P - *test object*
T - *voltage divider*
S - *shunt*
Tr - *voltage transformer*
U - *voltage measurement output*
I - *current measurement output*

of the total current. For small measuring uncertainties the temperature coefficient has to be taken into account. Fig. 12.7 shows an example [5].

The variation of the total resistance between - 15 °C and 50 °C is less than 0.05 %. Such a small deviation can only occur if the resistor elements are carefully selected, according to temperature coefficient, and combined in such a way that the temperature coefficient for a stack of resistors fulfils the requirement. Each resistor element may have a larger coefficient than that allowed for the complete chain.

The influence of the temperature coefficient and voltage coefficient can be compensated by a voltage divider, if the elements in the high voltage and low voltage arm have the same characteristic. Fig. 12.8 shows the simplified equivalent circuit diagram.

It should be noted that the measuring instrument needs a very high input impedance in order to prevent any influence on the transformation ratio. The direct voltage very often has a ripple. Depending on the measuring instrument the reading is the mean arithmetic value (moving-coil instrument) or the r.m.s. values (static voltmeter). This means that the instantaneous value of the high voltage can be much higher than the measured mean arithmetic value as shown in Fig. 12.9.

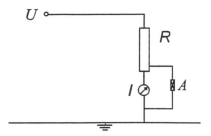

Fig. 12.6   Direct voltage measurement with high ohmic resistor

U - *high voltage*
R - *high ohmic resistor*
I - *current through the resistor*
A - *surge arrester*

A further problem of a resistive divider is the non-linear voltage distribution during a fast voltage change. A measure against this effect is the capacitive grading of the resistive divider. Fig. 12.10 shows an equivalent circuit diagram with the grading capacitors C_p and the stray capacitors C'_E. Depending on the ratio between C_p and C_E the distribution of transient voltages is non-linear as shown in Fig. 12.10.

12.3.2 Alternating voltage

A resistor or resistive divider can also be used for alternating voltage measurements, but an additional error is the phase shift due to the capacitance and inductance. Normally, a capacitor instead of a resistor will be used for alternating voltage measurements. Fig. 12.11 shows the circuit diagram with a capacitor and measuring instrument.

The measuring uncertainty depends strongly on the harmonic content. The r.m.s. value of the voltage is given by the equation:

$$U_{eff} = \left(U_1^2 + U_3^2 + U_5^2 + ... \right)^{\frac{1}{2}} \qquad (12.1)$$

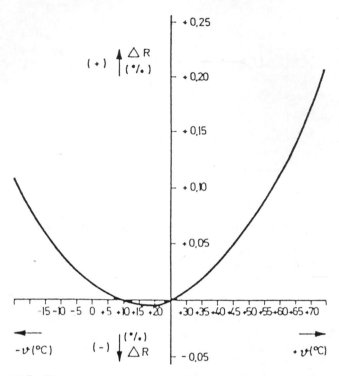

Fig. 12.7   Change of resistance as a function of the temperature

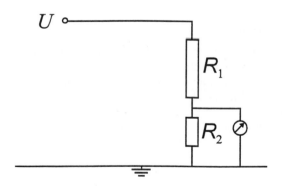

Fig. 12.8   Direct voltage measurement with a resistive voltage divider
   U   -   *high voltage*
   R_1 -   *resistor of high voltage arm*
   R_2 -   *resistor of low voltage arm*

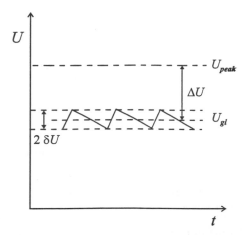

Fig. 12.9 *Direct voltage output with voltage drop ΔU and ripple δU*
U_{peak} — *peak voltage*
U_{gl} — *mean value of direct voltage*

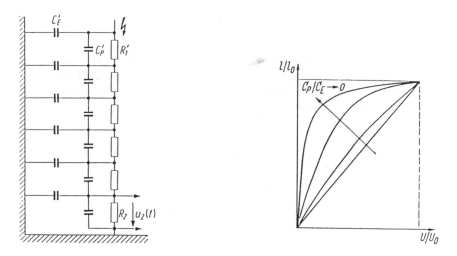

Fig. 12.10 *Schematic diagram of a resistive divider with capacitive grading*

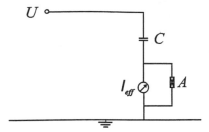

Fig. 12.11   Alternating voltage measurement with a capacitance
U -   *high voltage*
C -   *high voltage capacitor*
I_{eff} -   *current through the capacitor*
A -   *surge arrester*

The current is given by the formula:

$$I_{eff} = \omega \, C \, U_{eff} \qquad\qquad (12.2)$$

which means that, for example, the third harmonic drives a current which is three times higher than the basic frequency current and the calculated voltage is too high.

The capacitive voltage divider, shown in Fig. 12.12, is therefore one of the measuring systems for alternating voltage measurements.   The measuring instrument again needs a very high impedance, in order to prevent load on the divider, which leads to a frequency dependent transformation ratio.   The output voltage is given by the ratio of the capacitances:

$$U_2 = \frac{U_1 \, C_1}{\left(C_1 + C_2 + C_I\right)} \qquad\qquad (12.3)$$

where C_I is the capacitance of the measuring instrument.

In power networks both types of voltage transformers, capacitive and inductive, are very often used, because they also perform some other tasks besides the measurement of the r.m.s. value, e.g. network protection and measuring voltage distortions. For testing purposes, capacitive voltage dividers are preferred.

The dielectric strength of insulating material depends on the peak value of voltage, when the voltage stress is short, and the recommendations normally require the peak value divided by the factor $\sqrt{2}$ as a criterion. If the wave shape is not pure sinusoidal, the evaluation of the peak value from the r.m.s. measurements is not possible.

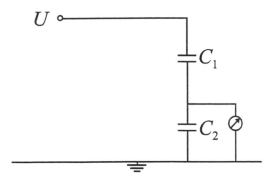

Fig. 12.12 Alternating voltage measurement with a capacitive divider
U -   high voltage
C_1 -   capacitor of high voltage arm
C_2 -   capacitor of low voltage arm

Fig. 12.13 shows a measuring circuit for peak voltage measurement according to Chubb and Fortescue [6]. Assuming a symmetrical shape of the voltage and only one maximum within a half period, the peak value of the voltage is given by the equation:

$$i_M = \left(\frac{1}{T}\right) \int_0^{T/2} i \; C(t) \; dt = 2 \; f \; C \; \hat{u} \qquad (12.4)$$

From Equation 12.4 it can be seen that the frequency directly influences measuring uncertainty, so measurement of the frequency is required.

Peak voltage measurement with a voltage divider is shown Fig. 12.14. The measuring capacitor C_M will be charged up to the peak value of $u_2(t)$ and the measuring instrument shows the voltage on C_M. The series resistance of the diode must be small. There are three main parameters which influence measuring uncertainty. The resistor R_E influences the transformation ratio. The resistor R_M discharges the measuring capacitance C_M. The capacitor C_M is in parallel with the capacitor C_2 during charging, which changes the capacitive ratio between the high and low voltage parts. The last two factors are frequency-dependent.

Fig. 12.15 shows a measuring circuit according to Rabus which compensates the measuring uncertainties. Due to its symmetry, the resistor R_E is not necessary [6].

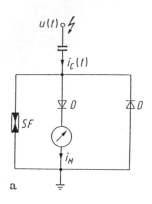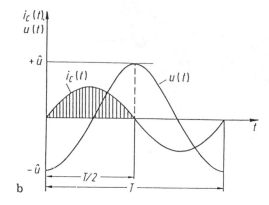

Fig. 12.13   Peak voltage measuring circuit

u(t)	-	*input voltage*
D	-	*diode*
SF	-	*surge arrester*
$i_{M(t)}$	-	*measuring current*
$i_{C(t)}$	-	*current through the capacitor*

The measuring uncertainty due to discharge of C_M is negligible, because the parallel capacitor C_S supports the voltage on C_M. The estimated measuring uncertainty is in the range of $\pm 1.5\%$ for frequency between 16⅔ and 100 Hz. A measure against charging and discharging errors is the use of an operation amplifier.

Another means of measuring the peak value is the use of an analogue-digital converter as a recording device. The evaluation of the recorded signal can be done by a built-in computer or by a host computer. Furthermore, the computer can make more calculations, e.g. a fast Fourier transformation, and can evaluate the amplitudes of the relevant harmonics. With the 12-bit resolution of a commercially available digital recorder, the measurement uncertainty of the recording device is negligible compared with the other components of the measuring system.

Finally, the sphere gap is also a device for peak voltage measurements. Its disadvantage is the relatively high number of tests, because in the tables of IEC Recommendation Publication 52, the mean value of the disruptive breakdown voltage is noted. The measurement uncertainty is $\pm 3\%$, but the device is very simple and the gap can be checked with a simple scale.

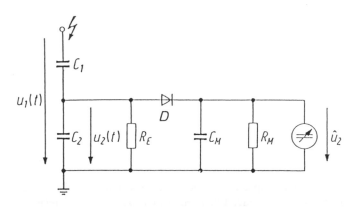

Fig. 12.14   Peak voltage measuring circuit with voltage divider

$u_1(t)$ - high voltage
u_2 - low voltage
D - diode
R_E - discharging resistor
C_M - measuring capacitor
R_M - discharging resistor

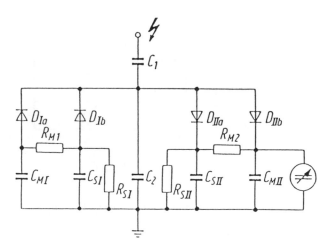

*Fig. 12.15   Peak voltage measuring circuit with additional capacitors
(for legend see Fig. 12.14)*

12.3.3 Impulse voltage

The measurement of impulse voltages requires a system which has a known scale factor and adequate dynamic behaviour. Transfer behaviour is characterised by its unit step response.

A resistive divider can be used for impulse measurements if the resistance is low enough that the transfer behaviour is not influenced by stray capacitances. A good estimation of the time constant of the unit step response of a resistive divider is given by the equation:

$$T = R \frac{C_E}{6} \tag{12.5}$$

where C_E is the stray capacitance and R the resistance of the divider. From Equation 12.5 it can be deduced that the stray capacitance and resistance should be as small as possible. Fig. 12.16 shows a unit step response of a resistive divider.

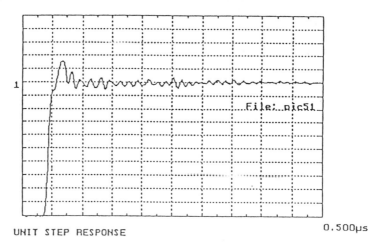

UNIT STEP RESPONSE   0.500µs

Fig. 12.16   Unit step response of a resistive divider

Another problem, which is widespread for measuring techniques, should be mentioned here. Due to the low resistance of the measuring system, the wave shape of the impulse will be influenced in such a way that no switching impulses can be measured with resistive dividers.

A further means of measuring impulse voltages is a resistive-capacitive divider: Fig. 12.17 shows the schematic diagram. This divider is also frequency-independent, but the inductance of the high voltage lead, the voltage source, the divider elements, the earth connection and the capacitance builds up a series resonance circuit.

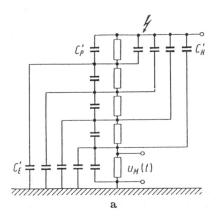 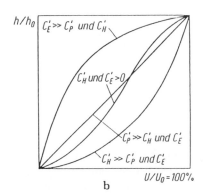

a   b

Fig. 12.17   Schematic diagram of a resistive-capacitive divider
C_P' -   parallel capacitor
C_H' -   stray capacitance
C_E' -   stray capacitance

Because high voltage dividers are normally large and consist of several identical parts, the oscillation should be damped by resistors. The damped capacitive divider is shown in Fig. 12.18.

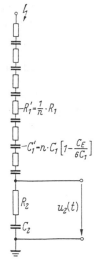

Fig. 12.18   Schematic diagram of a damped capacitive divider

It is clear that the time constant R_1C_1 should be equal to R_2C_2 in order to make the divider frequency-independent. Depending on the value of the damping resistor there are two types of damped capacitive dividers, the optimum damped and the slightly damped divider. Equation 12.6 shows the resistance of an optimum damped divider:

$$R \approx 3...4 \left(\frac{L}{C_E} \right)^{1/2} \tag{12.6}$$

This leads to a value of 400 to 800 Ω for a divider in the MV range. The slightly damped divider has a smaller resistance, which does not depend on the stray capacitance but only on the inductance of the measuring circuit. Therefore, it is possible to remove the resistor in the secondary part [7]. The resistance for such a divider is in the range of:

$$R \approx 0.2 \left(\frac{L}{C_E} \right)^{1/2} \tag{12.7}$$

Besides the requirements for design, good symmetry, low inductance and matching with the recording device are also very important.

For fast signals, the measuring cable should be matched with its characteristic impedance. With an oscilloscope as recording device, the input impedance is very high, if the cable is directly connected to the deflecting system. Therefore, the standard measuring circuit has the characteristic impedance at the beginning of the measuring cable. This system can also be used for a digital recorder as recording system and is shown in Fig. 12.19.

The measuring signal $u_2(t)$ will be divided by the factor 2 due to impedance Z and the characteristic impedance of the cable. The voltage will be doubled at the recording device, because the cable is open at this side. The wave travelling back has no reflection because the cable is matched with the impedance Z. For fast transients the transformation ratio is given by the equation:

$$\frac{u_1(t)}{u_2(t)} = \frac{(C_1 + C_2)}{C_1} = n \tag{12.8}$$

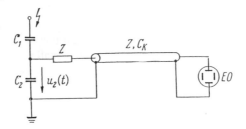

Fig. 12.19 Damped capacitive divider with cable matching resistor
  Z - *characteristic impedance of the measuring cable*
  C_K - *capacitance of the measuring cable*
  EO - *oscilloscope*

After twice the travel time, the ratio changes to:

$$n = \frac{(C_1 + C_2 + C_K)}{C_1} \qquad (12.9)$$

Depending on the cable length, the measuring error can be neglected or should be compensated. Fig. 12.20 shows such a measuring circuit, where the ratio is constant for high and low frequencies, if the condition $C_1 + C_2 = C_k + C_3$ is fulfilled.

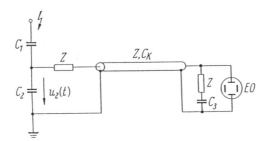

Fig. 12.20   Damped capacitive divider with matching network

The evaluation or measurement of the peak value can be made with an oscilloscope, a peak voltmeter or a digital recorder. Evaluation of the oscillogram permits determination of the mean curve through oscillations, if necessary, according to the relevant recommendations.

A peak voltmeter gives only the highest amplitude recorded, but this is in many cases sufficient, in particular if the wave shape is checked by an oscilloscope. The principle of the measuring system is the same as the peak measuring circuit for alternating voltage, but it should be taken into account that the impulse voltage or chopped impulse voltage is very short, which requires small values of capacitance.

A digital recorder can replace the oscilloscope and when used with a computer the peak value can be evaluated very easily.

12.3.4 Impulse current

Measurement of an impulse current can be carried out with measurement of a voltage across a defined resistor, but it is very important that the influence of the high magnetic field is considered. Fig. 12.21 shows the simplified equivalent circuit diagram together with an example of a typical current impulse.

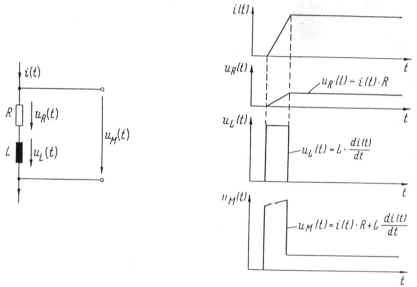

Fig. 12.21   *Simplified   equivalent   diagram   for   an   impulse   current measurement*

The voltage has two components, resistive and inductive. It is necessary to compensate the inductive part by, for example, a coaxial design of the resistor or shunt. A further point to note is the loss of power and the mechanical stress due to the action integral $\int i^2 dt$. Both parameters should be taken into account in the design of a current measuring system.

The magnetic field can also be used to measure the current. The simplified equivalent diagram is shown in Fig. 12.22, where the output voltage $u_1(t)$ is equivalent to the change of the current.

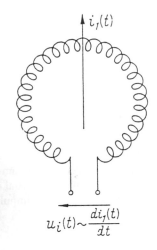

$$u_i(t) \sim \frac{di_1(t)}{dt}$$

Fig. 12.22   Rogowski coil

An integration of the measured signal is necessary to obtain a signal proportional to the current. This can be done by a simple RC circuit and then the measured voltage u_M is proportional to the current according to Equation 12.10.

$$u_M(t) = \left(\frac{1}{RC}\right) \int u_i(t)\,dt = \left(\frac{M}{RC}\right) i_1(t) \qquad (12.10)$$

M is the proportional factor given by the mutual inductance of the current path and the measuring coil. The advantage of such a so-called Rogowski coil is the potential separation between the test and measuring circuit.

12.4 Time parameters

The measurement of time parameters is mainly limited to impulse measurement. It is necessary to record the shape of the impulse and then to evaluate the required parameter. The evaluation method depends on the recording device. With an oscilloscope as recording device the evaluation can only be done manually using an oscillogram. With a digital recorder, the evaluation can be carried out automatically by a computer.

12.5 Measuring purposes

Measurement are carried out for two main reasons:

•   Check on the applied voltage and current in order to ensure that the required voltage and current wave form and amplitude have been applied;
•   Assessment of test results for compliance with a test requirement.

12.5.1 Dielectric tests

A dielectric test consists of applying a voltage up to a certain level and proving that no partial discharges or flashover occur. Measurement during these tests confirms the required wave form and amplitude. In addition, the test criterion, flashover or no flashover, will be recorded at the same time, but the measuring system is not necessary for this. In the case of a partial discharge test a measuring system is necessary. Fig. 12.23 shows a lightning impulse measured with a digital recorder.

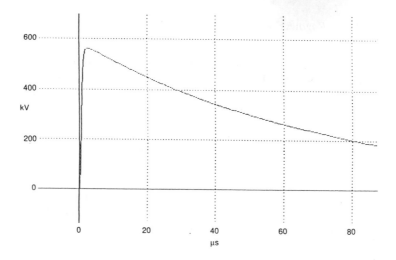

Fig. 12.23   Lightning impulse measured with a digital recorder

Another example is a test with chopped impulses. The record of the impulse should show the correct shape and the time to chopping, which is determined within certain limits in the relevant recommendations. Fig. 12.24 shows a voltage test for a metal oxide arrester. The upper curve shows the voltage chopped by the test object, the lower curve the current through the test object.

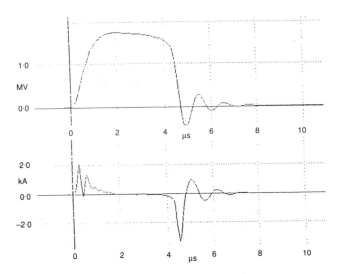

Fig. 12.24   Voltage and current of a metal oxide arrester

A switching impulse on a transformer is also determined by the test object itself and the test prehistory. Therefore, the measurement should not only detect a flashover but also record the wave shape for evaluation of the different time parameters.

12.5.2 Linearity tests

Pure dielectric linearity tests need to be carried out, in particular on transformers. The reason for such a test is to check the linear behaviour of the test object. The transformer will be stressed first with 50% of the voltage, later with 100% and the neutral current will be measured. The test will be successful if the comparison between the current and voltage at 50% and 100% levels shows no differences. In this case, the result of the measurement can be used as a diagnostic tool. With a digital recording system, the recorded data can be used for further evaluation and the diagnosis can be improved. Fig. 12.25 shows the two superimposed voltages at 50% and 100% test voltage level (a) and the calculated and by a factor 10 enlarged difference between the two impulses (b). The respective currents are shown in (c) and (d).

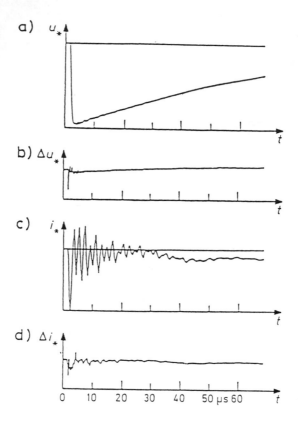

Fig. 12.25   Voltage and current of a transformer
(a) 50% and 100% superimposed impulse voltages
(b) calculated voltage difference enlarged by a factor of 10
(c), (d) respective currents

12.6 References

1   "Impulse voltage dividers", Bulletin E 142, (E. Haefely & Cie Ltd, Basel, Switzerland)

2   IEC CD 42 (Secretariat) 82, "High voltage test technique Part 2: Measuring systems", June 1992

3   MALEWSKI, R. and GOCKENBACH, E.: "Neue Möglichkeiten der Beurteilung von Stoßspannungsprüfungen an Transformatoren durch Einsatz eines digitalen Meßsystems". *ETZ-Archiv*, 1989, **11**, pp.179 -185

4   MODRUSAN, M. and GOCKENBACH, E.: "Eine kombinierte Prüfanlage für die Arbeitsprüfung von Metalloxidableitern", 19 International Blitzschutzkonferenz (ICLP), Graz, 1988, Paper 6.10

5   GOCKENBACH, E., *et al*.: "DC voltage divider with high precision for HVDC transmission", 4th International Conference on AC and DC power transmission, London 1985

6   SCHWAB, A.: "Hochspannungmeßtechnik", (Springer Verlag, Berlin) 1981

7   FESER, K.: "Heutiger Stand der Messung hoher Stoßspannungen", PTB Seminar 1982, Physikalisch-technische Bundesanstalt, Braunschweig

Chapter 13

Traceable measurements in high voltage tests

R.C. Hughes

13.1 Introduction

An acceptable form of stating the result of a voltage measurement in a high voltage test would be:-

1072 ± 16 kV (for an estimated confidence level not less than 95%).

A requirement for an accredited test laboratory is that the measurement shall be traceable through an uninterrupted chain of calibrations to national standards of measurement.

In the example, the value 1072 kV is the result of the calculation of the product of a low voltage instrument reading, corrected for any systematic error, and the value of the scale factor of the measuring system for the particular voltage shape.

Whilst traceable calibrations of low voltage d.c. and a.c. instruments are readily available, similar services for impulse peak voltmeters, oscilloscopes and digital recorders are not. Furthermore, calibration of the scale factor of high voltage measuring systems, particularly for impulse voltages, requires a different concept of traceability. In a calibration of a low voltage instrument traceability is shown to a National Standard of Measurement, e.g. unit of voltage or resistance. For the complete high voltage measuring system the demonstration of traceability, especially for impulses, is to a "consensus reference standard" obtained through international intercomparison of reference measuring systems by accredited calibration laboratories and some National Standard Laboratories.

Comparison against a reference measuring system is the preferred method of determining the assigned scale factor of an approved measuring system. The technique has been used successfully for a long time in many fields of measurement.

In a comparative measurement, the system to be calibrated is connected in parallel with a reference system for which the scale factor and its uncertainty are both known. The circuits are arranged for minimum mutual interference, usually a Y or T arrangement. At least 10 sets of simultaneous measurements of the outputs of the two systems are made for the same nominal input. The mean value of the difference in the relevant output is the error of the system to be approved with a stated overall uncertainty.

13.2 Direct voltage

The measurement of all electrical quantities is based on national standards of direct voltage and resistance traceable to the standards of other countries by comparisons performed periodically at the Bureaux Internationale Poids et Measures, BIPM. These standards are compared to within one part in 10^8 but the uncertainty in absolute value is estimated to be about one part in 10^6. They are presently evolving towards different, but more stable, devices such as those based on Josephson and quantum Hall effects.

The calibrations of d.c. voltmeters up to 1 kV are directly traceable to the primary voltage standards through self-calibrated voltage dividers with uncertainties of 10^{-5}.

Intercomparison measurements by several national laboratories [7] have resulted in voltage dividers for a few hundred kilovolts with uncertainties of 10^{-5} being available in a number of countries. Calibrations of reference dividers, with an uncertainty of 1 %, can be achieved with existing techniques.

13.3 Alternating voltage

Alternating voltages up to a few hundred volts, at power frequency, can be measured by comparison with direct voltages through a.c./d.c. transfer devices, with uncertainties of 10^{-4}. Traceable calibration at higher voltages is a well established procedure using voltage transformers or voltage dividers.

Intercomparison measurements by several National Standards Laboratories in Europe, on a voltage transformer and on a compressed gas capacitor, have resulted in National Reference Measuring Systems being available up to 100 kV with overall uncertainties of 10^{-5} [8,9].

Industrial testing laboratories regularly perform calibrations of voltage transformers with uncertainties in the range of 10^{-4}, for tariff metering equipment. Voltage transformers calibrated to an uncertainty of 10^{-3} are readily available, easily transportable and influenced very little by ambient conditions. They provide the best working reference standard for an Accredited Calibration Laboratory, up to several hundred kilovolts. Compressed gas capacitors, available for voltages up to 1 MV, are not influenced much by ambient conditions. The capacitance is usually independent of voltage, frequency and small temperature changes to within 1 part in 10^3 and they can be used in a testing laboratory after a low voltage check of capacitance. With special precautions during transport they have been used very successfully as transportable working references [10]. As with d.c., low voltage measurement in a.c. systems is by readily available digital voltmeters of sufficient accuracy. An overall uncertainty of 1 % specified for reference systems can be achieved by existing procedures.

13.4 Impulse voltage

Traceability to primary standards of voltage for impulse measurements is more difficult to achieve. Only a few National Standards Laboratories are engaged in impulse measurements. Calibration of impulse recorders is not yet readily available for uncertainties better than 1 % for voltage. Although the sampling frequency can be measured with traceability to the unit of time, the evaluation of the time parameters of an impulse requires the determination of the crossing of voltage levels at specified times, e.g. at 0.3 and 0.9 of the peak value. The uncertainty of the time parameters for full lightning impulses is estimated to be about 1 %.

The concept of international traceability for impulse voltage measurements is participation by National Standards and Accredited Calibration Laboratories in intercomparison tests, with the achievement of small deviations between the results from all laboratories.

Several intercomparisons of impulse measuring systems by national and industrial laboratories in Europe and North America, with full and front-chopped lightning impulses in the range 100 kV to 200 kV, have shown that uncertainties of less than one third of the limits for voltage and time errors required by the Draft Standard [2] are achievable [11,12]. Many laboratories agreed to within 0.2 % in the measurement of the peak value of full lightning impulses. As more experience is gained it is expected that Accredited Calibration Laboratories should achieve uncertainty values, in calibrations by comparative measurement techniques with full and front-chopped lightning impulses, of less than 0.5 %.

The evaluation of the front time and other time parameters can be affected by oscillations superimposed on the impulse shape. Several smoothing techniques, among them digital filtering, are known to work but they must be applied carefully.

Recommendations for data processing are now being studied by CIGRE and IEC working groups.

A measuring system for impulse voltages does not necessarily have a unique value of scale factor. Calibration over the complete range of impulse shapes to be measured is necessary and a correct measurement of the unit step response is a part of the traceability procedure.

The scale factor of a digital recorder must be measured for the relevant impulse shapes and for all its input ranges as deviations of several percent can occur.

13.5 Linearity test

The voltage rating of many systems is likely to be much higher than the voltage at which a comparative measurement with a reference system can be made to provide a traceable calibration of scale factor. A measurement of scale factor may be at a voltage as low as 20 % of the maximum voltage for which the system is to be approved and used. Additional tests must then be made to prove that the scale factor is linear up to the required value of voltage. There is no single test that will meet all situations. One or more of a number of tests must be performed to provide the necessary evidence of linearity.

Where another system of a higher voltage rating, which has already been qualified as an Approved Measuring System, is available, then the linearity test may be made by comparative measurement technique against the approved system. The scale factor, however, must be determined from the comparative test with the Reference System at the lower voltage.

A voltage divider consisting of several sections in series can sometimes be checked for linearity by testing separately each of the sections up to its proportion of the maximum voltage for which the complete divider is to be approved. Then the complete divider must be tested to check that no partial discharges occur at the highest voltage needed for approval. The linearity of a converting device may be affected by corona above a certain voltage resulting in a change of scale factor. The detection of d.c. and a.c. corona with commercially available instruments is a relatively simple test. However, for impulses, it may be necessary to record and compare the impulse waveshape at various voltages.

The relation between charging voltage of an impulse generator and its output has been shown to be constant within 1 % and can therefore be used to demonstrate linearity. Tests for linearity may be performed against an IEC Standard Measuring Device, a rod gap for d.c. and an irradiated sphere gap for impulse and a.c. It is necessary to show that each measured value is within 2 % of the corresponding IEC sparkover voltage.

Recently developed field sensors seem likely to provide the most useful technique to prove the linearity of a converting device.

13.6 Uncertainty of measurement

The uncertainty in a measurement [13] is an important factor in any statement about its traceability. All measuring systems are influenced to some degree by factors not necessarily directly related to the quantity being measured. Corrections can sometimes be made for parameters (e.g. ambient temperature, proximity effect) where these effects are constant and are known, and where screening of the system from them is not possible. But some uncontrollable influences remain.

If a measurement is repeated several times under apparently constant conditions there will be a spread in the observed results. If the tests can be repeated many times it is usually found that most of the results fall close to one central value and that the distribution of the results is Gaussian. The central value tends to become constant as the number of measurements increases and can be regarded as the mean value of the distribution.

Most high voltage tests are characterised by only a single or small number of measurements of voltage. A single measurement has a chance of taking any value in the expected distribution and the difference between this single measurement and the mean value of the distribution gives rise to a random component of uncertainty. In most measurements, the overall uncertainty will result from a combination of systematic and random uncertainty contributions.

Arithmetic summation of the contributions in an uncertainty budget is likely to give an unrealistic and pessimistic value. A more realistic method of combining uncertainty contributions is generally given by using a root sum of the squares method, which makes allowance for the probability that not all contributions will act simultaneously in the same direction. A statement of uncertainty should include the method of combination of the contributions and the level of confidence (e.g. 95 % confidence level) in the given value of overall uncertainty.

Examples of systematic uncertainty are:

— Calibration uncertainties;
— Errors in graduation of an instrument scale;
— Use of an instrument under constant conditions, but different from those of calibration.

Examples of random uncertainty are:

— Measurement of electrode separation in an air gap;
— Proximity effect;
— Voltage shape;
— Setting a pointer to a fiducial mark on a scale;
— Interpolation between marked points on an analogue scale;
— Digitising error;
— Interference;
— Fluctuation in any influence parameter, e.g. air temperature, humidity.

In the example, $1072 \pm 16\,\text{kV}$, the value of $16\,\text{kV}$ is the estimate of overall uncertainty, obtained by the method of root sum of the squares, for an estimated confidence level of 95 %.

13.7 Definitions related to accreditation

Accreditation
Formal recognition that a laboratory is competent to perform specific calibrations or tests.

Accredited laboratory
A laboratory that meets the general requirement for quality control specified by an accrediting body and the special requirements for high-voltage and impulse current measuring systems in the relevant standards.

Accrediting body
The official national authority responsible for the granting, renewal or termination of accreditation in respect of calibration or testing laboratories. It negotiates agreements on mutual recognition with other national schemes to obtain international acceptance of the competence of accredited laboratories. Many countries have completely separate organisations for accrediting calibration laboratories and for accrediting testing laboratories.

Approved measuring system
A measuring system which is shown in its record of performance to comply continuously with the requirements for accuracy and performance in the relevant standard.

Calibration
Specific type of measurement on measurement standards, measuring instruments and measuring systems to establish the relationship between the indicated and known values of a quantity. Calibration may be performed in a standards laboratory or in a testing laboratory.

Measuring system
A complete set of devices suitable for performing a high-voltage or impulse-current measurement.

Quality system
The organisational structure, responsibilities, procedures, processes and resources for implementing quality management, formalised in a maintained quality manual.

Record of performance of a measuring system
A working document, established by the user. It contains the complete history of the system and the latest measured values of the scale factor.

Reference measuring system
A measuring system having a calibration traceable to a national standard of measurement and used solely for the calibration or approval of other measuring systems by simultaneous comparative measurement.

Reference standard of measurement
A device of the highest meteorological quality at a given location, from which measurements at that location are derived.

Test
All types of objective measurement except calibration, performed in any location.

Traceability of measurement
The property of a result of a measurement by which the measurement can be related to the relevant national or international standard of measurement by an uninterrupted chain of comparisons.

13.8 Definitions related to uncertainty

Arithmetic summation of uncertainties
The most pessimistic method of combining uncertainty contributions.

viz $U_{\text{TOTAL}} = |U_1| + |U_2| . . + |U_n|$

where $|U_1|$ to $|U_n|$ are the moduli (i.e. the magnitudes are all taken as positive) of the contributions.

Confidence level
The probability that the true value (q.v.) will lie within a defined range of values. The rules governing its evaluation depend upon the assumed or measured distribution of values.

Control quantity
One of the quantities whose magnitude is specified as a reference condition for the test.

Conventional true value (of a quantity)
A value approximating to the true value of a quantity such that for the purpose for which that value is used, the difference between the two values can be neglected.

Correction
The value which, when added algebraically to an indicated or measured value, corrects for a known or assumed error.

Distribution
The frequency of occurrence of different magnitudes of value throughout a measured, or assumed, range of values.

Error of measurement
The result of a measurement minus the true value of the measured quantity. A term which cannot be quantified since the true value lies somewhere unknown within the range of the uncertainty.

Error of measuring instrument
The difference between the value indicated by an instrument and the most probable (or the conventional true) value of the measured quantity. Such an error is correctable by addition of a correction (equal to, but opposite in sign to the error).

Experimental standard deviation (s)
From a limited sample of measured values of a quantity, an estimate of the standard deviation in terms of the whole population is given by:

$$ s = \left[\frac{1}{n-1} \sum_{m=1}^{n} (x_m - x)^2 \right]^{\frac{1}{2}} $$

where n is the number of measurements
x_m are the measured values for $m = 1$ to n
x is the mean of the measured values.

Gaussian distribution
A distribution of a form shown graphically in Fig. 13.1(b) is alternatively known as a normal distribution.

Indicated value
(1) The indicated or recorded value of a measuring instrument.
(2) The nominal or stated value of a material measure.
(3) The set or nominal value of a supply device.

Limit of error and uncertainty
The sum of uncorrected error and uncertainty to give a maximum limit to the error of a measurement or a test. See also error of measurement and permissible limit of error and uncertainty.

Mean value
The average value of a set of readings or measurements of the same quantity. Denoted by a bar over the variable, e.g. $\bar{x}$.

Nominal value
The manufacturer's specified value of a component.

Permissible limit of error and uncertainty
A maximum value of the sum of uncorrected error and uncertainty in the measured value of a quantity permitted by contract, regulation or legislation. The permissible limit is obtained by the arithmetic summation of uncorrected error and

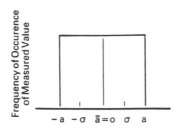

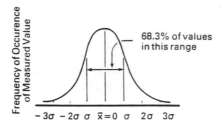

Fig. 13.1   *(a)*   *Value as difference from mean ($\bar{a}$); rectangular distribution*
          (b)   *Value as difference from mean ($\bar{x}$); Gaussian distribution*

uncertainty, the latter being the value at a confidence level of not less than 95%. See also limit of error and uncertainty.

Random error
Component of the error of measurement which, in the course of a number of measurements of the same measurand under the same measurement conditions, varies randomly with expectation zero.

Rectangular distribution
A particular form of distribution shown graphically in Fig. 13.1(a). It is characterised by the equal probability assigned to any value in its range.

Repeatability of measurements
The closeness of the agreement between the results of successive measurements of the same measurand carried out subject to all the following conditions:

—   The same method of measurement;
—   The same observer;
—   The same measuring instrument;
—   The same location;

—   The same conditions of use;
—   Repetition over a short period of time.

Note: Repeatability may be expressed quantitatively in terms of the dispersion of the results.

Reproducibility

The closeness of the agreement between the results of measurements of the same measurand, where the individual measurements are carried out, changing conditions such as:

—   Method of measurement;
—   Observer;
—   Measuring instrument;
—   Location;
—   Conditions of use;
—   Time.

Notes: Reproducibility may be expressed in terms of the dispersion of the results. A valid statement of reproducibility requires specification of the conditions changed.

Root sum of the squares (r.s.s.) of uncertainties

The most optimistic method of combining uncertainty contributions.

viz $U_{TOTAL} = (U_1^2 + U_2^2 \dots + U_n^2)^{1/2}$

where U_1 to U_n are the magnitudes of the contributions.

Standard deviation

σ, pronounced sigma, is a measure of the dispersion of a set of values. This symbol is generally associated with the population standard deviation. The experimental standard deviation (s) is usually used as a measure of a finite sample of values.

Systematic error

A component of the error of a measurement which, in the course of a number of measurements of the same measurand under the same measurement conditions, remains constant.

True value (of a quantity)

The value which characterises a quantity perfectly defined, in the conditions which exist when that quantity is considered.

Note: The true value is an ideal concept, and in general, cannot be known exactly.

Uncertainty of measurement
Result of the evaluation aimed at characterising the range within which the true value of a measurand is estimated to lie, generally with a given likelihood.

Note: Uncertainty of measurement comprises, in general, many components. Some of these components may be estimated on the basis of the statistical distribution of the results of series of measurements and can be characterised by experimental deviations. Estimates of other components can only be based on experience or other information.

13.9 References

1   IEC 60-1: "High voltage testing techniques", 1989

2   IEC 60-2: "High voltage testing techniques. Measuring systems"

3   ISO/IEC Guide 25: "General requirements for the technical competence of testing laboratories"

4   ISO/IEC Guide 38: "General requirements for the acceptance of testing laboratories"

5   IEC 790: "Oscilloscopes and peak voltmeters for impulse tests"

6   IEC 1083: "Digital recorders for measurement in high-voltage impulse tests Part 1: Requirements for digital recorders", 1991

7   DEACON, T. A.: "Intercomparison measurement of the ratios of a 100kV d.c. voltage divider". BCR Report EUR 10178 EN, Brussels, 1985

8   BRAUN, A., and RICHTER, H.: "Intercomparison of the calibration of a voltage transformer". BCR Report EUR 10193 EN, Brussels, 1985

9   SCHON, K. and LATZEL, H. G.: "Intercomparison measurements capacitance and loss factor at high voltage". BCR Report EUR 11453 EN, Brussels, 1988

10   RUNGIS, J. and BROWN, D. E.: "Mobile HV calibration unit". *The Austr. Physicist*, 1979 **16**, pp. 160-161

11   McCOMB, T. R., *et al.*: "International comparison of HV impulse measuring systems". *IEEE Trans. Power Delivery* (1988) **PWRD-4**, pp. 906-915

12   SCHON, K. and SCHULTE R.: "Intercomparison of HV impulse measurements." BCR Report, Brussels, 1991

13   "Uncertainties of measurement for NAMAS Electrical Product Testing Laboratories". NIS20.

Chapter 14

Fundamental aspects of air breakdown

N.L. Allen

14.1 Introduction

In almost all cases where we are dealing with the dielectric breakdown of air, we are concerned with electric fields which are highly non-uniform. It is characteristic of such breakdowns that the final sparkover is preceded by streamer formation, followed in many cases by a "leader" which develops prior to the breakdown itself. These processes are known to occur under impulse overvoltages, and under alternating and direct voltages. Before discussing the breakdown characteristics themselves, a brief explanation of the preceding processes is given, since the former are largely determined by the latter.

14.2 Pre-breakdown discharges

A description is best given, initially, in terms of an impulse overvoltage applied to an electrode system in which the electrodes have sharp radii, giving rise to large electric fields. Near the positive electrode, ionising electron collisions in the air may form avalanches sufficiently large to create a streamer system — this is a corona which spreads out like a fan from the electrode (Fig. 14.1). The streamers propagate with a velocity $> 10^5 \, \text{ms}^{-1}$ and it is known that the electric gradient in the streamer channel is of the order of $500 \, \text{kVm}^{-1}$. Once the corona is formed, further development is "choked off" for a period up to a few tens of microseconds by its own space charge which reduces the electric field near the electrode, but in the prevailing electric field some heating, which can be envisaged as ohmic heating (I^2R), takes place in the very resistive channel. This results in a rise in temperature, a decrease in gas density and increased ionisation, so fostering the transition to an arc-like leader channel which, starting at the electrode, extends across the gap at the relatively low velocity of $\approx 10^4 \, \text{ms}^{-1}$. This channel is highly conducting with a low electric gradient and, in a simple picture, can be regarded as an extension of the anode across the gap. More streamers thus propagate from the tip of the advancing leader. Fig. 14.1 shows a time-extended picture of these events, for an impulse reaching crest in, e.g., 250 μs.

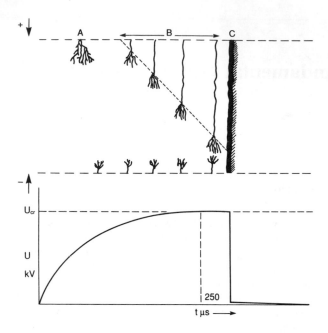

Fig. 14.1 *Time-extended picture of pre-breakdown corona and sparkover in a system with highly stressed positive and negative electrodes, e.g. the rod-rod gap. Impulse voltage reaching crest V_{Cr} at time $T_{Cr} = 250 \mu s$.*
A - Initial streamer corona; B - Leader growth phase; C - Breakdown

Avalanches, resulting in streamer formation, also occur at the cathode, but here electrons advance into regions of reducing field. The regions of ionisation are less extensive and require a higher electric gradient, of the order of 1000 kVm⁻¹ for streamer propagation. The negative corona is, in general, much less extensive than the positive corona.

Sparkover occurs after a conducting channel is established across the gap, that is, when the respective ionisation zones meet. The voltage at which breakdown occurs can then be quantified in the following way:

Let E_s^+, E_s^-, l_s^+, l_s^- be the gradients and lengths of positive and negative streamers respectively at the instant of breakdown. E_l^+, E_l^-, l_l^+, l_l^-, are the corresponding quantities for the positive and negative leaders. The voltage V_s at breakdown for a gap spacing d is:

$$V_s = E_l^+ l_l^+ + E_s^+ l_s^+ + E_l^- l_l^- + E_s^- l_s^- \qquad (14.1)$$

also

$$d - l_l^+ : l_s^+ : l_l^- + l_s^- \qquad (14.2)$$

In most electrode configurations, the lengths l are unknown, though the gradients are known or can be estimated. However, there is a useful simplification with the rod-plane electrode system where, since the negative electrode is only lightly stressed with a nearly-uniform electric field, $l_l^- = l_s^- = 0$. Thus, from equations (14.1) and (14.2), eliminating l_s^+:

$$V_s = E_s^+ d - l_l^+ (E_s^+ - E_l^+) \qquad (14.3)$$

Comparison of equations (14.1) and (14.3) makes it clear that where there is no negative corona, as in the rod-plane case, V_s is lower than it would have been in a gap where a significant negative corona occurred, as in the case where two highly stressed electrodes are used. Thus, the rod-plane gap exhibits the lowest sparkover gradient of any non-uniform field gap and this property forms a useful reference with which other gaps are compared.

14.3 The "U-curve"

From the foregoing description of the growth rates of streamers and leaders, it follows that the sparkover voltage of a gap depends upon the rate of rise of voltage in an impulse and on the electrode spacing. These dependences are manifest in the existence of the so-called "U-curves". For a fixed electrode spacing, the sparkover voltage, plotted as a function of impulse time-to-crest, passes through a clearly defined minimum. There is a different U-curve for each electrode spacing, with the minimum (known as the "critical time to crest" T_{Crit}) occurring at larger times-to-crest as the spacing increases. Examples for the rod-plane gaps are shown in Fig. 14.2 [1].

The characteristic profile of the U-curve arises because the leader, which advances at a velocity which varies little with voltage, is able to extend further into the gap as the time-to-crest voltage increases from a small value. For example, taking the rod-plane gap as the simplest case; where a lightning impulse is applied rising to crest in the order of $1\,\mu s$, there is little time for more than rudimentary leader growth after a streamer corona has occurred (usually on the wavefront) and in equation (14.3), the second term is negligible so that:

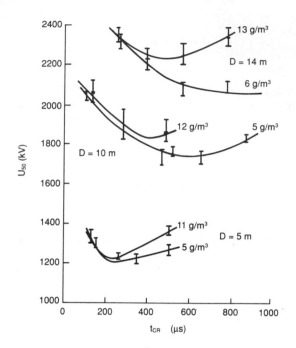

Fig. 14.2   U-curves obtained with impulse voltages of various times-to-crest applied to rod-plane gaps. Atmospheric humidity varied in these experiments (see Section 14.6.1 [1])

$$V_s = E_s^+ d \qquad\qquad (14.4)$$

Since E_s^+ has a constant value of approximately $500\,\mathrm{kVm^{-1}}$, regardless of streamer length, the positive sparkover voltage varies linearly with electrode spacing, a result that has been confirmed by experiments with electrode spacings up to $6\,\mathrm{m}$ [2], see Fig. 14.3.

For longer times-to-crest, the leader, with associated streamers, has more time to advance across the gap. Since its gradient E_l^+ is relatively small (and decreasing, in fact, with l_l^+), [3], inspection of equation (14.3) shows that $V_s < E_s^+ d$. More detailed investigation shows that V_s decreases with increasing time-to-crest and a minimum value is reached at T_{Crit}, when optimum leader development occurs.

Where the impulse has a time-to-crest $> T_{Crit}$, that is, with a low rate of rise of voltage, formation of the leader may be preceded by several streamer coronas, so

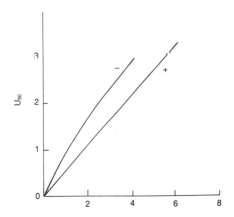

Fig. 14.3 *Positive and negative lightning impulse sparkover characteristics for rod-plane gaps. After [2]*

reducing the electric field near the positive electrode and inhibiting the ohmic heating of the streamer channels. Leader formation then requires higher total voltages, and the value of V_s tends to increase. Thus a minimum occurs in the V_s versus time-to-crest characteristic.

Further consideration of the effects summarised in equations (14.1) and (14.3) will show that the value of T_{Crit} increases with the electrode spacing d.

14.4 The gap factor

This is a rule-of-thumb used for estimating the sparkover voltages of gaps of various geometries, based on the sparkover voltage of a rod-plane gap of the same length. Again, this is discussed in terms of sparkover under positive impulse voltage in the first instance.

Equation (14.1) indicates that the sparkover voltage depends on the spatial extent of corona at the electrodes; this in turn depends on the nature of the electric field in the vicinity of the electrodes and hence on their geometry. This has given rise to the concept of the "gap factor" which is defined as the ratio of the sparkover voltage for a given electrode configuration to that for the rod-plane gap which, as we have seen, exhibits the smallest flashover voltage for a given electrode separation and can be taken as a reference. Tables 14.1 and 14.2, reproduced from the work of Hutzler *et al.* [4], show values of gap factor K that have been derived by experiment, together with empirical formulae for the calculation of their values.

Table 14.1   *Gap factor of a number of basic configurations. If $k > 1.6$ withstand in negative polarity is less than in positive*

Configuration		k
Rod-plane	d	1
Conductor-plane	d	1.1 to 1.15
Rod-rod	$H \quad d \quad H'$	$1 + 0.6 \dfrac{H'}{H}$
Conductor-rod	$H \quad d \quad H'$	$(1.1 \text{ to } 1.15) \exp\left(0.7 \dfrac{H'}{H}\right)$
Protuberances	$H \quad d \quad H'$	$k_0 \exp\left(\pm 0.7 \dfrac{H'}{H}\right)^{\bullet}$ $k \geqslant 1$

*Sign + for protuberances at the negative electrode
Sign − for protuberances at the positive electrode.

It is, however, necessary to use the concept with caution. Returning to the growth characteristics for streamers and leaders, we see that an electric field near a stressed electrode will, depending on its shape, determine the extent of the streamer growth and consequent leader initiation. Progress to breakdown thus depends on combined effects due to the time-to-crest voltage and the electric field distribution. The U-curve of the electrode system under consideration and that of the rod-plane gap will thus not be linked by a simple proportionality; in other words, the gap factor will vary with the waveshape.

As an example, the U-curves for 13-metre positive conductor-plane, sphere-plane and rod-plane gaps are presented in Fig. 14.4, together with gap factors calculated from the curves. These show significant variations with time-to-crest over a range of switching impulses [4].

Table 14.2   *Experimentally derived gap factor K [4]*

Sketch of the configuration	Formula	K
Sketch of the configuration «conductor-cross-arm»	$K = 1.45 + 0.015 \left(\dfrac{H}{D_1} - 6\right) + 0.35 \left(e^{-8\,S/D_1} - 0.2\right)$ $+ 0.135 \left(\dfrac{D_2}{D_1} - 1.5\right)$ Applicable in the range: $D_1 = 2 \div 10\,m$ $D_2/D_1 = 1 \div 2$ $S/D_1 = 0.1 \div 1$ $H/D_1 = 2 \div 10$	$K = 1.45$
Sketch of the configuration «conductor-window»	$K = 1.25 + 0.005 \left(\dfrac{H}{D} - 6\right) + 0.25\, e^{-8\,S/D - 0.2}$ Applicable in the range: $D = 2 \div 10\,m$ $S/D = 0.1 \div 1$ $H/D = 2 \div 10$	$K = 1.25$
Sketch of the configuration «conductor-lower structure»	$K = 1.15 + 0.81 \left(\dfrac{H'}{H}\right)^{1.167} + 0.02\,\dfrac{H'}{D}$ $- A\left[1.209 \left(\dfrac{H'}{H}\right)^{1.167} + 0.03 \left(\dfrac{H'}{H}\right)\right](0.67 - e^{-2\,S/D})$ where $A = 0$ if $S/D < 0.2$ and $A = 1$ if $S/D > 0.2$. Applicable in the range: $D = 2 \div 10\,m$ $S/D = 0 \div \infty$ $H'/H = 0 \div 1$	$K = 1.15$ for cond-plane to 1.5 or more
Sketch of the configuration «conductor-lateral structure»	$K = 1.45 + 0.024 \left(\dfrac{H'}{H}\right) - 6 + 0.35 \left(e^{-8\,S/D} - 0.2\right)$ Applicable in the range: $D = 2 \div 10\,m$ $S/D = 0.1 \div 1$ $H/D = 2 \div 10$	$K = 1.45$
Sketch of the configuration «rod-rod-structure» (open switchgear)	Horizontal rod-rod-structure $K_1 = 1.35 - 0.1\,\dfrac{H'}{H} - \left(\dfrac{D_1}{H} - 0.5\right)$ Rod-lower structure $K_2 = 1 + 0.6\,\dfrac{H'}{H} - A\,1.093\,\dfrac{H'}{H}(0.549 - e^{-3\,S/D_2})$ where $A = 0$ if $S/D_2 < 0.2$ and $A = 1$ if $S/D_2 > 0.2$. Applicable in the range: (K_1) $D_1 = 2 \div 10\,m$ $D_1/H = 0.1 \div 0.8$ $D_1 < D_2$ (K_2) $D_2 = 2 \div 10\,m$ $S/D_2 = 0 \div \infty$ $D_2 < D_1$	$K_1 = 1.3$ $K_2 = 1 + 0.6\,\dfrac{H'}{H}$

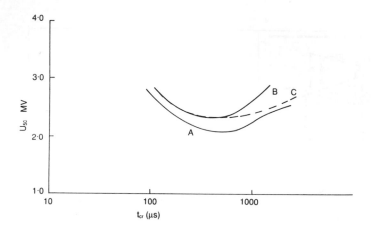

Fig. 14.4   U-curves   for   13 m   positive   rod-plane,   sphere-plane   and
conductor-plane gaps, after [4]. Gap factors as follows:

T_{Cr}	R-P (A)	S-P (B)	C-P (C)
100 μs	1.00	1.07	1.07
200 μs	1.00	1.10	1.10
500 μs	1.00	1.12	1.10
1000 μs	1.00	1.07	1.19
1100 μs	1.00	1.04	1.23

Again, for the specific example of the vertical rod-rod gap, the gap factor
depends upon the height of the tip of the "earthy" rod above the ground plane.
Fig. 14.5 shows results obtained from the sparkover characteristics of a rod-rod
gap of electrode spacing d in which the height above ground h of the tip of the
rod was progressively increased, while keeping the ratio h/d constant [5]. The gap
factor gradually increases as the height of the gap above ground level increases.
This is the result of changes in the disturbing effect of the ground plane upon the
electric field distribution in the gap. Under lightning impulse, gap factors tend to
be smaller than under switching impulse. This is a consequence of the small
amount of leader development [6]. Nevertheless, the proximity of earthed metal
surfaces can again exert a considerable effect upon the breakdown characteristics.
Fig. 14.6 presents the results of a study in a rod-rod gap in which an earthed metal
plane was placed parallel to the axis of a rod-rod system, at a distance 14 %
greater than the electrode separation [7]. The same study showed that the
proximity of such an earth plane to a rod-plane gap of similar range of dimensions
had a very much smaller effect than in the case of rod-rod gap.

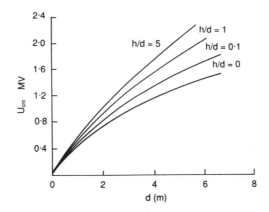

Fig. 14.5   *Rod-rod positive switching impulse sparkover characteristics as function of height h of lower rod above ground plane. Positive polarity to upper rod. After [5]*

h/d	Gap factor (d = 3 m)
0	1 (rod-plane)
0.1	1.15
1.0	1.32
5.0	1.46

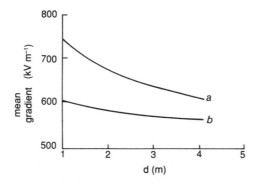

Fig. 14.6   *Rod-rod sparkover characteristics with earthed metal plane in proximity, expressed as mean gradient, kVm⁻¹*
a) = without plane, b) = with plane

14.5 Sparkover characteristics

14.5.1 Test procedures

The variability in the pre-breakdown processes summarised in Section 14.2 results in a variability in sparkover voltage, particularly when impulse voltages are used. It is necessary therefore to devise an average sparkover voltage in order to describe the strength of a particular gap. For this purpose the U_{50}, or 50 % sparkover voltage is used — this is the impulse crest voltage at which there is a probability of 0.5 that the gap will spark over. With impulse voltages, it can be determined in two ways: (a) the variable voltage method, and (b) the "up-and-down" method [7].

In method (a) an estimate of the likely value of U_{50} is made, either from experience or by means of a few trial shots and several sets of, say, 20 impulses each at constant crest voltage may then be applied to the gap. The voltage levels of the sets may differ from one another by the order of 1 % of U_{50}. Each set of 20 yields a probability of sparkover at the voltage used. The probabilities are plotted against voltage on normal probability paper and an approximate straight line is usually obtained. U_{50} is then immediately determined.

Method (b) is more economical in the number of impulses required. The voltage is raised, around the estimated U_{50}, in constant steps of 2 or 3 % of this value, until a sparkover takes place. It is then lowered by one step. If no sparkover occurs, it is raised again and this procedure is repeated, raising or lowering according to the result occurring in the preceding impulse. After a number of impulses, usually less than about 50, a voltage is reached which is very close to U_{50}. This technique can be made more precise if, after completion of the foregoing procedure, groups of 50 shots are tried at the estimated U_{10} and U_{90} conditions. If the results so obtained are plotted with the rest of the data, a value of U_{50} can be determined with very little uncertainty.

Each of these methods implicitly assumes a relation between the probability p of sparkover and a voltage U of the Gaussian form:

$$p = \frac{1}{(\sqrt{2}\,\pi\,S)} \int_0^U \exp \frac{(U - U_{50})^2}{S^2} \, dU \tag{14.5}$$

where S is the variation in U between U_{50} and the value of U at which p is either 0.16 or 0.84. It is this value of S that is commonly quoted in describing the deviation about U_{50} in practical tests. Values of the order of 5 % are common under switching impulses.

This assumption of a normal, or Gaussian, distribution has proved adequate for most test purposes, but more painstaking investigations [9] have shown that the probability distributions can sometimes be far from normal.

When alternating voltages are being measured, the sparkover voltage depends upon the time for which the voltage is applied; it can decrease by up to 10 % for an application of voltage to the test gap which is rising for periods up to 1000 μs. It is therefore necessary to specify, for a given test, the rate at which the voltage rise takes place.

Similar considerations apply in the case of direct voltages. Here, for instance, the IEC [7] lays down that the voltage shall be raised to the breakdown value in approximately one minute, with a rate of rise of ≈ 1 %/s in the final stages. The scatter in sparkover values, expressed as a standard deviation, is usually less than 1 %, where the more highly stressed electrode is positive.

14.5.2 Sparkover voltage characteristics

14.5.2.1 Impulse conditions: rod-plane gaps

The practical engineer is faced with a variety of possible electrode geometries, but we have seen that the rod-plane gap forms a valuable reference. Jones and Waters [10] have presented a summary showing the mean breakdown gradient U_{50}/d against gap length for this gap; it is reproduced here as Fig. 14.7. Results for alternating and direct voltages are also included here.

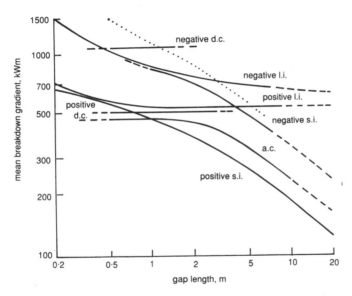

Fig. 14.7   *Mean breakdown gradient of rod-plane gaps under standard atmospheric conditions as a function of gap length for different testing waveforms [9]*

A further important reference condition becomes apparent. For the positive lightning impulse, the mean gradient remains constant, for gap lengths greater than about 1 m, at approximately 500 kVm⁻¹. This is approximately the same as the streamer gradient E_s^+ (Section 14.2 and equation (14.4)), a fact that is utilised in the IEC atmospheric correction procedures (Section 14.6). Thus, the sparkover voltage U_{50} increases linearly with gap length. For the negative lightning impulse, the gradient is higher but is not constant with gap; the sparkover voltage increases with electrode spacing less rapidly than linearly. No simple explanation can be offered here, if only because negative polarity has been much less studied than positive; the reason for this is that the higher gradients make negative impulses less dangerous to power systems than positive impulses.

Under positive switching impulse, the gradient declines with increasing gap length, due to the increasing growth of the low-gradient leader channel. The breakdown strength depends only on the gap length and is often expressed by the formula [4]:

$$U_{50(crit)} = \frac{3400}{(1 + 8/d)} \quad kV \qquad (14.6)$$

where $U_{50(crit)}$ is the value of U_{50} at the critical time to crest, that is, the minimum value. This expression holds over the range of gaps $2\,m < d < 15\,m$.

Figure 14.7 shows that the positive switching impulse characteristic becomes more linear for gap lengths $> 10\,m$ and a more appropriate relationship is the following:

$$U_{50(crit)} = 1400 + 55d \quad kV \qquad (14.7)$$

The following formula has also been proposed [4] as being reasonably accurate for all gaps up to $25\,m$:

$$U_{50(crit)} = 1080 \ln(0.46d + 1) \quad kV \qquad (14.8)$$

Finally, for the particular case of the IEC standard positive switching impulse, that is $250/2500\,\mu s$, the value of U_{50} (which is not now at a critical time-to-crest) is well described by:

$$U_{50} = 500 \, d^{0.6} \quad kV \qquad (14.9)$$

Under negative switching impulses, the following relationship holds with reasonable precision over the range $2\,m < d < 14\,m$ [4]:

$$U_{50(crit)} = 1180 \ d^{0.45} \qquad \text{kV} \qquad\qquad (14.10)$$

Fig. 14.8 compares practical values with some of these relationships [4].

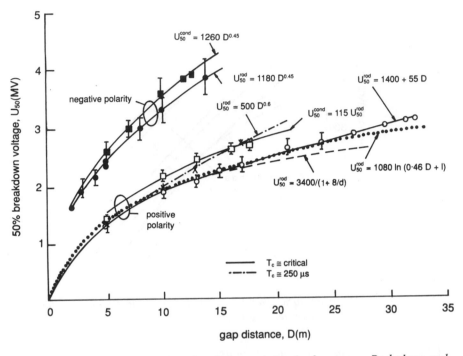

| *Fig 14.8* | *Switching impulse characteristics for long gaps. Rod-plane and phase conductor-plane. After [4]* |

14.5.2.2 Air gaps of other shapes

The expressions (14.6) and (14.7) can, in principle, be applied to gaps of more practical interest, provided the RHS of each is multiplied by the gap factor K. This statement must, of course, be qualified by the reservations already discussed in Section 14.4.

Fig. 14.9, reproduced from [10], summarises the available information on the minimum flashover mean gradients, that is, at critical time-to-crest, for a variety of gaps. Again, it will be noted that gradients are higher when the highly stressed electrode is negative, rather than positive.

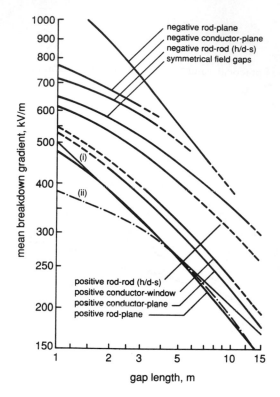

Fig. 14.9   *Minimum 50% sparkover gradient as a function of gap length for various reference geometries [9]*

14.5.2.3 Sparkover under alternating voltages

Here again, the rod-plane gap exhibits the lowest sparkover voltage and the 50% breakdown condition is given by [11]:

$$U_{50} = 750 \ln(1 + 0.55d^{1.2}) \qquad kV \qquad (14.11)$$

for values of d greater than 2 m. To give perspective, it may be noted that the peak values of U_{50} are of the order 20% higher than those for switching impulses of critical time-to-crest for the same gap.

For estimation of the flashover voltage of gaps other than the rod-plane, it has become customary to use the gap-factor K for switching impulses. This procedure is clearly subject to the same limitations as have been outlined already, but the following equation has been given [11]:

$$U_{50} = U_{50}(rp) \ (13.5K - 0.35K^2) \qquad \text{kV r.m.s} \qquad (14.12)$$

For gaps in the range $2\,m < d < 6\,m$ this formula holds to within $\pm 10\%$ for most of the following gaps: conductor-plane, conductor-structure (underneath), conductor-structure (lateral), rod-rod, conductor-rod, conductor-rope [12]. As noted earlier, flashover voltages are influenced by the rate of increase of voltage, with a decrease of up to 5% for slow rates of rise of voltage towards breakdown of the order of 1 hour.

14.5.2.4 Sparkover under direct voltages

In positive rod-plane gaps in the range $0.5\,m < d < 5\,m$, direct voltage sparkover voltages increase linearly with gap length, with a mean gradient of the order of $500\,kVm^{-1}$ [13, 14]. This characteristic is similar to that of the rod-plane gap under lightning impulse. In both cases, linearity is due to the dominance of the streamer pre-discharge, but in the positive direct voltage case, the reproducibility of sparkover voltage is greater, the standard deviation being less than 1%.

For the negative rod-plane gap, the gradient is higher, of the order of $1000\,kVm^{-1}$ with a slight tendency for the flashover voltage to increase less rapidly than linearly in gaps larger than $1\,m$ in length [15]. Gap factors for gaps other than rod-plane, and derived from direct voltage measurements are given in Table 14.3.

Table 14.3   Gap factors under direct voltages (data from [11])

Configuration	Gap d (m)	Sparkover voltage (kV)	Gap factor
Rod-plane	1	450	1
	2	910	1
Rod-rod*	1	520	1.16
	2	1050	1.15
Single conductor-structure	1	600	1.33
	2	1150	1.26
Bundled (4) conductor-structure	1	460	1.02
	2	950	1.04

* Tip of lower-rod 1.5 m above ground plane

The rod-rod gap has proved of particular use as a tool for the measurement of direct voltage, since it retains the property of linearity up to $d = 3\,m$ [8] and, provided it is mounted well above the ground plane, possesses the same characteristics whatever the polarity of the voltage being used. It has now been adopted by the IEC as a secondary standard for the measurement of direct voltages.

14.5.2.5 Flashover across insulator surfaces in air

The weakest aspect of the insulation of high-voltage systems is frequently the dielectric strength of the surface of the solid insulator which is used to separate the conductors. Knowledge of the basic processes involved is still relatively fragmented, but it is known that streamers, once initiated, may travel more rapidly across certain insulator surfaces in air than they do in the air in the absence of the insulator, and it is known that charges deposited on the surface by corona may so alter the electric field distribution as to facilitate breakdown. However, in practice, other factors may likewise cause field distortions and encourage breakdown: unequal distribution of capacitance between insulators; stray capacitances to ground; field concentration at the "triple junction" between insulator, conductor and air; layers of pollution; rain and deposition of moisture — all of these can contribute to weakness in the dielectric strength of a surface.

The tendency of a surface to reduce the insulation strength is minimised by the adoption of "sheds" to increase the total length of path between conductors. The "cap and pin" insulator takes this concept to the limit and almost all power lines operating at above 100 kV use strings of insulators of this type. The deep sheds ensure that discharges must cover most of their path in the air and tests have shown [16] that the loss of dielectric strength, compared with that in air is not usually more than about 3 %, under impulse voltages. This result is true for conductor-tower cross arm situations, for both polarities, but larger polarity differences may arise when a configuration such as rod-plane, with insulators, is tested. Indeed, when negative impulse voltages are applied to the highly-stressed electrode, results which are very sensitive to the electrode profile may be obtained; this is because high field concentrations at the "triple junction" may set up a cathode mechanism which encourages the initiation of discharges.

14.6 Atmospheric effects

All that has been written in the preceding sections must be qualified by the realisation that in air, the temperature, pressure and humidity change in an uncontrolled way. It is necessary, therefore to establish procedures by which (a) it is possible to estimate flashover voltages at any atmospheric condition from those measured at another condition, and (b) measurements can be corrected to a standard atmospheric condition for the purposes of comparison of results obtained in different laboratories at different times.

14.6.1 Density effects

Temperature and pressure together determine the density of the air. The ionisation processes involved in corona and breakdown depend primarily upon the air density and not, to any significant degree, specifically upon temperature except insofar as it changes density. Moreover, many experiments have shown that over limited ranges of density, breakdown voltages vary linearly with density. Thus, it has proved a relatively straightforward matter to define a set of conditions that can be regarded as characteristic of a standard atmosphere, and to relate other conditions of pressure and density to the standard atmosphere, assuming the simple gas laws to be correct. A standard atmosphere is defined for these purposes as at 1013 mbar (760 torr) and 293 K (20 °C). Then at any temperature T and pressure p, the density has changed from that at the standard atmosphere by:

$$\delta = \frac{p}{1013} \cdot \frac{293}{T} \qquad (14.13)$$

and clearly the value of δ is the ratio of the density at (p, T) to that at the standard condition. δ is called the relative air density; $\delta = 1$ at the standard condition.

If now, the value of U_{50} at relative air density $\delta = 1$ is denoted by U_0, then the value U at (p, T) is given by:

$$U = U_0 \cdot \delta \qquad (14.14)$$

It should be observed here that this discussion has ignored any effects due to change of humidity. These effects will be discussed below, but it should be noted that the standard atmospheric condition assumes an absolute humidity of 11 g moisture content per cubic metre of air.

Over the normal extreme range of temperatures encountered in practice, that is, between 313 K and 243 K approximately, the density correction procedure has been judged satisfactory. Values and curves of flashover voltages presented in the literature are usually described as uncorrected or alternatively corrected to $\delta = 1$, in the latter case so enabling comparisons to be made. The procedure is used for impulse, alternating and direct voltage measurements.

The procedure has also proved satisfactory for normal variations of pressure at altitudes up to 2000 m, but evidence has accumulated showing that at greater altitudes (lower pressures) the linearity breaks down and a modified procedure must be used [17].

14.6.2 Humidity effects

Atmospheric humidity can change from a moisture content of less than $1 \, gm^{-3}$ in very cold countries to the order of $30 \, gm^{-3}$ in the tropics. The effects of humidity on sparkover are quite complex, but the following general points can be made:

(a)   Humidity has its strongest influence on the positive pre-breakdown discharge. In particular, the streamer gradient E_s^+ (equation (14.1)) increases at the rate of roughly 1 % per gm^{-3} increase in moisture content;

(b)   Humidity has no significant effect on the leader gradients E_l^+ and E_l^- though it does have the effect of increasing the leader velocity;

(c)   Humidity has no significant effect on the negative sparkover under lightning impulse and under direct voltages.

A rough general rule therefore follows from (a), (b) and (c): where a sparkover is preceded only by positive corona, there is a significant humidity influence, but where it is preceded only by negative corona, there is no significant effect. Thus, for instance, under lightning impulse, humidity shows a maximum effect on sparkover in the positive rod-plane gap but no effect in the negative rod-plane gap.

The effects of humidity upon sparkover have been widely studied and have led to a correction procedure which uses the IEC curves shown in Fig. 14.10. Here k is a factor by which a sparkover voltage at a standard humidity h of $11 \, gm^{-3}$ must be multiplied in order to estimate what its value would be at any other humidity. Note that the abscissa here is in fact h/δ where δ is the relative air density; this usage arises from the fact that a given partial pressure of water vapour makes up a varying proportion of the air moisture as the density changes. A brief account of practical cases is now given.

Under positive lightning impulse, we have seen that sparkover is determined substantially by the streamer gradient E_s^+ and this is reflected in the expression for k, given in Fig. 14.10 where, for impulse voltage the rate of change with h/δ is 0.01, or 1 % per gm^{-3}. The correction procedure holds very well in this case. Where a switching impulse is used, Fig. 14.10 must be used with care, for substantial leader growth occurs and we have seen in (b) above that humidity has little influence on the leader gradient. Thus the rate of change of k with h/δ falls appreciably below 0.01 when the leader has traversed a significant part of the gap prior to sparkover. Also, since the leader velocity increases with humidity it is found that the minimum of the U-curve shifts towards lower times-to-crest; this is shown clearly in Fig. 14.2. In such cases, it is best to determine humidity influence by practical tests where possible.

Under direct voltages, sparkover is again determined substantially by streamer growth and the correction procedure holds very well with a rate of change of k with h/δ of 0.014 or 1.4 % per gm^{-3}. There is an important limitation, however, for gaps of the order of $1 \, m$ or greater and for humidities greater than $\approx 13 \, gm^{-3}$. Here the linearity between sparkover voltage and electrode spacing breaks down and, again, it is advisable to check humidity influence by practical tests where possible.

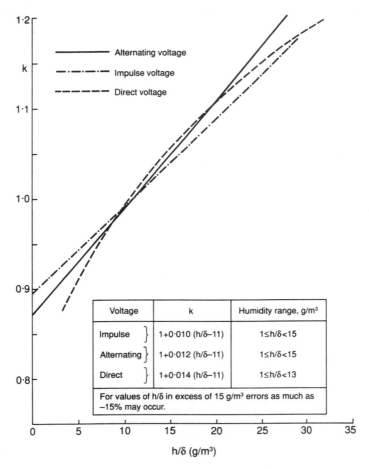

The table within the figure:

Voltage	k	Humidity range, g/m³
Impulse }	1+0·010 (h/δ−11)	1≤h/δ<15
Alternating }	1+0·012 (h/δ−11)	1≤h/δ<15
Direct }	1+0·014 (h/δ−11)	1≤h/δ<13

For values of h/δ in excess of 15 g/m³ errors as much as −15% may occur.

Fig. 14.10   IEC curves and expressions for humidity correction procedure

Under alternating voltages, application of the correction procedure is very difficult and it has been shown that at moderate humidities, around 13 gm⁻³, the sparkover voltage can become very variable [18]. The effects of humidity upon alternating voltage sparkover have been studied least of all and there are many uncertainties that remain to be resolved.

A full account of humidity effects and problems is given in [19].

14.6.3 Other atmospheric effects

High voltage power systems must contend with atmospheric pollution, rain, ice and snow and occasionally forest fires beneath overhead lines. The first three of these have their greatest effects upon insulators, as indeed does the fourth, by deposition of soot, etc. The last mentioned also has drastic local effects upon air density, which may result in sparkovers.

Detailed discussion of these topics is beyond the scope of this chapter; further information can be found in references [20, 21 and 22].

14.7 New developments

There is interest in the subject of "line compaction" of overhead lines in large-scale power systems, in order to reduce both costs and the visual impact on the environment. This will require a re-appraisal and an extension of measurements of sparkover characteristics in gaps of various geometries, with emphasis on the low-level probabilities of sparkover. Live-line working is also an issue at the present time, and studies of the influence of perturbing factors such as tools, floating objects and men working on lines on sparkover voltages and clearances between conductors are being undertaken in several countries. Finally, the efficacy of non-ceramic and composite insulators, as alternatives to the traditional ceramic or glass insulators, is under discussion and their flashover characteristics require study.

14.8 References

1   BÜSCH, W.: "Air humidity, an important factor of UHV design". *IEEE Trans.* (1977), **PAS-91** (6), pp.2086-2093

2   PARIS, L. and CORTINA, R.: "Switching and lightning impulse discharge characteristics for large air gaps and long insulator strings". *IEEE Trans.* (1968), **PAS-98**, pp.947-957

3   Les Renardières Group: "UHV air insulation: physical and engineering research". *Proc. IEE* (1986), **133-A**, pp.438-468

4   HUTZLER, B., GARBAGNATI, E., LEMKE, E., and PIGINI, A.: "Strength under switching overvoltages in reference ambient conditions", in: CIGRÉ Working Group 33.07, "Guidelines for the evaluation of the dielectric strength of external insulation". (CIGRÉ, Paris, 1993)

5   DIESENDORF, W.: "Insulation coordination in high voltage electric power systems". (Butterworth, London, 1974)

6   GALLIMBERTI, I. and REA, M.: "Development of spark discharges in long rod-plane gaps under positive impulse voltages". *Alta Freq.* (1973) **42**, pp.264-275

7   STANDRING, W.G., BROWNING, D.N., HUGHES, R.C. and ROBERTS, W.J.: "Impulse flashover of air gaps and insulators in the voltage range 1-2.5 MV". *Proc. IEE* (1963), **110** (6), pp.1082-1088

8   IEC 60-1 (1990): "High voltage test techniques, Part 1: General definitions and requirements"

9   MENEMENLIS, C. and HARBEC, G.: "Particularities of air insulation behaviour". *IEEE Trans.* (1976), **PAS-95**, pp.1814-1821

10   JONES, B. and WATERS, R.T.: "Air insulation at large spacings." *Proc. IEE*, (1978), **125** (11R), pp.1152-1176

11   PIGINI, A., THIONE, L. and RIZK, F.: "Dielectric strength under AC and DC voltages," in: CIGRÉ Working Group 33.07 "Guidelines for the evaluation of the dielectric strength of external insulation". (CIGRÉ, Paris 1993)

12   CORTINA, R., MARRONE, G., PIGINI, A., THIONE, L., PETRUSCH and VERMA M.P.: "Study of the dielectric strength of external insulation of HVDC systems and application to design and testing." CIGRÉ, 1984, Paper 33-12

13   FESER, K. and HUGHES, R.C.: "Measurements of direct voltages by rod-rod gap," *Electra* (1988), **64** (117), pp.23-35

14   BOUTLEND, J. M., ALLEN, N.L., LIGHTFOOT, H.A. and NEVILLE, R.B.: "Transitions in flashover characteristics of rod-plane gaps under positive direct voltage in humid air." Proc. 7th Int. Symp. on High Voltage Engineering, Dresden, Paper 42.17, 1992

15   KNUDSEN, N. and ILICETO, F.: "Flashover tests on large air gaps with DC voltage and with switching surges superimposed on DC voltage." *IEEE Trans.* (1970), **PAS-89**, pp.781-788

16   HUTZLER, B. and RIU, J.P.: "Behaviour of long insulator strings in dry conditions." *IEEE Trans.* (1979), **PAS-98** (3), pp.982-991

17   MORENO, M, PIGINI, A. and RIZK, F.: "Influence of air density on the dielectric strength of air insulation," in :CIGRÉ Working Group 33.07 "Guidelines for the evaluation of the dielectric strength of external insulation", (CIGRÉ, Paris, 1993)

18   FESER, K.: "Influence of humidity on the breakdown characteristics under AC voltage," E.T.Z. A Bd 91 No.10, 1970

19   ALLEN, N.L., FONSECA, J., GELDENHUYS, H.J. and ZHENG, J.C.: "Humidity influence on non-uniform field breakdown in air." *Electra*, (1991), **134**, pp.63-90

20   LEMKE, E., FRACCHIA, A., GARBAGNATI, E. and PIGINI, A.: "Performance of contaminated insulators under transient overvoltages. Survey of experimental data," in: CIGRÉ Working Group 33.07 "Guidelines for the evaluation of the dielectric strength of external insulation" (CIGRÉ, Paris, 1993)

21   KAWAMURA, T. and NAITO, K.: "Influence of ice and snow" in: CIGRÉ Working Group 33.07 "Guidelines for the evaluation of the dielectric strength of external insulation" (CIGRÉ, Paris, 1993)

22   FONSECA, J., SADURSKY, K., BRITTEN, A., MORENO, M., and VAN NAME, J.: "Influence of high temperature and combustion particles (Presence of fires)", in: CIGRÉ Working Group 33.07 "Guidelines for the evaluation of the dielectric strength of external insulation" (CIGRÉ, Paris, 1993)

Chapter 15

Application of new high voltage measurement technology

E. Gockenbach

15.1 Introduction

New technology, particularly in electronics, is now changing high voltage testing and measurement due to the reduced electromagnetic sensitivity of the electronic components to HV fields. Use of electronics is therefore possible in high voltage and high power laboratories. The two main fields in which electronics have many advantages are automation and measurement and diagnostic techniques.

15.2 Automation

Automation is directly connected with economy; this means that investment in automation will only be made if the return on the investment is reasonable. Another aspect of automation is the replacement of human by machine in many applications. The main cases for automation in high voltage tests are now discussed.

High voltage tests can be separated into three main parts; routine test during the manufacturing procedure as part of the manufacturing quality assurance and type test as part of research and development quality assurance. A further part concerns on-site tests, but for these the potential for automation is probably less. However, the same automation used for routine or type tests can also be applied to on-site tests [1].

The routine test is the most frequent test and therefore its automation brings the most benefit. The use of a computer to control the test procedure and test generators leads to a reduction in staff, testing time and error probability. Fig. 15.1 shows remote computer control of a high voltage impulse test.

The operator can input the data for the test object and the computer selects the correct test program. Depending on the program the relevant parameters can be calculated, e.g. correction of the test voltage according to instantaneous air

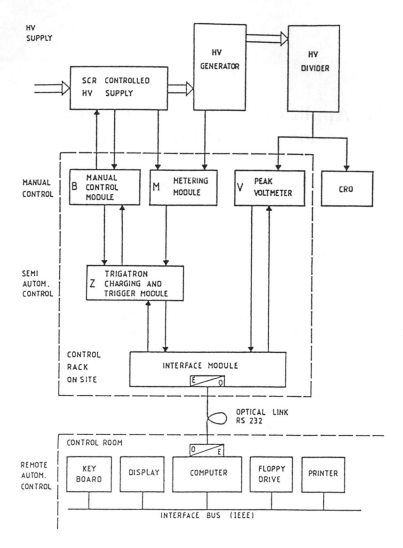

Fig. 15.1   Remote control system for an impulse voltage generator

pressure, temperature and humidity if necessary. The next step can be automatic adjustment of the test generator parameters.

Control of the output voltage is not a problem for direct or alternating voltage. However, for impulse voltage generators, control of the generator is via the charging voltage depending on the generator type, test object and test configuration. The output voltage, measured by a peak voltmeter, an oscilloscope or a digital recorder, should be available for the computer control system in order to adjust the charging voltage to the required output voltage. This feedback system can be used to check the proper function of the test and control system. After the first reference impulse the efficiency factor will be calculated and confirmed by the operator. This guarantees a first check of the whole system by the operator and later an automatic check by the computer control system itself, because the efficiency should be constant over the whole charging voltage range.

A further step in automation is the recording of the whole test sequence. This requires a reasonable storage means, particularly if all impulse data are to be stored. A complete test report can then be written by the computer after the test run. A test report, for immediate use by the customer for a transformer test where not only the peak of the test voltages but also its time parameters are shown, is seen in Fig. 15.2 [2].

Digital recording of test results has a further advantage for automation. Test results can be transmitted to a host computer, which collects all the data for a customer order, so that high voltage tests can also be a part of computer integrated manufacturing (CIM).

15.3 Digital recorders for impulse measurements

15.3.1 Performance of the digital recorder

The main difference between an analogue recording device and a digital recorder is that the signal is measured only at discrete sampling points and no information is available between two adjacent sampling points. Therefore, it is necessary for the resolution of the impulse digital recorder to be as good as, or better than, that of the best available impulse oscilloscopes. Besides the non-linearity of time base and amplitude, which is also important for analogue recording devices, the resolution and sampling rate of digital recorders are the main parameters of the performance which are not independent of each other [3].

The resolution mainly determines the accuracy of the peak value, and the sampling rate mainly the time parameter. The standards require a resolution of 0.4 % of the full scale deflection or better for tests where only the impulse parameters are to be evaluated. For tests which require comparison of records, a resolution of 0.2 % of the full scale deflection or better are recommended. This leads to digital recorders with a 8 bit or 9 bit analogue digital converter [4].

The sampling rate is expressed in terms of the time interval to be measured and varies between 32 and 60 Msample/s according to the equation:

$$\text{sampling rate} \geq \frac{30}{T_x} \qquad (15.1)$$

where T_x is the time to be measured.

```
HAEFELY      test  protocol

IMPULSE VOLTAGE TEST                           Serialnumber: 421000
Ordernumber  : WO 609047
Test object  : 3Ph Trafo 40MVA 110kV
Standard     : VDE 0532
Reportnumber :
Test leader  :
Performed by : H.P.Haeusler
Inspector    : E.Klaus

Last measure : 11:36:57 25 Feb 1967
```

```
LIGHTNING IMPULSE VOLTAGE TEST on terminal 1U.
Nominal test voltage -550kV. Tap changer position 1.
```

MEASURED VALUES type	amplitude kV		impulse shape us	oscillogram number
Lightning imp.	62.5%	-345	1.16/58	2
Lightning imp.	62.5%	-346	1.16/58	3
Lightning imp.	80%	-440	1.16/58	4

```
CHOP.LIGHTNING  VOLTAGE TEST on terminal 1U.
Nominal test voltage -630kV. Tap changer position 1.
```

MEASURED VALUES type	amplitude kV		impulse shape us	oscillogram number
Chop.lightning	62.5%	-395	1.16/3.94	5
Chop.lightning	100%	-626	1.17/3.9	6
Chop.lightning	100%	-632	1.18/3.89	7

```
LIGHTNING IMPULSE VOLTAGE TEST on terminal 1U.
Nominal test voltage -550kV. Tap changer position 1.
```

MEASURED VALUES type	amplitude kV		impulse shape us	oscillogram number
Lightning imp.	100%	-545	1.17/58	9
Lightning imp.	100%	-545	1.17/58	10
Lightning imp.	62.5%	-346	1.16/58	11

```
THE TEST OBJECT PASSED THE SPEC. TEST.    SIGN:
```

Fig. 15.2   Test report

Investigations of the influence of sampling rate on the accuracy of front time of a standard lightning impulse [5] have shown that this requirement on the sampling rate may be too stringent.

Digital recorders can be separated into two groups on the basis of performance:

• Digital recorders for tests where only the impulse parameters for standard impulses are to be evaluated, with a resolution of 0.4 % (8 bit) and a minimum sampling rate of 40 Msample/s;

- Digital recorders for tests that require comparison of records including tests with front-chopped waves, with a resolution of 0.2 % (9 bit) and a minimum sampling rate of 60 Msample/s.

The influence of individual errors, such as noise, non-linearity of time base and amplitude, on the evaluation of peak value and time parameters is so complex that simulations of the various parameters are necessary to check the requirements of the recommendations and evaluation procedures [6].

15.3.2 Evaluation of peak value

The influence of recorder performance on peak value evaluation can be simulated. The limits of this individual error are 0.4 % or 0.5 % depending on the standards for wave form parameter measurements. This leads to noise of 1 bit for an 8 bit digital recorder or 0.65 % of the peak value assuming an impulse amplitude of about 60 % of full scale deflection. For comparative measurements, the required individual error is 0.1 %. Fig. 15.3 demonstrates the influence of noise on the peak value.

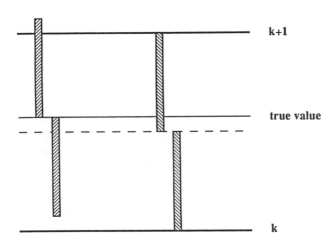

Fig. 15.3   Influence of noise on peak value error

The left part of Fig. 15.3 shows the sample at the true peak value with possible deviation due to noise. The level $k+1$ can be the recorded level, but with a relative low probability. The right part shows a sample after the true peak value, whereas the impulse has an actual value represented by the dotted line. The dotted line is the mean value between level k and $k+1$, so that all samples have the level k independent of the noise.

The non-linearity of time base has no influence on peak value measurement. The non-linearity of the amplitude, particularly static integral non-linearity, influences the measuring uncertainty of the peak value. Fig. 15.4 shows the maximum peak error as a function of sampling rate with 0.5 % noise and maximum static integral and differential non-linearity of amplitude. No additional error due to reading methods will be expected here.

The evaluation is valid for smoothed impulses, but for lightning impulses with oscillation in the peak area the evaluation is more complicated and will be described later in context with the definition and determination of overshoot.

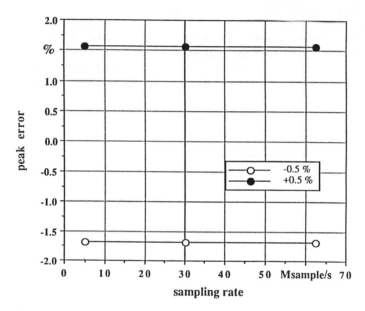

Fig. 15.4   Peak error as a function of sampling rate for full lightning impulses with 0.5 % noise and maximum static integral and differential non-linearity

15.3.3 Evaluation of time parameter for lightning impulses

The recommendations state that the uncertainty caused by the method of reading the record shall be less than 1 % of the value measured.

The performance of a digital recorder contributes to the error in time parameter evaluation, even for a smooth impulse. Assuming a linear interpolation between the two samples nearest to the relevant level, the front time error is shown in Fig. 15.5 as a function of sampling rate.

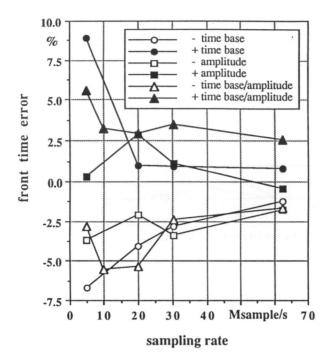

Fig. 15.5   *Front time error as a function of sampling rate with static and differential non-linearity of time base and amplitude*

Simulations show that the errors due to non-linearity are of the same order for the time base and the amplitude, but they will not add to each other. Therefore, the error values given in Fig. 15.5 are valid for both non-linearities. It can also be clearly seen that the required limits for non-linearity in the standards are quite reasonable. But it should be mentioned here that the evaluation is made using an interpolation which reduces the error markedly.

IEEE Standard 1122 does not refer to the evaluation of front time from the time difference between 30% and 90% of the peak value of a full lightning impulse. IEC Publication 1083-1 includes an algorithm reading time parameters. The time to, for example, 30% is determined as the time of the sample nearest to the respective level. If the samples in the vicinity of this level increase or decrease monotonically a linear interpolation may improve the accuracy. If the record in this vicinity is non-monotonic a curve fitting may improve the accuracy of the measured time.

The decision as to which sample is the nearest to the respective level (sample i or sample $i + 1$) is made by a linear interpolation between these two samples and results in sample i of Fig. 15.6, which also represents the respective time.

Assuming a curve shape shown in Fig. 15.6, the true value is given by the crossing point between the real curve and the respective level. The difference between the true value and the measured value is 75% of a sampling interval, the difference between the linear interpolation point and the true value is only 30% [7].

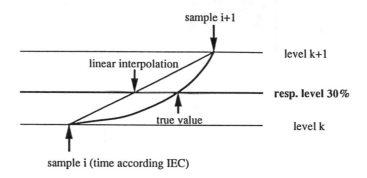

Fig. 15.6   *Evaluation of the time at 30% peak value with and without linear interpolation*

The second example is typical for evaluation of the time at 90% of a full lightning impulse.   Fig. 15.7 shows the sampling procedure.   Assuming a monotonic increasing of the real curve, the nearest sample to the respective level exists twice, sample $i + 2$ and sample $i + 3$.

Depending on the starting point of the evaluation procedure, the front time differs with at least with one sampling interval. This is equivalent to 3.3% at a sampling rate of 60 Msample/s and a time to be measured of 500 ns. The proposed linear interpolation between sample $i + 2$ and $i + 5$ gives, in this particular case, the same result as the simple linear interpolation between sample $i + 3$ and sample $i + 4$, because the number of samples per level is identical for level k and level $k + 1$ and the respective level of 90% is very close to the mean value between level k and level $k + 1$.

The problem will be greater when the real curve has an oscillation in the prospective area. Fig. 15.8 shows a simple example of an oscillation near the 30 % level. All samples at level k (sample $i + 1$, sample $i + 3$, sample $i + 4$) can be used for the front time determination, if level k is nearer to the respective level of 30%.

Both standards recommend a so-called mean curve. However, the smoothing procedure or the generation of a mean curve is not clearly defined in the standards. Fig. 15.8 also shows a mean curve made by a test engineer according to the relevant standards. The proposed procedure is linear interpolation between the first sample below the relevant level (sample $i + 1$) and the first sample above the relevant level (sample $i + 5$). This procedure is identical with the procedure

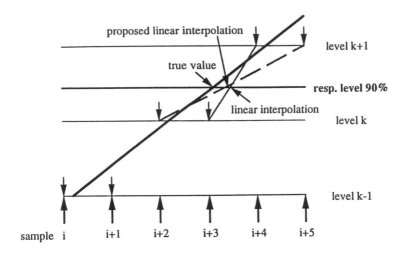

Fig. 15.7   Evaluation of the time at 90% peak value (two samples on the same level) with and without linear interpolation

used in Fig. 15.6 and therefore only one linear interpolation procedure can be used for the evaluation of front time for impulse with more than one sample at the same level, or with oscillations. The comparison between true value and evaluated value by linear interpolation shows a very good agreement. This may not be the case for all possible kinds of oscillations. However, the advantages of this procedure are its simplicity, traceability and reproducibility.

15.3.4 Evaluation of overshoot for lightning impulses

The evaluation of overshoot is combined with evaluation of the peak value and time parameters. Fig. 15.9 shows relevant examples of lightning impulse with oscillations or overshoot according to existing standards [8].

There are many possibilities to solve this problem, but from a practical point of view it seems to be necessary to use a simple, short and traceable procedure to evaluate the overshoot. The recorded data can be filtered or the computer can calculate the best fit curve, e.g. a double exponential curve. All these procedures have the disadvantage that the shape of the curve will be changed or that the judgement for fitting should be made by an experienced test engineer. Therefore, another proposal will be described. The maximum value of the record is always available. Taking all the samples between the maximum sample and the 50% value into account a calculation of the tail in the form:

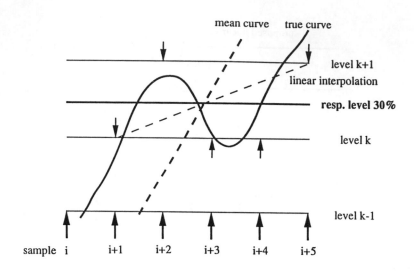

Fig. 15.8   Evaluation of the time at 30 % peak value (superimposed oscillations) with and without linear interpolation

$$U^* = U_0 \, \exp\left(\frac{-t}{T}\right) \qquad (15.2)$$

is possible, where T is the time constant of the tail.

If the impulse has no oscillations and no overshoot, the measured curve and the calculated curve U^* are identical on the tail but have no crossing points. If oscillations exist as in (a) and (c) in Fig. 15.9, crossing points exists. The frequency can be calculated easily and the decision can be taken according to relevant standards. Furthermore, the overshoot can be calculated in the same way, using U^* as the peak value of the mean curve. It might be reasonable first to calculate the overshoot according to the procedure described.

A little more difficult is the decision for overshoot without oscillations as in (b) and (d) in Fig. 15.9. But again the function U^* has two crossing points with the measured curve, whereby the crossing point in the tail should be defined as a point where the difference between U^* and the measured curve is less than 1 %. Using these two points, the time can be calculated where the measured and true curves deviate and this again leads to determination of the peak value and value of overshoot according to the existing standards. Fig. 15.10 shows a simple example.

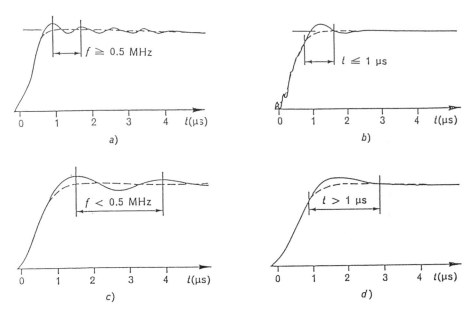

Fig. 15.9   *Examples of lightning impulses with oscillations or overshoot.*
(a,b) *The value of test voltage is determined by a mean curve (broken line).*
(c,d) *The value of test voltage is determined by the peak value (full line)*

The deviation between the calculated overshoot and that evaluated by experienced test engineers is about 1 %. The disadvantage of this procedure is the small deviation from the existing procedure, in particular for impulses with overshoot without oscillations, but the great advantages are the simplicity, reproducibility and traceability [9].

15.3.5 Evaluation of time to peak for switching impulse

The problems of evaluation of the time to peak are partly similar to front time evaluation of lightning impulses. The beginning of the impulse can be evaluated by the same procedure of linear interpolation between the sample on or below and above the calculated zero level.

The more crucial problem is the evaluation of the sample, which represents the peak of the curve. Fig. 15.11 shows the number of samples, which have an amplitude in the range of 100 % - 0.4 % (fsv) as a function of the sampling rate for two different switching impulses (200/1000 µs; 300/4000 µs) and for two measuring ranges (100 % and 60 %).

The error of time to peak can theoretically be ±25 samples at an impulse waveshape 200/1000 µs and a sampling rate of 1000 ksample/s. This is equivalent

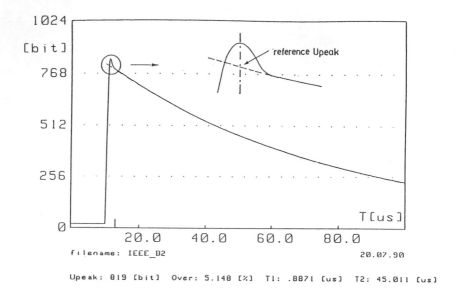

Fig. 15.10   *Evaluation of overshoot (proposal)*

to ±12.5%, and agrees very well with measurements of switching impulses.

An increase in resolution from 0.4% (8 bit) to 0.025% (12 bit) reduces the number of samples in the peak area, but Fig. 15.12 shows very clearly that at a sampling rate of 500 ksample/s 6 samples exist which can represent the time to peak. Assuming the true time to peak is the mean value of these 6 samples, the error in front time evaluation caused by the evaluation procedure can be ±3 samples or ±6 μs.

A simple and traceable method to reduce evaluation error is interpolation between the first and last sample on the highest level and the definition of the time to peak as the mean value between these two samples. Fig. 15.13 shows the time to peak evaluated by the first and last samples at the highest level and the mean value of the first and last sample, as already proposed. It can be seen clearly that even with a large deviation in time to peak evaluated by only one respective sample, evaluation by the mean value gives very good results. The sampling rate for these measurements was 1000 ksample/s [10].

15.3.6 Calibration procedures

It is beyond the scope of this chapter to describe the whole calibration procedure for impulse digitisers. Table 15.1 shows the performance of an impulse calibrator

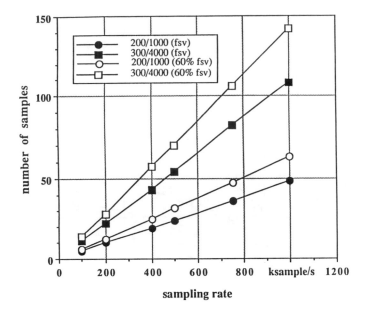

Fig. 15.11   Number of samples for switching impulses as function of sampling rate with different measuring ranges

in respect of amplitude and time parameter accuracy. A problem in high voltage measuring technique, but not uncommon in many fields of metrology, is that the uncertainty of the calibration device has the same order of magnitude as the uncertainty of the device to be calibrated.

15.4 Diagnostic techniques using digital measuring systems

The diagnostic test in high voltage techniques is the determination of a variation of a condition and the prediction of its consequences. The variation can occur during a test or during the service time of the apparatus. A typical example is the high voltage impulse test of a transformer. Fig. 15.14 shows the voltage and current shape and its differences in a test with full lightning impulse voltages [11].

Calculation of the difference is only possible if the data are available in a digital recording system. Furthermore, to stress the statement, the difference can be amplified by a certain factor. A further step in the diagnostic technique is the calculation of the transfer function. This function will be created by two steps. First the voltage as input signal and current as output signal will be transferred from the time domain to the frequency domain. The ratio between current and voltage, as a function of the frequency, is the transfer function of the test object.

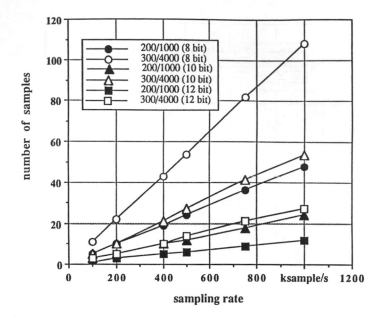

Fig. 15.12 *Number of samples for switching impulses as function of sampling rate with different resolutions*

Fig. 15.15 shows a transfer function for a transformer.

If a test circuit is linear, the transfer functions at reduced and full impulse voltage level should be identical. The small difference in Fig. 15.15 between the transfer function of the reduced and full levels of a lightning impulse test is caused by partial discharges at full test level and not by an unacceptable failure within the test object. This can be confirmed easily with a digital measuring technique. A comparison of two oscillograms according to convention can lead to a difficult and time-consuming discussion. A failure inside the transformer leads to a variation of the poles, as is shown in Fig. 15.16 [12].

Another example of a diagnostic technique is a partial discharge measurement under noisy conditions. The noise can be separated into three groups: the continuous sinusoidal signal (from radio stations), the impulse-shaped phase-constant signal from thyristors and the random (also impulse-shaped) signals. The digital measuring technique allows the suppression of the noise by digital filtering [13]. Fig. 15.17 shows the frequency content of the noise at a bushing of a transformer. The main noise can be suppressed with about ten filters. The phase-constant noise can be suppressed by a window method, but recording and comparison of some cycles, which leads to a suppression of the noise signal by calculation of the difference, seems to be a better solution. The third group of noise is difficult to separate from true partial discharge signals. Fig. 15.18 shows

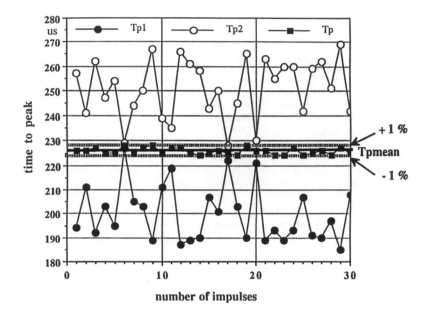

Fig. 15.13   Time to peak evaluation of switching impulses

T_p1 *time to peak evaluated by the first sample*
T_p2 *time to peak evaluated by the last sample*
T_p *time to peak evaluated by the mean value of T_p1 and T_p2*

the signal with and without adaptive filtering on a high voltage cable. The system uses the fact that the partial discharge signal will be reflected on the open end of the cable and that the travelling time is known.

A final example is the voltage monitoring of a transformer, which itself can also be a type of diagnostic technique. The voltage stress of a transformer and the neutral current will be recorded continuously by a digital recorder. If an overvoltage due to a lightning stroke occurs, the digital measuring device records voltage and current. By calculating the transfer function the stress can be evaluated, and compared with the stress during its acceptance tests. Fig. 15.19 shows a frequency spectrum of test voltages and the voltage stress during a lightning stroke. The two curves at a frequency of several kHz are very close, which gives some information to the operator in order to take care of the transformer [14].

Table 15.1   Requirements for pulse generator [4]

Impulse type	Parameter being measured	Value	Long term accuracy (1)	Short-term accuracy (2)
	Time to half value	60 µs	±2%	±0.2%
Full and standard chopped lightning	Front time	0.84 µs	±2%	±0.5%
	Peak voltage	To produce 40% to 100% full-scale deflection	±1%	±0.2%
	Time to chopping	500 ns	±2%	±1%
Front chopped lightning impulse	Peak voltage	To produce 40% to 100% full-scale deflection	±1%	±0.2%
	Time to peak	20 µs	±2%	±0.2%
Switching impulse	Time to half value	4000 µs	±2%	±0.2%
	Peak voltage	To produce 40% to 100% full-scale deflection	±1%	±0.2%

(1)   The accuracy is determined from the mean value of a sequence of at least 20 pulses

(2)   The stability is the standard deviation of a sequence of at least 20 pulses

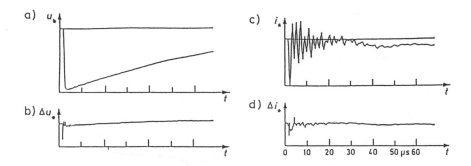

Fig. 15.14   Voltage and current shape for a transformer impulse test

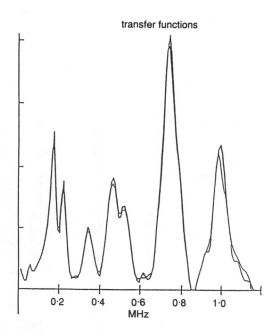

Fig. 15.15   Transfer function for a transformer

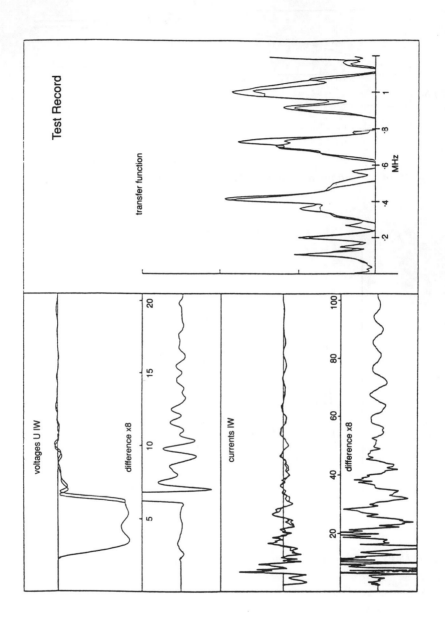

Fig. 15.16 Transfer function of two chopped impulses with a failure

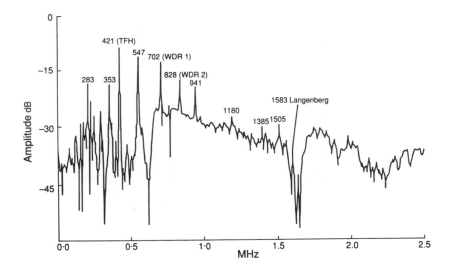

Fig. 15.17   Noise signal measured on a transformer bushing

15.5 Field measuring device

Electrical field strength is one of the most important parameters for the dimensions of high voltage apparatus. Optical transmission of a signal allows the use of a potential-free sensor to measure the electrical field at any point. Under certain circumstances the relationship between field strength and applied voltage provides measurements of voltage or voltage shape. The negligible feedback of the generating circuit is advantageous for comparative measurements of impulse voltages.

15.5.1 Measuring system

The measuring system consists of the electrical field sensor with a fibre optic transmission cable, receiver and recording instrument. The sensor is a sphere of 80 mm diameter, which can measure the electrical field components in two orthogonal directions. The measured signal is transmitted by two fibre optic cables, one for each direction, and then converted into an electrical signal by the

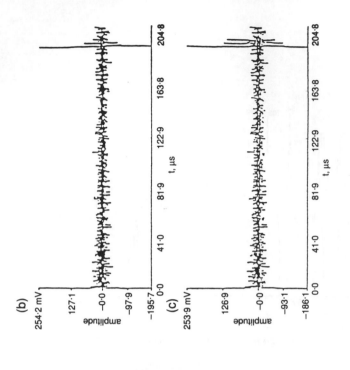

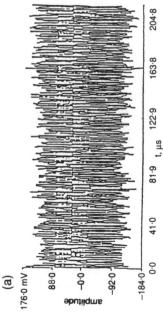

Fig. 15.18   PD and noise signal;
(a) original signal,
(b) signal with adapted filtering,
(c) signal with partly-adapted filtering

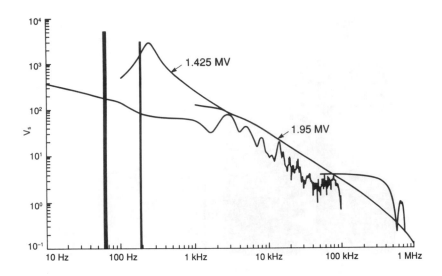

Fig. 15.19   Frequency spectrum of test voltages and lightning stroke

receiver. The type of recording instrument depends on the measuring purpose, e.g. the peak value of an electrical field can be recorded by a digital peak voltmeter, but the shape of an impulse voltage should be recorded by an oscilloscope or a digital recorder. Fig. 15.20 shows the complete measuring system with electrical field sensor, fibre optic link and receiver [15].

Generally an electric field sensor consists of a measuring electrode connected via a measuring impedance to a reference electrode often at earth potential. The principle of all capacitive field sensors can be deduced from Maxwell's first equation:

$$\int_H Hdl = \int_A gdA + \varepsilon \int_A \left(\frac{dE}{dt}\right) dA \qquad (15.3)$$

In an equivalent circuit the integral on the left hand side of Equation (15.3) can be replaced by a current source $i_o(t)$. The conduction current integral is normally negligible in this application and therefore the first part on the right hand side of Equation (15.3) can be neglected. With these assumptions the equivalent simplified circuit is given in Fig. 15.21. The equation for this circuit is:

Fig. 15.20   Electrical field measuring system

$$\varepsilon \int_A \left(\frac{dE}{dt} \right) dA = i_o(t) = C_2 \left(\frac{dU}{dt} \right) + \frac{U}{R} = i_C + i_R \qquad (15.4)$$

According to Fig. 15.21 two different measuring methods exists for the electric field measurements.

The sensor is terminated with the characteristic (low) impedance of the measuring cable, e.g. $R = 50\,\Omega$. In this case the capacitive current i_C is much lower than the ohmic current i_R and therefore:

$$u_M = R\,\varepsilon\,A_{eff}\,\frac{dE}{dt} \qquad (15.5)$$

The measuring voltage u_M is proportional to the derivative of the applied field. To obtain the electric field, or, in a calibrated configuration, the voltage, the measuring signal has to be integrated.

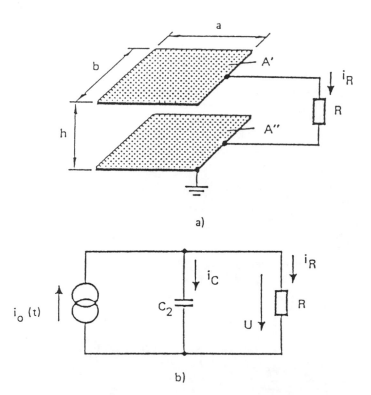

Fig. 15.21 *Principal diagram of electrical field sensor;*
(a) arrangement;
(b) equivalent simplified circuit

The sensor is terminated with a high impedance ($R > 1\,\mathrm{M\Omega}$). In this case the capacitive current i_C is much higher than the ohmic current i_R and the measuring voltage is given by the equation:

$$u_M = \frac{A_{\mathit{eff}}}{C_2}\, \varepsilon\, E \qquad (15.6)$$

The measured voltage is proportional to the applied electric field or, with a calibration to the applied voltage.

Both principles are used in practical applications. In most cases the high ohmic termination is used with modifications such as enlarged sensor areas, termination networks or an impedance converter to allow the connection of a long measuring cable. The construction of the field sensor is normally adapted to the measuring problem and therefore numerous designs exist.

15.5.2 Practical field measurement

Fig. 15.22 shows a typical high voltage arrangement of a porcelain insulator with a toroidal electrode on the top. The electrical field near the electrode can be measured and the influence of the field sensor is very small assuming that the distance between sensor and electrode is larger than the diameter of the sphere.

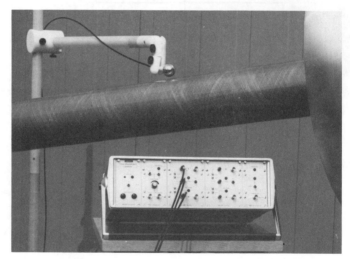

Fig. 15.22 Electrical field measurement

As long as no predischarges occur at the electrode, the sensor can also be used for voltage measurements. The voltage amplitude can be determined if the relationship between output voltage of the sensor and applied voltage is known for a defined position of the sensor [16].

Comparative measurement with a voltage divider and an electrical field probe is a step forward to solving the problem of linearity checking, because transfer behaviour can be checked in the final position of the voltage divider and with the impulse shape, for which the divider will be used. The following measurements were made with a damped capacitive divider for 3200 kV and an electrical field sensor, fixed near to the flange between the first and second high voltage units of the divider. The recording instruments were a high voltage impulse oscilloscope for the divider and an 8-bit digital impulses analysing system for the field sensor. Fig. 15.23 shows a lightning impulse voltage with a front time of 1.64 μs and a time to half value of 67.4 μs.

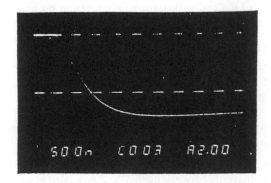

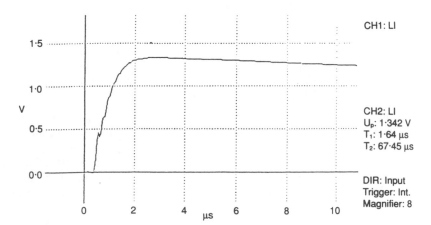

Fig. 15.23   Lightning impulse voltage measured with a damped capacitive divider and an electrical field probe

An analogue recording system was used for two reasons; one is the well-known electromagnetic compatibility of the high voltage impulse oscilloscope and the other is to show the differences between analogue and digital recording systems and their oscillograms.

15.6 Very fast transient measuring devices in GIS

To measure fast transients or very fast transients (VFTs), usual'external voltage dividers are inadequate and specially designed probes have to be included in the metallic enclosure of the SF_6 test vessel. The probe must be placed close to the point where the transient voltage is to be measured, since the voltage may differ along the length of the test vessel due to propagation phenomena.

The basic design of such a probe consists of a plane metallic electrode (live electrode) close to the grounded enclosure, but insulated from it, and facing the high voltage electrode. It constitutes a voltage divider, the high voltage arm of which is the stray capacitance (around 1 pF) between the live electrode and the high voltage electrode. In order not to impair the dielectric strength of the test object, the probe must be placed in a weak field region, for instance inside an access port.

According to the nature of low voltage capacitance, two kinds of field probes are distinguished:

- The low voltage capacitance is made up of a set of elementary components mounted coaxially between the live and grounded electrodes as for external capacitive dividers. A small ceramic capacitor without terminals is directly welded between two brass discs in order to achieve a minimal inductance of connections and a low voltage capacitance of some tens of nF is obtained.

- The low voltage arm is made of a thin layer of insulating material inserted between the opposite flat surfaces of the live and grounded electrodes. Layers of different materials (polyethylene, polypropylene, etc.) and thicknesses (10 μm to 100 μm) can be used.

The output of the probe is connected to the recording instrument by a coaxial cable matched on both sides. Furthermore, the length of the coaxial cable must be as short as possible, as its own capacitance must be kept small with respect to the capacitance of the probe. Given the characteristic value of 100 pF/m, the maximum length is 1 m (or a few meters, depending on the type of the probe). Fig. 15.24 shows two typical designs of measuring devices for very fast transients in GIS.

Calibration of such a measuring device is difficult because it should be done in the real arrangement. It is of great advantage to use the power frequency voltage for calibration, but it should be taken into account that the scale factor of a capacitive divider with an ohmic load on the secondary side is frequency-dependent. It should be ascertained whether the transformation ratio is also valid for high frequencies [17].

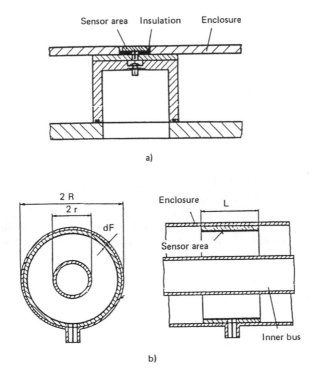

Fig. 15.24   Typical design of a measuring system for VFTs

15.7 References

1   "Computer aided control system for impulse test plant", Bulletin E 137, (E. Haefely & Cie. Ltd., Basel, Switzerland)

2   HIAS - "High Resolution Impulse Analysing System", Bulletin E 147 (E. Haefely & Cie. Ltd., Basel, Switzerland)

3   GOCKENBACH, E. and HÄUSLER, H.P.: "Some aspects of the evaluation of high voltage impulse measured with an impulse digitizer", 6th International Symposium on High Voltage Engineering, New Orleans, 1989, Paper 50.03

4   IEC Publication 1083 - 1: "Digital recorder measurements in high voltage impulse tests Part 1: Requirements for digital recorders", 1991

5 HÄUSLER, H. P. *et al.*: "Transient recorder for HV impulse tests" 5th International Symposium on High Voltage Engineering, Braunschweig, 1987, Paper 70.02

6 GOCKENBACH, E.: "Influence of digitizer performance and evaluation procedures on errors in high voltage impulse measurements", 7th International Symposium on High Voltage Engineering, Dresden, 1991, Paper 62.01

7 GOCKENBACH, E.: "Auswerteverfahren bei der digitalen Messung von Hochspannungsimpulsen". *Elektrie*, 1992, **46**, pp. 382 - 390

8 IEC Publication 60: "High voltage test technique Part 1: Definitions and test requirements", 1989

9 GOCKENBACH, E.: "A simple and robust evaluation procedure for high voltage impulses," International Symposium on Digital Techniques in High Voltage Measurements, Toronto, 1991, Paper 3-1

10 GOCKENBACH, E. and CLAUDI, A.: "High voltage impulse parameter evaluation using relevant standards for digital recorders", IEEE Summer Meeting 1991, San Diego, Panel Discussion

11 MALEWSKI, R. and POULIN, B.: "Digital monitoring system for HV impulse tests". *IEEE Trans.* 1985, **PAS-104**, pp. 3108 - 3116

12 GOCKENBACH, E. *et al.*: "High voltage impulse tests on power transformers using a digital measuring system" 5th International Symposium on High Voltage Engineering, Braunschweig, 1987, Paper 72.05

13 HARTJE, M.: "Erfassung von Teilentladungen an Transformatoren im Netzbetrieb" (Dissertation) (VDI Verlag Nr. 61, Hannover, 1989)

14 MALEWSKI, R. *et al.*: "Digital recording and processing of switching transients in Hydro Quebec's EHV power transmission system", 5th International Symposium on High Voltage Engineering, Braunschweig, 1987, Paper 72.03

15 "Field measuring system", Bulletin E 142, (E. Haefely & Cie. Ltd., Basel, Switzerland)

16 FESER, K. *et al.*: "Distortion-free measurements of high voltage impulses". *IEEE Trans.* 1988, **PD-3** pp.857 - 866

17 "Monograph on GIS Very Fast Transients" CIGRE, 1989

15.8 Case studies

15.8.1 Lightning impulse measurement

Task 1

Evaluation of peak value, front time T_1 and overshoot of the digital recorded signal in Fig. 15.25.

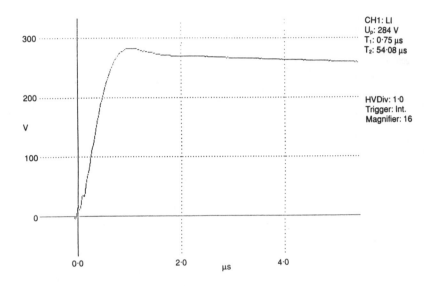

Fig. 15.25   Measured lightning impulse voltage with overshoot

Procedure

Draw a mean curve according IEC Publication 60 High Voltage Test Technique (see Fig. 15.9).

Determine the value of the test voltage.

Determine the 30% and 90% levels and calculate the front time T_1.

Calculate the overshoot as the maximum difference between the mean curve and the measured curve.

Results

Calculations by three engineers, as well as evaluation by a computer, according to the proposed method in Fig. 15.10, show that the value of the test voltage is determined by the mean curve (the value U_p in Fig. 15.25 is the peak value, because the proposed method is not yet incorporated in the evaluation program of the digital recording system). The difference between manual calculation and computer evaluation is about 0.5%. The overshoot is 4% for manual calculation and 3.3% by computer evaluation. The front time T_1 is 0.75 μs in both cases.

Task 2
Evaluation of peak value and front time T_1 of the digital recorded signal in Fig. 15.26.

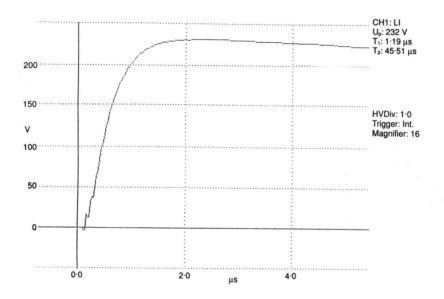

Fig. 15.26   Measured lightning impulse voltage

Procedure
Determine the value of the test voltage.
Determine the 30% and 90% levels and calculate the front time T_1.

Results
The value of the test voltage is given by the peak value and evaluation by hand or computer gives the same peak value with a difference of 0.5%. The front time is 1.19 μs evaluated by the computer and 1.20 μs evaluated by an engineer. The steps in the record are the problem for computer evaluation. Samples represent a time interval of 75 ns at a sampling rate of 40 Msample/s. This means a deviation in the front time of about 10% depending on which sample will be used for the front time evaluation. The proposed procedure according to Fig. 15.7

reduces the deviation by a factor of 2 and gives reproducible results, which are also less sensitive to recorder behaviour.

15.8.2 Switching impulse measurement

Task
Evaluation of the time to peak of the digital recorded signal in Fig. 15.27.

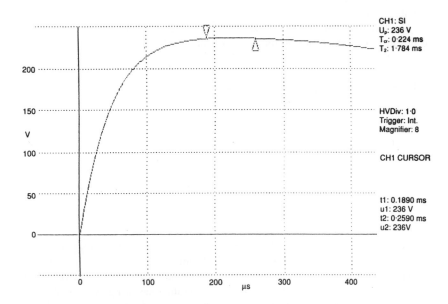

Fig. 15.27   Measured switching impulse voltage

Procedure
Determine the highest sample.
Calculate the time to peak.

Results
The two cursor positions show samples between 189 µs and 259 µs at the same level. A simple procedure, calculating the mean value, gives a time to peak of 224 µs. An evaluation by an engineer would give more or less the same result. Fig. 15.13 confirms the validity of the proposed procedure.

15.8.3 Impulse tests on a bushing

Task
Check of linearity with chopped and full lightning impulses with a digital recorder.

Procedure
Compare voltage and current oscillograms at different voltage levels or for different impulses, as shown in Fig. 15.28.

Results
The oscillogram of full lightning impulse shows no difference between the impulse voltage before (Imp. 9) and after (Imp. 10) tests with chopped impulses. The currents show a small difference, particularly with an enlargement by a factor of 10. The transfer function clearly shows a deviation, which indicates a defect of the bushing, generated during the test with chopped impulses.

15.8.4 Field distribution measurement

Task
Measurement of electrical field strength on a cable termination; peak value and distribution.

Procedure
Measure electrical field according to Fig. 15.29 with a potential-free field sensor at a constant voltage.

Result
The measured field strength is shown in Fig. 15.30. The distribution is not linear, but the maximum field strength is, at 2.3 kV/cm, very low. The measured and calculated values show good agreement. The small size of the sensor and the optical link ensure a negligible field distortion by the measuring system. A simple check of the measurement is to integrate the measured field strength along the tube, which should be equal to the applied voltage according the equation:

$$\int E \, ds = U$$

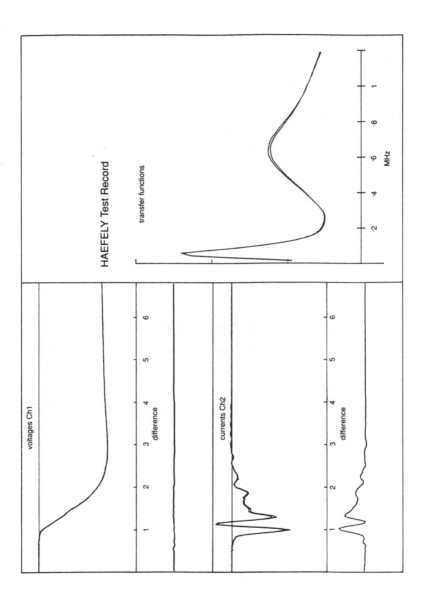

Fig. 15.28 Measured lightning impulse voltage and current and calculated transfer function

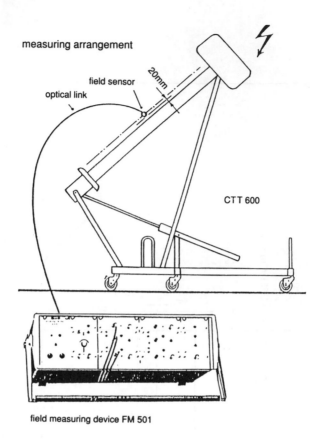

Fig. 15.29 *Arrangement for the measurement of electrical field strength*

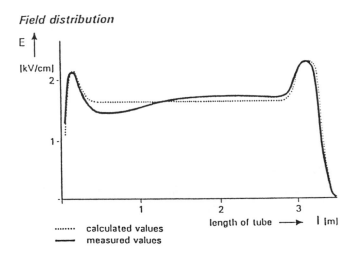

Fig. 15.30   *Measured and calculated field strength distribution*

Chapter 16[1]

Optical fibre based monitoring of high voltage power equipment

G.R. Jones

16.1 Introduction

Optical fibre usage has largely been restricted to the domain of the telecommunications industry until recently, but is now emerging in the fields of data communications, community antenna television (CATV) and control applications, where the high inherent bandwidth available and relative immunity to electromagnetic interference (EMI) is becoming increasingly advantageous. Fibre, especially in digital communications, overcomes, for example, the problems of differential earth potentials between locations, and is largely immune to lightning strikes and other electromagnetic interference. As the data transfer rates of local area networks increase, fibre becomes increasingly attractive and more economically viable. In terms of pence per metre per unit bandwidth, fibre as a transmission medium is the obvious choice for an increasing number of applications.

In the domain of electrical power systems optical fibre technology (with the above advantages) has the scope of being used for a variety of purposes. The adoption of telecommunications technology is already well-advanced and the realisation of reliable optical data transmission for protection and control is being convincingly demonstrated. Furthermore, research into optical fibre-based parameter sensing has reached the stage that properly engineered systems are becoming available whose integration into an optical fibre network should eventually lead to a purely optical monitoring and transmission system.

---

[1] This chapter is based upon part of Chapter 27 of the *Electrical Engineer's Reference Book, 15th edition* (edited by Jones, Laughton and Say), published by Butterworth-Heinemann, 1993 and appears with permission.

16.2 Optical fibre fundamentals

16.2.1 Optical propagation in fibres: ray theory

Light propagation through an optical fibre depends upon total internal reflection at the interface between two transparent materials with high and low refractive indices. Fig. 16.1 shows that as a ray approaches a boundary within a transparent medium, it can be totally internally reflected at the high-low refractive index interface, and is guided along the high refractive index medium. As the angle at which the approaching ray increases a critical value θ_2 may be reached beyond which light "leaks" out of both media.

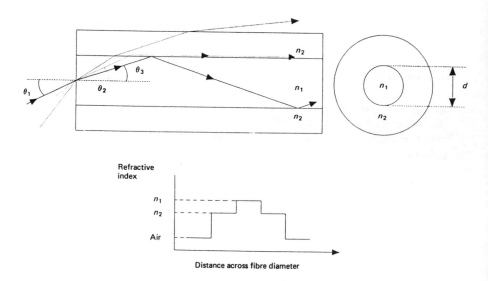

Fig. 16.1   *Ray diagram for multimode step index fibre*

The most basic type of optical fibre can be developed from the above principle by using a cylindrical geometry: rays beyond the critical angle are trapped within the fibre core and travel down the fibre.

16.2.2 Acceptance angle and numerical aperture

Consider the fibre illustrated in Fig. 16.1, with a circular core of diameter d, and a uniform refractive index n_1, surrounded by a cladding layer of uniform refractive index n_2. Light launched into the core at angles θ_1 will be propagated within the core at angles θ_3 up to a maximum value θ_2 to the axis. Light at angles greater than θ_2 will not be internally reflected and will be refracted into the cladding. The maximum launch or acceptance angle θ_1 can be expressed as a function of the numerical aperture (*NA*), where:

$$NA = (n_1^2 - n_2^2)^{\frac{1}{2}} = \sin\theta_1 = n_1 \sin\theta_2 \qquad (16.1)$$

Note that reciprocity dictates that what is true for light entering the core of a fibre is also true for light exiting. The fibre core diameter (*d*), the *NA*, and the operating wavelength (λ) are often used together in a single parameter, known as the normalised frequency, waveguide parameter, or fibre parameter (*V*), which is of importance in characterising all fibres:

$$V = \frac{\pi d}{\lambda} . NA \qquad (16.2)$$

16.2.3 Basic fibre types, modes, mode conversion and bandwidth

Waveguide theory shows that the total number of modes which can be sustained in a step index fibre is given by $N \approx V^2/2$. Modes can be visualised as rays propagating at differing angles to the fibre core. Discrete and definable modes only can propagate because of the geometric constraints of the fibre, and are analogous to modes in hollow metallic waveguides used at microwave frequencies (ca. 1-100 GHz). Typical multimode fibres, with core diameters of 50-200 µm propagate 100 - 1000 modes. It is not necessary for the general user of fibres to understand the mathematics of propagation in detail.

In a multimode step index fibre (Fig. 16.2(a)), however careful one is to launch a single mode, conversion between modes or ray angles is inevitable because of bending and fibre imperfections. This is a great drawback for such fibres, because different modes travel at different speeds down the fibre, causing different arrival times at the receiving end. The difference in transit time between the extreme ray paths for a multimode step index fibre of length *L* is given by:

$$\Delta T_{intermodal} = \frac{L}{c} (n_1 - n_2) \qquad (16.3)$$

where *c* is the velocity of light. The difference in transit times causes a sharp pulse of launched light to become spread at the distant end, limiting the bandwidth of a system. Typically, for an all-silica based fibre of $NA \approx 0.2$, pulse spreading is of the order of 50 ns/km, and is inversely proportional to the length of the system; the longer the system, the lower the bandwidth.

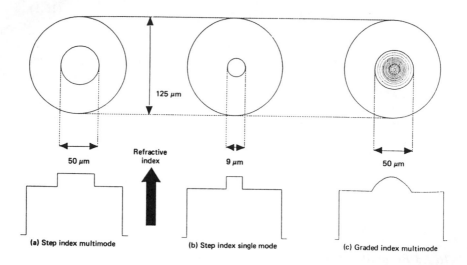

125 μm

Refractive index

50 μm   9 μm   50 μm

(a) Step index multimode   (b) Step index single mode   (c) Graded index multimode

Fig. 16.2   Refractive index profiles (typical dimensions shown)

As V is reduced, less guided ray or modes can be supported, and when $V < 2.405$, only a single waveguide mode can propagate. Such single mode fibres have a core diameter which is comparable with the wavelength of light (d is commonly 8-10 μm for telecommunications fibres — see Fig. 16.2(b)), making fibre-fibre and fibre-device interconnection more difficult and generally less efficient than for multimode fibres. Single- or mono-mode fibres have, however, become the predominant type for most telecommunications links over 1 km or so, and are generally used at 1300 nm and 1550 nm wavelengths where attenuation is low and sources and detectors are available. Intermodal dispersion does not occur and bandwidths can be very high.

Note that fibre which is single mode at, say, 1300 nm will not necessarily be single mode at 850 nm or below. V increases as the wavelength decreases, and will generally be greater than 2.405 at 850 nm. A wavelength known as the cut-off wavelength is an important manufacturing parameter defining the onset of multimode behaviour.

Simple ray optics do not describe energy through a single mode fibre; it is difficult to depict a single ray being guided. Mathematical modelling of single- and multi-mode fibre is achieved by solving Maxwell's equations with boundary conditions defined by fibre geometry and wavelength. This involves Bessel functions and is beyond the scope of this introduction. One of the important results of mathematical modelling is that optical power is not confined to the core alone, but extends appreciably into the cladding region — the power distribution is approximately Gaussian. The extent of cladding penetration is dependent primarily on refractive index difference and wavelength.

A fibre-type intermediate between step index multimode and step index single-mode was developed to overcome bandwidth and connection difficulties; this is the graded index fibre. The principle is qualitatively described in Fig. 16.2(c). Here, the refractive index profile is graded, and ray paths are curved as the rays are continually refracted (Fig. 16.3). Rays which travel nearer the core-cladding boundary are in a region of lower refractive index, and travel faster than those in the denser central core area. The overall effect, given an appropriate refractive index profile, is that rays travelling along different paths arrive at the far end at approximately the same time. The exact index profile to minimise dispersion effects is dependent on the composition of the fibre and the operating wavelength, but is approximately parabolic.

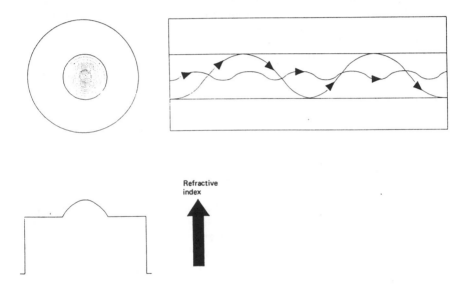

Fig. 16.3   Typical ray paths for graded index fibre

Graded index fibres dominated the telecommunications market in the early 1980s until single-mode technology was perfected. Single-mode fibre is used almost universally in telecommunications, but multimode fibre has realised a new lease of life in emerging high speed (ca. 100 megabits per second (MBps)) local area networks, and 62.5/125 (core/cladding diameters in μm) fibre has been specified as the first choice for the ANSI X3T9.5 committee FDDI (Fibre Distributed Data Interface) standard.

One other fundamental dispersion mechanism operates in all fibres. This is material dispersion, which is the result of light of different wavelengths (from an

LED and even a narrower linewidth laser source) travelling at different velocities down the fibre, and which predominates in single-mode fibres. The speed of light propagating in the fibre is inversely proportional to the refractive index of the propagating medium, and the refractive index of silica drops from 1.46 to 1.44 between 600 and 200 nm, approximately. The variation of the material dispersion parameter $M(\lambda)$ with respect to wavelength is given by:

$$M(\lambda) = -\frac{\lambda}{c}\frac{d^2 n_1}{d\lambda^2} \quad \text{(ps/nm/km)} \tag{16.4}$$

and is shown for silica in Fig. 16.4. Pulse broadening in a particular case can be calculated by multiplying the value of $M(\lambda)$ by both the length of fibre in question and the linewidth of the source in nanometres. An LED source, for example, operating at 850 nm and with a linewidth of 40 nm will give pulse spreading of some 4 ns/km, and this spreading may be significantly reduced by using a laser source with much reduced linewidth of typically 4 nm or less. The bandwidth-length product is generally specified for multimode fibre in the region of 850 nm, called the first window (where sources and detectors were available, and fibre losses were acceptable at approximately 3 dB/km or greater), but second window systems operate at 1300 nm and can exploit the dispersion zero to give high bandwidths. Also, at 1300 nm, losses can be substantially below 1 dB/km, giving far greater transmission distances before regeneration is required. A third window at 1550 nm, at which attenuation can be less than 0.2 dB/km, is now in common telecommunications use. Fig. 16.5 shows nominal attenuation versus wavelength for all-silica fibre, and indicates operating windows.

By modifying the chemical composition of single-mode fibre, and the geometry of the core and cladding, the zero of $M(\lambda)$ can be moved to 1550 nm, but this is achieved only at the expense of an attenuation penalty and is not common practice. (High bandwidth systems are generally achieved by using very narrow spectral linewidth lasers.)

The above treatise generally refers to all-silica fibres — these are by far the most widely used, as they are deployed in the telecommunications industry. The two other common types are plastic-clad-silica (PCS) and all-plastic fibres. PCS fibres have an all-silica core, and polymer-based cladding, commonly a silicone resin which also serves as a protective layer. They are generally less expensive to manufacture, but are characterised by higher attenuation and lower bandwidth (as they have a step index) than all-silica fibres, but are used for relatively short data links. All-plastic fibres, generally manufactured from polymethylmethacrylate (PMMA), are the least expensive type. They currently have the highest attenuation of commonly available fibres, and are generally step index. They have applications as "light pipes" over short distances (metres).

Table 16.1 summarises fibre types and usage.

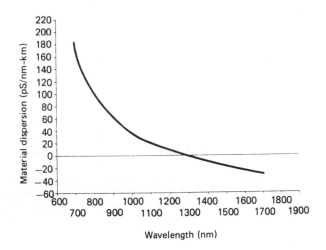

Fig. 16.4   *Material dispersion versus wavelength for silica*

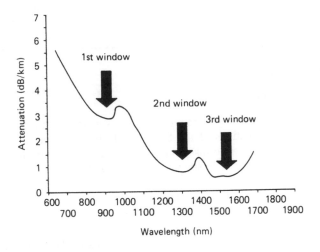

Fig. 16.5   *Nominal attenuation versus wavelength for silica fibre (showing operating windows)*

Table 16.1 Fibre types

Type	Core/cladding diameter (µm)	Typical attenuation (dB/km)	Typical bandwidth (MHz/km)	Applications
All-silica				
Step index multimode	50/125-200/300	3-10 at 850nm	20	Data links
Graded index multimode	50/125-100/140	3 at 850nm <1 at 1300nm	200-1000	Telecommunications, data links
Single mode	5/125-10/125	<0.5 at 1300 nm <0.25 at 1550 nm	>1000	Telecommunications, high speed data links
Other				
PCS	50/125-200/300	5-50 at 850nm	20	Data links
All plastic	50/100-500/1000	>100	<20	Light pipes, electrical isolation, short data links

16.2.4 Fibre protection

The attenuation of a fibre, single- or multi-mode, is increased if it is subjected to bends. Repeated perturbations of the fibre couple light between modes and, if severed, can cause leakage into lossy, radiative modes (those which couple light into the cladding and beyond). In simple terms, excess bending of the fibre causes a light ray on the outside radius of a bend to approach or exceed the critical angle, and light may be lost through the cladding. Such losses are strongly influenced by core and cladding diameters, and *NA*. A guide to the susceptibility to microbending of a particular multimode fibre type can be made by using the following "figure of merit" (that is based on step-index analysis but can be used as a general guide):

$$\gamma = K \ \frac{d_{core}^4}{NA^6 \ d_{cladding}^6} \tag{16.5}$$

An optical fibre must therefore be protected against radial forces. This is accomplished by mechanically decoupling the fibre from its immediate surroundings, and is commonly achieved by surrounding the fibre with a very low elastic modulus material such as silicone rubber followed by an extruded polymer layer (termed a "tight" packaged fibre), or by encapsulating the fibre loosely within a polymer tube ("loose" packing). The type of protection depends upon the particular application. Note that the pristine surface of an all-silica fibre is always protected during manufacture by a thin ultraviolet (UV) radiation or heat cured polymer layer.

16.3 Power equipment monitoring with optical fibre sensors

16.3.1 Introduction

Although early forms of optical fibre sensing systems have illustrated the potential of such methods for power system monitoring, there have been practical deficiencies which have detracted from their widespread use. The original systems were cumbersome, unreliable, costly and involved much unfamiliar optical processing and interfacing. It may be argued that insufficient attention was given to the real needs of power system monitoring. The problem has been exacerbated because the power industry has been unsure of the modes in which such novel technology might best be used whilst simultaneously a major thrust of optical fibre sensors research has been for methods which are over-sophisticated and costly for power system applications.

However, the many fundamental advantages of optical fibre sensing systems remain attractive. For instance, a major advantage is for the condition monitoring of a circuit-breaker during fault current interruption when signatures of impending faults may be more distinguishable. A further implication is that such an approach minimises supply disruption since such monitoring would be undertaken live, a capability which emerges because of the system's electromagnetic immunity and its inherent electrical insulation properties.

This section examines the advantages of optical fibre monitoring for power equipment applications, reasons for limited progress in implementing such technology and possible strategies which are evolving for the site testing and evaluation of these systems.

16.3.2 Technology implementation difficulties

One objective of the realisation of optical fibre sensing in the power industry is for the implementation of an optically controlled substation system. Such an objective involves producing optical fibre sensors for monitoring a range of parameters governing the condition and operation of power equipment such as circuit breakers and transformers. It entails interlinking the various monitoring systems with an optical fibre system and connecting these to various data stations and control units (Fig. 16.6). The attraction of optical fibre sensors for such monitoring is that their intrinsic properties offer many advantages in hostile and challenging environments. For high-voltage power apparatus applications these properties include:

(1)   Inherent isolation of electronic instruments;
(2)   Immunity from electromagnetic interference;
(3)   Geometric flexibility;
(4)   Inherent electrical insulation;
(5)   No electrical shock hazard;
(6)   Corrosion resistance;
(7)   Compact and light weight.

Despite such attractive advantages, the uptake of the fibre sensing technology has been retarded because of general market penetration problems. These include:

(1)   Uncompetitive prices;
(2)   End-user unfamiliarity;
(3)   Conservative attitude of large industries;
(4)   Range of different optical systems for monitoring various parameters;
(5)   Failure of early, immature systems;
(6)   Absence of established markets.

Recent developments in optical fibre sensing are addressing these commercial problems. In order to appreciate the significance of the developments it is necessary to review briefly the methods available for modulating an optical signal.

The basic modulation methods involve modifying either the amplitude, phase or polarisation of the optical signal (Fig. 16.7). Absolute amplitude monitoring is unreliable for sensing purposes because a number of external (e.g. fibre bending) and intrinsic (e.g. source variability) factors in addition to the sensor may modify the amplitude. Phase monitoring is based upon optical interferometry which involves complex instrumentation and because of its digital nature is not so attractive for situations in which the instrument power supply may be interrupted. Methods based upon monitoring changes in the plane of polarisation of an optical signal have hitherto relied upon even more complex optical systems.

The range of parameters which are measurable, in principle, with these basic modulation methods is given in Table 16.2. Attempts to overcome the deficiencies of the amplitude modulation method by referencing a modulated signal at one wavelength (λ_1) with respect to an unmodulated signal at a second wavelength (λ_2) have met with only limited success because the referencing is not completely

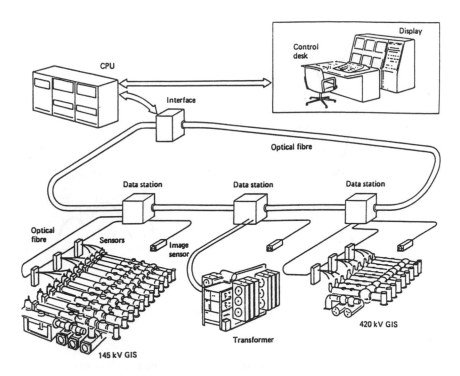

Fig. 16.6   *Optically monitored substation concept (after Mitsubishi)*

reliable and because of power-budget limitations.

The problem with these basic methods is that the intention with any measurement system is to seek an output (V) which is proportional to the measurand (M) via a constant which is the sensitivity S:

$$V = S.M \qquad\qquad 16.6$$

With an optical fibre system the relationship (equation 16.6) between the output V and modulation $M_1(\lambda)$ is complicated and depends upon parameters such as source power $P(\lambda)$, fibre transmission $T(\lambda)$, and fibre perturbation $M_2(\lambda)$ which can be induced by external influences or age:

$$V = q[\int_{\lambda} (P(\lambda)[\int_{l} T(\lambda)M_2(\lambda)dl)] \, R(\lambda)M_1(\lambda)d\lambda]^P \qquad (16.7)$$

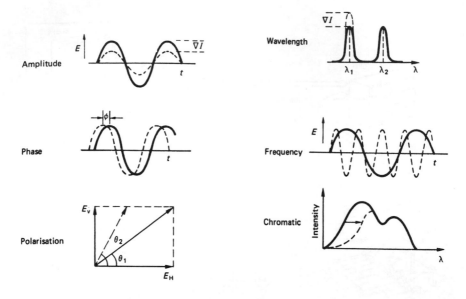

Fig. 16.7   Basic optical modulation methods

It is therefore necessary to seek modulation methods which are less susceptible to these effects. Two main methods are currently emerging. The first relies upon frequency modulation which is the solution adopted by the telecommunication industry. The difficulty with such an approach for sensing is in realising practical forms of modulators.

The second method relies upon chromatic modulation whereby the spectral content of the optical signal is varied and the optical signal is monitored over its centre spectral range by several detectors each having a different but overlapping spectral response.

Both these methods are intensity-independent and lead to practical, cost-effective fibre sensing systems. The chromatic approach has the added bonus of using common instrumentation for monitoring a range of different parameters so avoiding the complicated situation depicted in Table 16.2.

16.3.3 Evolving fibre sensing systems — an example

An example of fibre sensing evolution which respects the real needs of the power industry is the emergence of new forms of current and voltage monitors. Two main methods may be identified corresponding to direct optical transduction of the

Table 16.2   *Fibre optic sensor systems and associated optical phenomena (Mitsubishi Electric)*

Measured physical quantity	Optical modulation	Optical phenomenon used	System*
Current, magnetic field	Polarisation	Faraday effect	DO, FF
	Phase shift	Interference (magnetic distortion)	FF
Voltage, electric field	Polarisation	Pockels effect	DO
	Phase shift	Interference (electric distortion)	FF
Temperature	Light intensity	Light shading by a plate	DO
		Changes in light absorption of semiconductors	DO
		Fluorescent radiation	PP
	Light intensity, spectrum	Radiation from a heated body	PP
	Polarisation	Double refraction	DO
Angular velocity	Phase shift	Sagnac phenomenon	FF
Velocity, liquid velocity	Frequency	Doppler effect	PP
Vibration, acceleration, pressure	Light intensity	Loss due to microbend	FF
		Light shading by a plate	DO
		Change of reflection intensity by a diaphragm	DO
	Polarisation	Photoelasticity	DO
	Phase shift	Interference (photoelasticity)	FF
	Frequency	Doppler effect	PP

*DO = Discrete optical sensor, PP = Pick-up-probe sensor, FF = functional fibre sensors

current/voltage signal and to sensing via intermediate electronic transduction (hybrid transduction).

Initially, the attraction was for purely optical monitoring relying on Faraday rotation and Pockels effect for current and voltage transduction, respectively. Such methods may be further subdivided into intrinsic systems (which utilise the transmitting fibre itself as the sensor) and extrinsic systems (which utilise separate, localised sensing elements); see Fig. 16.8(a) and (b), respectively.

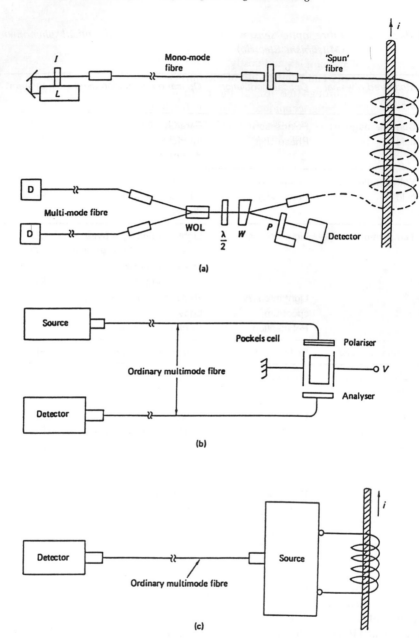

Fig. 16.8   Evolution of optical fibre current/voltage monitoring system:
(a) Intrinsic electro-optic sensing;
(b) Extrinsic electro-optic sensing;
(c) Hybrid sensing

The penalties associated with the intrinsic approach include the poor optical stability and cumbersome size of the sensor elements and the need for complex and delicate optical processing. Attempts to stabilise the systems by compensation are cumbersome, complex and costly.

The complexity and unreliability of the intrinsic approach may be alleviated to a large extent through the use of the extrinsic approach in conjunction with chromatic monitoring techniques. The benefits in system simplicity (and hence cost and reliability) which occur are apparent from a comparison of Figs. 16.8(a) and (b). The need for special fibres and intermediate optical components (Wallaston prisms, fibre splitters etc.) is removed through the use of the chromatic extrinsic system. Nonetheless, the optical sensing elements of such systems remain susceptible to temperature and vibrational instabilities.

The simplicity of the optical sensing system may be further reduced and the difficulties associated with the stability of the electro-optic elements overcome through the use of the hybrid system. Such systems utilise current-voltage transformers for intermediate transduction into a suitable electronic form for optical modulation. The systems have the added advantages, therefore, of being retrofittable on existing transformers already installed on power systems. They also provide a number of additional benefits when used in a non-retrofittable mode. For instance, current and voltage may be monitored from a common transformer unit, and the current and voltage signals may be multiplexed via a single fibre.

Problems of variable transformer burden with electrical transformer connections which are produced by transmission over different lengths of wire are removed due to the low, variable burden of the fibre-energising electronics. There is a consequent reduction in size and cost of the current transformer. Because of the inherent insulation nature of the optical fibres, transformer insulation requirements may be relaxed leading to further reductions in cost and size.

The replacement of wire by fibre connections avoids electromagnetic pick-up between signals from different power system phases, provides weight reduction advantages and provides automatic electronic isolation. Finally, the optically linked current transformer is more compatible in producing digital inputs for microprocessors which are increasingly being used for controlling protective functions on power equipment.

A typical example of a combined current/voltage hybrid is shown in Fig. 16.9. This type of unit may be operated either in frequency modulation or chromatic modulation modes. The performance characteristics of such units are typically: high linearity; a resolution of 1 in 10^4; a frequency response in the kilohertz range; a range from 0.3 A minimum; transmission distance at least several hundred metres; and output as required. Such units are currently being site evaluated both in the UK and Europe. Electro-optic monitoring systems are being evaluated in France and for GIS (gas-insulated system) use in Japan.

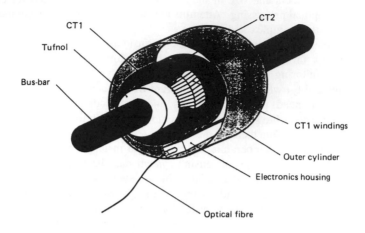

Fig. 16.9   General arrangement of an optical fibre current and voltage sensor

16.3.4 Monitoring non-electrical parameters

Notwithstanding the importance of monitoring electrical parameters such as current and voltage, there is also a major requirement for monitoring mechanical and thermal parameters on power equipment since failure often occurs in these respects. For instance, the majority of circuit-breaker faults are mechanical in nature.

Optical fibre-based systems for monitoring parameters such as the gas pressure and temperature of gas-filled equipment (circuit-breakers, and gas insulated bus-bar chambers), mechanical linkage on switchgear and the occurrence of arcing are available and have already undergone site testing.

The mass density of gas in SF_6-filled equipment may be monitored with a system produced by Kent Instruments (Fig. 16.10). This is an example of a frequency-modulated system whereby the natural frequency of vibration of a wire attached to a pressure-sensitive diaphragm is modulated by the gas density. The vibration of the wire is excited by pulsed optical power transmission via a fibre link and the vibrations are detected by the shuttering action of the wire on a fibre data link. However, this sensor system cannot distinguish between gas-density variations caused by pressure and temperature changes, so that distinction cannot be made between gas leakage and gas liquefaction effects. Separate monitoring of pressure and temperature is required if such a differentiation is to be made.

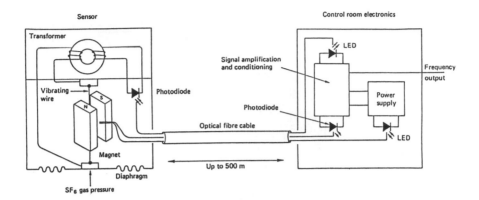

Fig. 16.10   Schematic diagram of the Kent P901 gas density sensor

Chromatically-based fibre sensors for monitoring pressure and temperature independently have already been evaluated during internal arcing tests on a high-voltage disconnect switch developed by Reyrolle and the University of Liverpool (Fig. 16.11). These tests were for r.m.s. fault currents of 65 kA flowing for 0.1 s. The temperature of the tank wall and internal gas pressure were monitored to obtain information about the likelihood of arcing to the switch wall causing burn through. The test results have shown how the arc heating causes the pressure to rise sharply whilst the wall heating is delayed and localised to the arc root position (Fig. 16.12). The optical fibre monitoring system survived the severe operating conditions better than a back-up electronic system. Such tests have established the robustness of fibre-based systems under arduous industrial conditions.

Site tests using a distributed fibre temperature system (GEC Cossor), relying on time-domain reflectometry, for monitoring temperature variations along lengths of several kilometres of underground cable are on-going. However, such systems are expensive so that their normal use on power systems is likely to be limited.

The monitoring of the movement of mechanical linkages of power switchgear is another area where optical fibre monitoring has been shown to operate well on site. A comparison between records obtained with a conventional rheostat-based travel recorder and an optical fibre equivalent using chromatic modulation is shown in Fig. 16.13. The optical fibre system has been shown to operate reliably over an extended period and during live testing of the circuit breaker.

Another example of mechanical-parameter measurement is the monitoring of the mechanical vibration signatures of various parts of power plant structures during the operation of separating circuit-breaker contacts to interrupt a fault current. Such signatures could lead to early diagnosis of impending mechanical faults, although much effort to distinguish fault-related signatures would be needed before

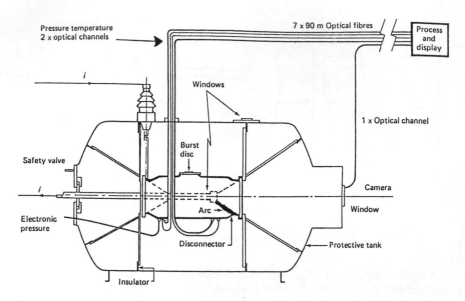

Fig. 16.11   *Disconnect switch-fault arc test (50 kA r.m.s., 15 ms duration)*

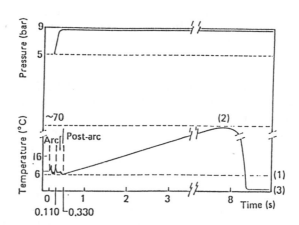

Fig. 16.12   *Gas pressure and wall temperature variations during internal arcing tests on a GIS disconnect switch*

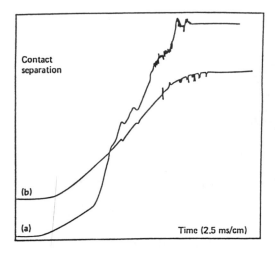

Fig. 16.13   Comparison of outputs from:
(a) Potentiometric and;
(b) Optical fibre travel recorders

such a method could be regarded as reliable. An example of such an optical fibre based method using a distribution fibre sensor is shown in Fig. 16.14(a). This utilises a length of fibre which has been specially prepared to respond to small stresses via induced microbending. The fibre is simply wrapped around the various parts of the circuit breaker structure to be examined and the vibration signature monitored by recording changes produced by the vibration-induced microbending in the spectral content of the transmitted light (Fig. 16.14(b)).

16.3.5 Prognosis for routine in-service monitoring

Optical fibre monitoring has evolved beyond scientific demonstrations to the engineering optimisation and commercial viability phase. The frequency and chromatic modulation techniques have not only provided routes for overcoming the deficiencies of simple intensity, phase and polarisation methods but are also leading to versatile cost-effective instrumentation which is suitable for high-volume sensing applications. Because these two new approaches are less sensitive to fibre connectors, the routine connection and disconnection of the optical sensing elements may be installed in a dormant state and only resurrected for addressing with portable instrumentation during servicing or fault diagnosis.

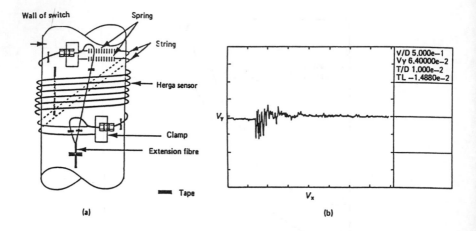

Fig. 16.14   Mechanical vibration monitoring using a distributed fibre sensor.
(a) Connection of optical fibre sensor to switchgear wall;
(b) Typical output waveform of the vibration sensor

Of course, the inherent electrical insulation nature of the optical fibres allows, in principle, such connection and disconnection of the monitoring instrumentation to be made on live equipment without supply disruption. Such use of optical fibre sensing systems has hitherto not been feasible on account of the intolerance of intensity, phase and polarisation sensing to any fibre link disruption.

The goal of a common electronic instrument for monitoring several parameters is being gradually realised. One such unit is already being manufactured by Optical Sensing Ltd., a member of the Lucas Engineering group. The availability of such a unit with the versatility of the chromatic approach is allowing engineering decisions to be addressed on their own merit as to the most appropriate form of sensing element for the measurement of particular parameters on different pieces of power equipment under various field conditions. The diverse range of mechanical parameters requiring monitoring, as described in Section 16.3.4 (pressure, temperature, mechanical movement, and vibration), are all, in principle, capable of being monitored with such a unit.

Index